REASON AND BEING

BOSTON STUDIES IN THE PHILOSOPHY OF SCIENCE

EDITED BY ROBERT S. COHEN AND MARX W. WARTOFSKY

VOLUME 17

B. G. KUZNETSOV

REASON AND BEING

Edited by

CAROLYN R. FAWCETT AND ROBERT S. COHEN

D. REIDEL PUBLISHING COMPANY

A MEMBER OF THE KLUWER ACADEMIC PUBLISHERS GROUP

DORDRECHT / BOSTON / LANCASTER / TOKYO

Library of Congress Cataloging-in-Publication Data

CIP

Kuznetsov, B. G. (Boris Grigor'evich), 1903–
Reason and being.

(Boston studies in the philosophy of science; v. 17) Translation, by Lynn Visson, of: Razum i bytie. 1972.
Bibliography: p.
Includes index.
1. Rationalism—History. 2. Ontology—History. I. Fawcett, Carolyn R. II. Cohen, Robert Sonné. III. Title. IV. Series.
Q174.B67 vol. 17 001'.01 s 86–20405
[BD181] [149'.7]
ISBN-13:978-94-010-8539-7 e-ISBN-13: 978-94-009-4590-6
DOI: 10.1007/978-94-009-4590-6

Published by D. Reidel Publishing Company,
P.O. Box 17, 3300 AA Dordrecht, Holland.

Sold and distributed in the U.S.A. and Canada
by Kluwer Academic Publishers
101 Philip Drive, Norwell, MA 02061, U.S.A.

In all other countries, sold and distributed
by Kluwer Academic Publishers Group,
P.O. Box 322, 3300 AH Dordrecht, Holland.

The translation, by Lynn Visson, of Boris G. Kuznetsov's *Razum i bytie* (Moscow, Nauka, 1972), has been revised by Carolyn R. Fawcett. Chapter 12 and the Afterword were added by the author for this English edition.

TABLE OF CONTENTS

Raphael: 'The School of Athens', detail (Stanza della Segnatura, Vatican); central figures of Plato and Aristotle. Alinari/Art Resource 3498.

B. G. KUZNETSOV

EDITORIAL PREFACE

Boris Kuznetsov was a scientist among humanists, a philosopher among scientists, a historian for those who look to the future, an optimist in an age of sadness. He was steeped in classical European culture, from earliest times to the latest *avant-garde*, and he roamed through the ages, an inveterate time-traveller, chatting and arguing with Aristotle and Descartes, Heine and Dante, among many others. Kuznetsov was also, in his intelligent and thoughtful way, a Marxist scholar and a practical engineer, a patriotic Russian Jew of the first sixty years of the Soviet Union. Above all he meditated upon the revolutionary developments of the natural sciences, throughout history to be sure but particularly in his own time, the time of what he called 'non-classical science', and of his beloved and noblest hero, Albert Einstein.

Kuznetsov was born in Dnepropetrovsk on October 5, 1903 (then Yekaterinoslav). By early years he had begun to teach, first in 1921 at an institute of mining engineering and then at other technological institutions. By 1933 he had received a scientific post within the Academy of Science of the U.S.S.R., and then at the end of the Second World War he joined several colleagues at the new Institute of the History of Science and Technology. For more than 40 years he worked there until his death two years ago. He was educated as economist and engineer, becoming a model student of the idealized polytechnical education of the early Soviet years. During the late 1920s he wrote on some specific problems of political economy; then he turned toward practical urgencies of his own engineering specialty and joined in work on the massive electrification program. By 1931 he had prepared one of the first detailed technical plans for a unified high-voltage electric power transmission network for the U.S.S.R., drawing upon the latest inventions and international standards. He was a responsible leader in the All-Union Institute of Power Engineering and Electrification, and during the 1930s concentrated upon mapping power grids for several regions of the Soviet Union. In the mid-thirties, he was awarded his doctorate in economic science. His interests in the history of scientific ideas were developing alongside of his efforts at solving urgent technical problems of economic importance. Thus, by 1936 he had published a substantial essay on the history of electrical engineering, and then in 1937 a treatise both on historical development and on his own proposal for new principles to provide the physical and engineering foundations of 'energetics'. Kuznetsov concluded that historical understanding of the changes in scientific principles, indeed a critical comprehension of such changes, would be required for any accurate mastery of current revolutions in energy technology.

From the start Kuznetsov was an imaginative as well as a disciplined thinker, the speculative philosopher as well as the scientifically precise historian; that is, he was a philosopher of science. Already by 1931 he had published a shorter work called *Dialectics, Science and Technological Reconstruction*, uniting predictions about the potentialities of science and technology with retrospective philosophical reflections on the sciences. He had undertaken detailed studies of modern physics by the mid-thirties, and also he had begun his investigations of the lives and works of both Russian and other European scientists. By 1945 he had prepared and published his pioneering book of essays on *Lomonosov, Lobachevsky, Mendeleev*, and then (1946–48) a series of interpretive papers on problems of space and motion in Newton, Leibniz and especially Euler. In that decade, Kuznetsov worked mainly on 17th and 18th century studies (Galileo, Descartes, Huyghens), and then turned to later scientists (Faraday, Lagrange, Maxwell). He also began his intensive effort to understand the fundamental nature of relativity and quantum mechanics, and he initiated his friendly discussions with theoretical physicists, among them the philosophically sensitive Ya. I. Frenkel.

During the following years, Kuznetsov devoted his writings to epistemological issues of the sciences, taking these issues in their widest as well as the most technical senses. He worked on the logic of both relativistic and quantum physics, on the scientific influences upon philosophy throughout its classical and modern periods, upon his ever interesting and creative investigations of the scientific development of economic theory, upon the future and its forecasts, upon the realities of cultural life in the literature, arts and buildings of Europe in the West and in his own land. There was no problem of two cultures in his spirit; C. P. Snow would have recognized a comrade in this cultivated, committed, and learned engineer who had become historian of science and philosophical essayist.

I believe it is not disturbing to call Kuznetsov an essayist. He was surely a modest thinker, meditating on his scientific and humanist heroes, digging deeply where they had ploughed and planted, interpreting and imagining what they might have said when he questioned them. He told me he was always a student, and his writings were often those of an enthusiastic youthful mind, indeed delightfully young to the end, eager to explain to his fellows just what some fascinating idea really meant, or was likely to mean. He also said he was a teacher whose expository comments on the texts of our civilization would, he hoped, bring interest and information and hope to his readers. Of course the readers must have numbered in the many hundreds of thousands: his splendid biography and the later cultural commentary on Einstein's life and work and times were published in editions in the tens of thousands, repeatedly; his *Philosophy of Optimism* (1972) likewise; and the imaginative and thoughtful *Journey Through Epochs: Memoirs of Count Cagliostro* (1975) in a first printing of 100 000.

Even the book before us is an essay, not fully a scientific treatise. Kuznetsov was intrigued by the tremendous liberation promised by the brilliant achievements of those who discovered non-Euclidean geometry. Reasoning was preserved in the break with Euclid's monopoly, but an ancient fear of necessity and closure without freedom might now be confronted with optimism as well as courageous understanding. Again and again, in this and in his other historical and philosophical works, Kuznetsov explored the 'non-classical' style of the age of Einstein and Bohr, and recurs in his enlightening retrospective sketches to the linkages between heroic figures of the past and our present crises of opportunity and horror. To his eye, the 'non-classical' suffuses the modern period, threatening the optimistic rationalism of the 18th century with the statistical explanations of the 19th and the ambiguous lessons of relativistic classicism and non-causal quantum uncertainty in the 20th. Has it only been genuinely a radical change of *style*? Or is it a change of form along with ontological content?

Kuznetsov dances about the reader in his essay on Einstein and Mozart where he so aptly cites Einstein's remarks about George Bernard Shaw which Kuznetsov applied to Einstein too:

> I once said of him that his plays remind me of Mozart. There is not one superfluous note in Mozart's music. The one in the medium of language, the other in the medium of melody, expresses perfectly with almost superhuman precision, the message of his art and his soul. (*Cosmic Religion*, 1931, p. 109)

But perhaps he would have been pleased with Pais's comment that the quality of Mozart in Einstein's special theory of relativity of 1905 was succeeded by the new style of the general relativity developed during 1907–1915, "reminiscent of Beethoven," says Pais. And Pais quotes the motto of the last movement of Beethoven's Opus 135: "Must it be? It must be" ('*Subtle is the Lord. . .*', 1983, p. 183).

Nevertheless whether it 'must be' cannot rest with either perfection or necessity, not in Kuznetsov's anguished and resonant reading of Dostoyevsky, nor in his response to Einstein. After all, Einstein said "Dostoyevsky gives me more than any other thinker, more than Gauss!". Was it in the portrayal of the devastating social reality of our earth "saturated with human tears from crust to center" that Einstein received so much from Dostoyevsky? Or in the novelist's desire for harmony? And then in his terrible descriptions of innocent suffering especially of children at the hands of human beings in 'normal' social circumstances? What kind of higher rationality can there be if it is the rationality of an overall whole which has such suffering individual beings within it? Ivan Karamazov says

> I don't want harmony . . . out of the love I bear mankind . . . Besides, too high a price has been placed on harmony. We cannot afford to pay so much for admission. And therefore I hasten to return my ticket of admission. (*The Brothers Karamazov*, tr. Magarshack, 1964 ed., p. 257)

Kuznetsov wrote that this

> is the most piercing note in that craving for harmony which has distinguished the history of human culture, and which remains as a question addressed to the 20th century.

He added:

> Whatever one may think of Dostoyevsky's public views, however erroneous or reactionary his political and social ideas may have been, the essence of his art is an appeal addressed to the twentieth century; man needs a social and moral harmony that does not ignore local disharmonies and that refuses to accept the individual sufferings of any man, a harmony that leaves no room for coercion, oppression or contempt for the weak. (*Einstein and Dostoyevsky*, 1972, p. 12)

Later he wonders whether Dostoyevsky imagined a 'complementarity' of individual, local 'inimitable' existence with a rational general scheme of a world harmony . . . "a harmony that would not ignore individual fates"? Yes, he thinks, but it remained vague for Dostoyevsky.

Kuznetsov's parallel of moral and scientific cognition goes perhaps too far, but the disharmony between science and humanity was insistent for Kuznetsov. At a later point he wrote

> Modern science cannot treat ultra-microscopic processes as real without introducing macroscopic concepts, without defining a particle's macroscopic behavior. Modern notions of moral harmony require that an individual existence be determined by its importance to the collective destiny. (*Ibid.*, p. 106)

Finally, then, we must see Einstein as perceived by Kuznetsov, a seeker of rational 'non-classical' harmony, rejecting classical determinism, inheritor of an Epicurean hope for a genuine 'swerve' and of Dostoyevsky's problem of the micro- and the macro-. Einstein saw a time beyond Dostoyevsky, a time of Auschwitz, Hiroshima, and more. How did Kuznetsov remain so optimistic? Characteristically his last letter to me from Paris was from his room at the Hotel de l'Avenir.

Boris Kuznetsov was a dear friend who could argue with energy, hope, and sympathy. We met in Boston and Montreal, in Dubrovnik and Paris, and in Moscow. He survived some of the gravest times; in living through more than the first half-century of the Soviet Union, he tried to understand and to share that understanding, to create but also to conserve. He was pleased when he learned that this book on his central theme of 'reason and being' was to appear in English in the *Boston Studies*. He helped time and again with questions of translation, and of references and citations.

I have appended to this Editorial Preface a selected bibliography of the published writings of Kuznetsov, taken from his more than 200 publications.

Carolyn Fawcett was the principal editor of the draft translation of this book. The reader will be grateful, as I am, for her careful, graceful and scholarly achievement.

November 1986

ROBERT S. COHEN
Center for Philosophy and History of Science
Boston University

SELECTED PUBLICATIONS OF BORIS KUZNETSOV

Note: Publication is in the Russian language unless indicated otherwise; all Russian publications listed here were published in Moscow.

1931 'Unified High-voltage Net'.
1936 *Essays on the History of Electrotechnology.*
1937 *History of Energy Technology.*
1945 *Lomonosov, Lobachevsky, Mendeleev: Essays of Lives and Outlooks.*
1955 *The Evolution of the Scientific Picture of the World in the Physics of the 17th and 18th Centuries* (Revised 1961).
1958 *Principles of Classical Physics.*
1959 'On the Foundations of Quantum-relativistic Logic', *Logical Research Studies* (pub. Acad. Sci. U.S.S.R.), pp. 99–112.
'Paths of Development of Quantum-relativistic Logic', *Papers of the Institute of History of Science and Technology* **22**, 69–105.
The Principle of Relativity in Ancient, Classical and Quantum Physics.
1960 'Die Lehre von Aristoteles von der relativen und der absoluten Bewegung im Lichte der modernen Physik', *Sowjetische Beiträge zur Geschichte der Naturwissenschaft*, ed. G. Harig (Berlin), 27–61.
1962 'Infinity and Relativity', in *Einstein and the Development of Physical and Mathematical Thought*, pp. 137–200.
Einstein (447 pp). (Revised several times. Thus, *Einstein: Life, Death, Immortality*, 1972, 608 pp.; again 1979, 680 pp.). English trans. Moscow 1965; reprint New York 1970. French trans. Brussels 1965, Paris 1967; *Einstein, sa vie, sa pensée, ses théories.* German trans. Berlin 1977: *Einstein: Leben, Tod, Unsterblichkeit.*
1963 *Development of Physical Ideas from Galileo to Einstein* (2nd ed., 1966). German trans. Berlin 1970: *Von Galileo bis Einstein: Entwicklung der physikalischen Ideen.*
Development of the Fundamental Ideas of Electrodynamics (238 pp. with a 54 pp. bibliography of literature from 1600 to 1962).
1964 *Physics and Logic.*
Galileo. French trans. Moscow 1973: *Galilée* (revised by Michel Rouzé).
'Moderne Logik, Physik und Geschichte der Physik', *Beihefte zur Schriftenreihe für Geschichte der Naturwissenschaften, Technik und Medizin: Für 60. Geburtstag Gerhard Harigs*, ed. I. Struve und H. Wussing (Leipzig), pp. 131–136.
1965 *Studies in Einstein* (rev. ed. 1970). French trans., rev. and enlarged (Moscow 1971): *Essais sur la relativité*, selected from this and related works by Kuznetsov.
1966 'Complementarity and Relativity', *Philosophy of Science* **33**, 199–209.
1967 *Physics and Economics.*
'Spinoza et Einstein', *Revue de synthèse* **3**, 31–52 (see also ch. 6 of *Reason and Being*).
1968 'Physik und Volkswirtschaft', *Ideen des exakten Wissens (Wissenschaft und Technik in der Sowjetunion)* 12/68, pp. 110ff.
Principle of Complementarity.
Paths of Physical Thought.
1969 'Einstein et le principe de Mach', *Organon* (Warsaw) **6**, 265–277.
Science in the Year 2000 (revised and incorporated within Kuznetsov's *Philosophy of Optimism* of 1972).
Relativity. 'Die Information "Know-How" ', *Ideen der exakten Wissens (Wiss. und Technik in der Sowjetunion)* 2/69, pp. 80ff.
'Relativität und Komplementarität', *ibid.* 12/69, pp. 779–785.

1970 'Einstein und der Machse Prinzip', *ibid.* 1/70, pp. 49–56 (trans. from same Russian ms. as the French above 1969).

'Le problème de la théorie du champs en XIXe siècle et le rapport de la physique aux mathématique', *Physis* **12**, 353–362.

'L'infinité du macrocosme et du microcosme chez Galilée', *L'Univers à la Renaissance: Microcosmos et Macrocosmos*, which is volume 14 of *Trav. de l'Inst. pour l'étude de la Renaissance et de l'Humanisme*, pp. 55–67.

'The Rationalism of Leonardo da Vinci and the Dawn of Classical Science', Diogenes no. 69, pp. 1–11.

Giordano Bruno and the Genesis of Classical Science.

1971 'Der wirtschaftliche Nutzen nichtklassischer Wissenschaften', *Ideen der exakten Wissens (Wiss. und Technik in der Sowjetunion)* 3/71, pp. 159–166.

'The Economic Effect of Non-Classical Science and Methods of Economic Forecasting', *Problems of Science of Science* (Warsaw), pp. 63–74 (special issue of *Zagadnienia Naukoznawstwa*).

'Quantum-relativistic Retrospection and the History of Classical Physics: Classical Rationalism and Nonclassical Science', *Historical Studies in the Physical Sciences* **3**, 117–135 (originally prepared for the Introduction to *Reason and Being*.

1972 *Reason and Being.*

Philosophy of Optimism (revised and enlarged ed. 1977). French trans. Brussels 1972; major selection in *La science en l'an 2000*, Paris 1972. English trans. of 1977 edition, Moscow 1977, in three sections: 'Epistemological Optimism', pp. 7–73; 'Science in the Year 2000', pp. 74–268; 'Economic Conception of Optimism', pp. 269–343.

Einstein and Dostoyevsky (London). Subtitled: 'A Study of the Relation of Modern Physics to the Main Ethical and Aesthetic Problems of the Nineteenth Century'.

'Einstein und Mozart', *Ideen der exakten Wissens (Wiss. und Technik in der Sowjetunion)* 12/72, 783–788.

'Einstein and Epicurus' *Diogenes* No. 81, 44–69.

'Lénine, Langevin et la préhistoire de la théorie de la relativité', *La Pensée* No. 161 2/72, pp. 43–47 (trans. from *Voprosy Filosofii* of 1969 for documentary article on 'Paul Langevin et Albert Einstein d'après une correspondance et des documents inédits', *ibid.* 3–40, edited by Luce Langevin).

1973 'Paris Traduit', *Change* No. 14, pp. 19–31.

1974 *History of Philosophy for Physicists and Mathematicians.*

1975 *The Value of Knowledge.*

A Journey Through Epochs: Memoirs of Count Cagliostro and Records of His Discourses with Aristotle, Dante, Pushkin, Einstein and Many Other Contemporaries.

1976 The Value of Scientific Errors and the Irreversibility of Science', *Diogenes* No. 97, 103–123.

1978 'The Poetry of Logic and the Logic of Poetry', *Organon* **15**, 9–21.

'The Tragedy of Knowledge at the Time of the Renaissance', *Diogenes* No. 104, 66–92.

1979 'Einstein and Classical Science', *Einstein and the Philosophical Problems of 20th Century Physics* (pp. 61–121 in the English translation, Moscow 1983).

Ideas and Images of the Renaissance..

1981 *Modern Science and Philosophy.*

Studies in Megascience.

'The Aesthetics of Nonclassical Science', *Diogenes* No. 115, 87–110.

PREFACE TO THE ENGLISH EDITION

The ideas of classical rationalism are now completing a journey that is somewhat analogous to the voyage of the *Mayflower*, only this journey is not in space, but in time, not across an ocean, but across the centuries, from the times of Galileo, Descartes and Spinoza to the present. These ideas, spanning four centuries, convey to our contemporaries the spiritual legacy of the cinquecento, the Baroque, and the Age of Reason, and embody the spirit of epistemological optimism as the basis of economic, social, and moral optimism. Modern nontraditional, nonclassical science embodies this spirit but in a new form. This history of rationalism in the light of nonclassical science has brought us to an understanding of its present value. Realization of this value (economic, ecological, social, moral) — fulfillment of the potentialities of the rationalist impulse of modern science — this is the most important task of mankind as it approaches a new millennium.

The actual and predictive conclusions drawn from the history of rationalism from a modern nonclassical perspective is emphasized more sharply in this edition than in the first edition. In *Reason and Being*, the initial paradox of being is demonstrated as a very old, but still fundamental conception: the past is no more, the future is not yet, and the present is a boundary of zero duration between the nonexistent past and the nonexistent future. Being signifies that the past and the future enter into the present. As time derivatives enter the concept of being, being becomes differential and dynamic. The dynamic *ratio* is a necessary component of being. This conception of 'the principle of being' was set forth in the first edition of *Reason and Being*. Now I should like to express it in another form as well, namely in an analysis of the concept of 'nothing' to be found in a new chapter (Chapter 12), 'Nothing and the Vacuum'. In this chapter the connection between social philosophy, the philosophy of modern physics, and the history of philosophy is shown through the concept of alienation. This connection is considered in a generalized form (including 'the alienation of nature') and is examined as a rupture between localized, individual, here-and-now existence and the integral harmony of being. Modern nonclassical science, in making a unity of the cosmos and the microcosm, of cosmology and the theory of elementary particles, opposes 'the alienation of nature'. In transforming the burdens and character of labor, the practical application of modern science has thereby eliminated the historical denial of the 'here-and-now existence', of the individual destinies of which Dostoyevsky once wrote. Thus, in this new edition, I am returning to my previous reflections on the analogy between the ideas of Einstein and the work of Dostoyevsky (see Kuznetsov 1972b). I do so in order to attempt to connect this problem of the individual and the cosmos with the most fundamental confrontations of reason and being.

I am profoundly and sincerely grateful to my American colleagues, to the translator of *Reason and Being*, Dr. Lynn Visson, and especially to the editors of the English edition, Carolyn R. Fawcett and Robert S. Cohen. Through their assistance I have been able to bring my thoughts on being, and on reason — its past, present, and future — before a new audience.

INTRODUCTION

INTRODUCTION

1. A NONCLASSICAL HISTORICAL PERSPECTIVE

Science — I should say, it is precisely science that — has imparted to contemporary civilization its unprecedented dynamism. What is the nature of the link between this dynamism, with its assumptions concerning the transformation of the most fundamental notions about the world, and rationalism? How does this rationalism evolve and what features of classical rationalism become more distinct in the light of contemporary science?

It would seem that the answers to these questions cannot avoid relying on certain predictions because it is in the nature of contemporary science to push forward. Therefore, these answers must say something about the velocity, acceleration and direction of that forward-moving process. When the contemporary physicist first observes a rapidly moving particle, he cannot identify it nor relate it to something already known, he cannot speculate on its being a new type of particle without first describing its behavior, namely its velocity, its acceleration, the curvature and length of its trajectory in some field, and therefore its lifetime. A localized ultramicroscopic event does not have a sensible physical existence without a space-time world-line. Analogously, it is impossible to answer questions about the content, meaning and significance of contemporary tendencies in science without at least some speculative sketch of the 'world-line' of science, without scientific prediction about science itself. In fact, without such prediction it is impossible to describe the nature, style and ideals of contemporary science, or even to answer the question about the relationship of contemporary science and rationalism.

It is also impossible to answer this question without adopting an historical perspective, without an analysis of rationalism itself. The history of science and philosophy has appropriated for itself the right which people denied the gods: the right to change the past. An historical perspective is not absolute, it defines the meaning and significance of discoveries and ideas within a certain system of coordinates.

Sometimes history does this without taking into account the relativity of its judgments and therefore takes them to be absolute. Indeed, it sometimes appears that history has juxtaposed ideas of natural science of the past with the definitive truth of the latest [scientific] discovery. In our dynamic age such pretensions would be an anachronism.

This book is an attempt at a *nonclassical* re-evaluation of *classical* thought. Our goal is to examine the classical rationalism of the seventeenth century and the evolution of dialectical thought during the eighteenth and nineteenth centuries, through the prism of contemporary nonclassical science. In other words, we will look back at classical thought through the prism of the contemporary tendencies of nonclassical science.

These tendencies demand that an historical perspective encompass not only classical rationalism but classical science as well. Classical science is the starting point and, in addition, a concrete embodiment of the idea of evolving reason. Neither present nor future science can be defined without historical digressions, which we will find especially necessary for defining the meaning and style of fundamental discoveries and generalizations. When new fundamental principles are discovered in science and philosophy, it is impossible to evaluate them or even to designate them without associating, likening and juxtaposing them to the old principles. Thus new principles and representations become the yardstick for the old, and sometimes the old even plays the role of standard for the new. What Niels Bohr said of science in our era — that to be correct, physical theory must be sufficiently mad — is to a certain degree characteristic of every era: scientific progress then removes the aura of paradox and 'madness' from an individual conception, deducing it from new, more general assumptions, in a natural and logically irreproachable manner. The more general assumptions are those which radically alter traditional views, and which revise ideas possessing still deeper historical roots. The more radical the new principles, the thicker the historical strata they uncover and the deeper they penetrate into history, generalizing the heritage of the past, and making it more concrete. Therefore, scientific progress is impossible without reviewing, re-evaluating and reinterpreting not only the arsenal [i.e., the working concepts] of physics, but also its pantheon. In any event, without this review, the scientific revolution is impossible. By this we mean that such a revolution comes about not only through continual and progressive change in its practice or, at any rate, through continual doubting and experimentally testing the conclusions and applications of science, but also, and all the more importantly, through doubting the initial and fundamental premises of science.

Nonclassical science has conducted just such a review of its own integral and evident components. Contemporary science may be described as nonclassical for two reasons. First, it has generalized and relativized the Newtonian axioms which previously had claimed the role of immutable canons of truth, much like the classical canons of beauty embodied in the art of antiquity. Second, contemporary science is also nonclassical in its style. Its canons are dynamic, and the ideals of scientific explanation change within a single generation. The dynamism of science, which is the alteration of its fundamental principles, has become evident.

In the light of contemporary science, we are discovering that classical science also was dynamic. This is the first result of nonclassical retrospection. It allows us to see that the most fundamental principles of classical science were fluid and changeable and that the Victorian era in physics was not all that Victorian. But the fluidity and changeability of the fundamental principles of physics during the seventeenth through the nineteenth centuries remind us of the movement of a minute hand, and even an hour hand. A nonclassical physics in its turn included in its conception of cognition something of an analogy to a second hand in that it made plain the dynamics of its fundamental principles. Once, on seeing a second hand for the first time, a person soon suspected that time involved movement. But now the idea of a moving mechanism becomes his immediate impression, the result of a single, specific observation. If we extend such an impression retrospectively to classical science, we begin to feel the dynamic value of its stages; we consider not only the positive [factual] results of science, but also the meaning of each new discovery and each generalization for the tempo and direction of subsequent scientific progress. Together with the criterion of truth, the criteria of conceptual velocity and acceleration entered into science in the comprehension of truth.

In contemporary nonclassical science, such criteria are constantly applied. The value of scientific discoveries and generalizations is measured not only by new information, which is subject to experimental test, but also by new methods, new questions and, new predictions — by everything that defines the velocity and acceleration of science. Thus the practical value of contemporary science is inseparable from its epistemological value: the more the new representations, methods and ideals prove to be fundamental and influential in the tempo of science, the more radical their influence on production. Moreover, this influence is measured by the differential coefficients of the velocity and acceleration of the technical process.

Philosophy long ago arrived at the idea that its fundamental categories are fluid and changeable. This idea grew out of generalizing the *development* of science. On the other hand, natural philosophy, which is a generalization of the positive, established *results* of science, constructed rigid and immutable matrices which it claimed had an *a priori* character and into which it attempted to line up the developing content of science. In this respect, natural philosophy was a continuation of the *identifying* tendency of rationalism, which found constant, invariant, identical relations within the uninterrupted flow of being. Such a tendency sometimes turned out to be cut off from the opposite, less official tendency, from the search for non-identity, irreducibility, paradox. Such a split was to a certain degree legitimate so long as the mutability of the fundamental principles of science was not apparent, just as the movement of the hour hand is not immediately obvious.

Nonclassical physics inaugurated an epoch of practically uninterrupted and obvious transformation of fundamental concepts. Correspondingly, science recognized in its past the less than official tendency just mentioned: the negation of identity, the transformation of the immutable matrices themselves into which new facts were laid out. This was the *questioning* tendency. Scientific thought now concentrated on what was not identical, repetitive, invariant, and it searched for new generalizing schemata. It did not find answers to the questions arising from this search, and it addressed these questions to the future, trying to penetrate into that future; it has, in this sense, been profoundly predictive in its style.

The development of science could not occur within the framework of rationalist thought without a synthesis of the basic 'official' identifying melody and the sensual, 'questioning' accompaniment which was undermining the established identity. Comprehension of *being* – about which more will follow – consists of a synthesis of universals with local, particular impressions. When rationalist thought attempted to explain not only the dislocation and behavior of bodies but their being, their nature as well, not only the structure of the world but also its substance, it turned to a world in which it had not always found identities and invariant correlations. At this point, rationalist thought inevitably broke away from the positive [and accepted] knowledge of its time, and it is precisely here that rationalism came closest to the future. When the future became the present, when the problem of being became the central problem of positive science and, having become a problem of experiment, acquired physical contours, then the history of science saw in the evolution of rationalism continual movement, the counter-identifying and individualizing accompaniment that lay outside 'official' limits.

The philosophy of Spinoza was the loftiest flight of rationalist thought, moving beyond the traditional framework. Descartes approached its limits but did not overstep them. For Descartes, the rationalist understanding of the world embraced the *behavior* of substance, but not its being. Both the being of matter and the initial laws of its movement transcend the limits of Cartesian physics. Within those limits, being is dissolved in behavior. Descartes described how bodies move, but could not answer the question of *what* moves, what bodies are, and how they are distinguished from space. Both being itself and the fundamental laws of being, like the very existence of matter, are *a priori* for the physics of Descartes. Spinoza moved beyond this framework; he transferred the reason for the existence of nature from the world beyond nature into nature itself; he viewed nature as the reason for its own being (*causa sui*) and thus created the rationalist conception of *being.* Not only during the seventeenth century, but also in all of classical science, Spinoza's conception did not find evident physical equivalents. Such equivalents finally did appear in nonclassical science.

Comparing nonclassical science with classical rationalism leads not only to a new understanding of the past but also furthers an understanding of contemporary ideas, of contemporary tendencies in science, of its moving forces and perspectives. We cannot easily say whether we are trying the seventeenth century before the court of the present or whether we are submitting the present to the verdict of classical rationalism. Probably both. In any event, both the composition of the court and the *corpus delicti* are changing during the court hearing. Classical rationalism turns its 'questioning' side towards us, its drive toward rational comprehension of being. In the forefront of contemporary science are its unresolved problems. Historical resonance does not unite the ideal of classical ideas with the ideal of contemporary science; here, contradictory, sometimes dissonant, occasionally tragic chords resound.

Without the 'questioning' side, without a sense of the incompleteness of existing theories, without the predictions and a differential approach to the value of contemporary science, our science could not arrive at a new historical perspective, a new evaluation of the past. This becomes obvious as soon as we try to examine a little more closely the mechanism of the historical perspective mentioned above. Let us begin with what may be called 'relativistic retrospection,' that re-evaluation of classical physics undertaken by Einstein when he was moving toward the theory of relativity.

2. THE CRITERIA OF 'EXTERNAL CONFIRMATION' AND 'INNER PERFECTION'

Both an historical perspective and prediction were necessary components in the intellectual surge, the brainstorm that led Einstein to the theory of relativity. Einstein spoke of the two criteria in the selection of a physical theory: 'external confirmation' and 'inner perfection.' The former consists of the correspondence of the theory's conclusions with empirical observations. The latter is the naturalness of the theory, its deduction from the broadest possible principles with maximum exclusions of *ad hoc* assumptions, i.e., for the explanation of only one given observation or series of observations (Einstein in Schilpp 1949, p. 25). In this book I shall be obliged to return more than once to these fixed, fundamental criteria for adequate knowledge of the world. During the formulation of the theory of relativity, 'external confirmation' demanded that the theory be consistent with the experiments that stumped, as well as with those that contradicted, classical science. Most important, the theory of relativity had to be consistent with the results of Michelson's experiment which demonstrated the paradox of the constancy of the velocity of light in relation to bodies with differing motions, both bodies which the light overtakes and those moving toward the light. But 'external confirmation' also demanded that under certain conditions (low energies), when the velocity of the bodies is not approaching the velocity of light, the theory of relativity approach the relations of classical mechanics, which, in such conditions, have received very substantial empirical corroboration. In this way classical mechanics gains a greater 'inner perfection' from the theory of relativity: it is shown to be the natural outcome of a more general and precise theory and, in addition, acquires historical being, loses its absolute and *a priori* character, and is proven to be connected with a definite stage in the development of experimentation, production and culture. This perspective permits us to see several internal contradictions in classical theory and to view classical theory not only as a collection of answers to questions which were asked of nature, but also as the sum of questions which were unanswered and were therefore readdressed to the future. The history of [actual] science (as distinguished from the history of a closed, 'idealized' system of knowledge), in analyzing a given period of the development of science, must also examine the questions inherited from the past as well as the questions readdressed to the future — retrospection and prediction.

In returning to a relativistic historical perspective, it must be noted that Einstein's theory allowed us to look differently at classical science, to see in it its 'questioning' side, because this selfsame science already included a 'questioning' component: prediction, searching, and the sense of its own incompleteness. Einstein's theory of relativity revealed the incompleteness of classical mechanics because Einstein found in the latter a certain limited, approximate modification of more general principles, such as the principles of the relativity of motion and the homogeneity of space. Einstein called the classical relations of dynamic variables in transition from one inertial system of reference to another, 'Galilean transformations.' This historical recollection is not simply a rapproachement of Galileo's views and those of the most up-to-date science; it is a presentation of the views of Galileo as an early, incomplete, future-oriented form of a very general principle.

In his *Dialogue Concerning the Two Chief World Systems*, Galileo proved that the uniform motion of a system does not cause changes in the operation of the internal processes of this system: whether a boat is at rest or in uniform motion, all the processes within its cabin take place in the same manner. It is impossible to form a judgment about the uniform motion of a system from its internal processes: motion is reduced to a change in the distances between a given body and the reference bodies, so that we may just as well consider the given body to be in motion and the bodies of reference to be at rest as the opposite. The first system of reference (in which the other bodies are at rest) and the second system (in which the given body is at rest) have equal rights: neither one can claim the title of absolute 'true' motion or absolute 'true' rest.

In the light of Einstein's theory, the classical idea of the relativity of motion does not seem general enough. Uniformly moving bodies are governed by this principle, but in addition to these bodies, in classical science there figured the continuous medium transmitting their gravitational interaction and electromagnetic oscillations — light. It is here that we discover the 'questioning', and moreover, identity-exploding, tradition-defying character of the Faraday-Maxwell concept of the field in classical science. Describing the field as a continuous medium makes it possible for us to speak about the absolute motion of a body, relative not to other bodies but to this medium. Such absolute motion, corresponding to the classical law of addition of velocities, must become apparent when we compare the differences in the velocity of the emission of light when a body moves toward the light and when the light overtakes the body. But such a conception does not have sufficient 'external confirmation.' Light moves with the same velocity in any medium, whether this medium moves toward or away from the source of light.

Einstein's theory arose from such a confrontation, having rejected the old, classical conception of a continuous medium in which the bodies were submerged. From the point of view of Einstein's theory, the classical theory of relativity did not seem general enough, as it was limited only to mechanical phenomena. Having demonstrated that in bodies moving at different speeds, not only the mechanical processes, but also the propagation of light occur in an identical manner, Einstein's theory generalized the principle of relativity. In this light the classical principle appears to be a question to which the science of the seventeenth through the nineteenth centuries did not supply a general answer, a question which has been readdressed to the twentieth century and given an appropriate answer.

The historical reinterpretation of classical relativity demanded that Einstein's theory itself possess a 'questioning', predictive component directed toward the future. In fact, Einstein's relativity theory does possess such a component. Einstein divides physical objects into two classes. One class is the moving body, the other class includes systems of reference, those measuring rods and clocks which permit the establishment of the location and time of each event. In accordance with the theory of relativity, the length of the moving body is measured differently when referred to the length of the measuring rod in a system in which the measuring rod is at rest, as compared with the length of a measuring rod in a system in which the rod is in motion. This does not mean that while in motion the length decreases in comparison with some 'normal' 'true' length of the measuring rod which is at rest. There is no absolute rest, and in the comparison of two measuring rods we can consider each at rest (and each length as not decreased) or each as moving (and then each length is decreased). By analogy, a clock slows its motion when it moves; this is not in comparison with a 'true' clock at rest, but, depending upon the particular systems of coordinates, one in which the clock is at rest and another in which it moves at one or another velocity, each of these systems of coordinates is equally correct, each of the clock movements, equally valid.

This conception has made the explanation of a great many observations possible; it possesses solid 'external confirmation'. The explanation rests not on *ad hoc* assumptions, but rather on very general assumptions, so that the 'inner perfection' of the theory is also great. But from Einstein's point of view this perfection was not great enough and we can discover in the theory of relativity a particular illogical premise which does not follow from more general principles. This premise consists of the following. The behavior of bodies seems to vary, depending upon the behavior of the reference bodies — the measuring rods and clocks — as in the contraction of the length of the measuring rods and the slowing of the motion of the clock. But how can the behavior of the reference bodies be explained? These bodies — both the measuring rods and the clocks — consist of moving parts. Do laws of motion exist for these parts, laws from which it is possible to deduce the behavior of the macroscopic reference bodies, the measuring rods and clocks? No; the theory of relativity postulates the behavior of reference bodies as independent of their movement and structure. "This, in a certain sense, is inconsistent," Einstein wrote.

Strictly speaking measuring rods and clocks would have to be represented as solutions of the basic equations (objects consisting of moving atomic configurations), not, as it were, as theoretically self-sufficient entities. However, the procedure justifies itself because it was clear from the very beginning that the postulates of the theory are not strong enough to deduce from them sufficiently complete equations for physical events sufficiently free from arbitrariness, in order to base upon such a foundation a theory of measuring rods and clocks (Einstein in Schilpp 1949, p. 59).

This statement of its own inadequacy is typical of the theory of relativity, and of nonclassical science as a whole. Nonclassical science is not only unable to become classical in the sense of idealized completeness, but in the course of time its very incompleteness, its very nonclassicality, acquires increasing importance as well. The evolution of nonclassical science increasingly includes fundamental problems within its 'questioning', in the number of [its own] unresolved problems (as distinguished from those *obviously* unresolved problems of the past). Accordingly, historical reflection becomes more and more thoroughgoing and far-reaching, and contemporary science revises concepts at an increasingly fundamental level, transferring them from a logical or even *a priori* class to an historical and approximate one. Such a re-evaluation goes even further, all the way back to classical antiquity, when the most general concepts of philosophy were discovered (or rather, guessed at for the first time). We may note in passing that contemporary science, fully armed with its new and unprecedentedly powerful experimental and theoretical means and methods, in some respects stands closer to the spirit of classical antiquity, which was very nonclassical in its style. Classical antiquity posited basic problems of being in the form of *aporia*, in the form of searches and approaches which V. I. Lenin saw in the *Metaphysics* of Aristotle, later the most dogmatized thinker of antiquity.[1]

Thus an historical perspective, and also prediction are explicitly or implicitly characteristic of nonclassical science. The incompleteness Einstein noted in the theory of relativity means in principle the possibility of a more general conception — more general not only in relation to the theory of relativity but also in relation to quantum mechanics.

The theory of relativity represents the world as a kind of space-time continuum. *The World as Space and Time* is the title of A. A. Fridman's (1965) exposition of the theory of relativity. In the old peripatetic vision of the world, space and time were examined separately. In the cosmology and physics of Aristotle, there is a purely spatial scheme of being, unchanged by time, taken at one moment: a world with a center, boundaries and natural location of bodies. On the other hand, Aristotle examines the nonspatial, purely temporal processes of qualitative change, the origin and destruction of bodies. However, according to the atomists (especially Lucretius), and in the mechanical view of the world, time lost its independent reality; processes occurring in time were viewed as occurring in space as well, and qualitative changes were reduced to spatial displacements. Newton gave space an independent being: the propagation of interactions and of forces. Since any movement can be momentary, it followed in principle that a process occurring in no time at all could be real, and the concept of simultaneity dominated the thinking of physicists.

3. THE WORLD-LINE

Einstein rejected such physics. His world consists not of purely spatial 'momentary' trajectories, but of four-dimensional *world-lines*, each point of which (a world-point) has four coordinates: three spatial and one temporal. (The momentary trajectories are retained as an approximate representation, valid for bodies moving at small velocities and for which the velocity of light and the velocity of the propagation of interactions can be considered infinite.)

If the world-line of the particle is straight, then the particle is moving in a uniformly straight line. It does not change its direction in space, which means it does not curve in a three-dimensional spatial projection of a four-dimensional world. It also does not undergo absolute acceleration, which means it does not curve on a plane formed by the spatial axes and the temporal axis. The theory of special relativity is a theory of inertial motion which is uniform and in a straight line. It portrays the harmony of the universe as a four-dimensional framework of straight world-lines. The geometry of such a world is analogous to Euclid's, in which a straight line, regardless of how far it is extended, remains straight, and does not intersect other lines parallel to it. The geometry of the world as it is portrayed by the theory of special relativity is a four-dimensional analogue to Einstein's geometry. It is called pseudo-Euclidean.

But the real world is composed of accelerated movements. How does one represent it as a four-dimensional framework of world-lines?

Einstein solved the problem in the following manner.

As is generally known, in addition to Euclidean geometry, there exist other, non-Euclidean geometries. In one of them, Riemann's, the shortest distance between points — what in Euclidean geometry is called a straight line — intersects with other very short lines parallel to it if they are extended sufficiently. In this geometry two perpendiculars to one and the same straight line meet, the sum of the angles in a triangle is greater that two right angles, and, in general, a number of extremely paradoxical relations appear. They cease to be paradoxical if two-dimensional Riemannian geometry is viewed as a geometry of a curved two-dimensional space, a geometry of the surface of a sphere. On the sphere, the role of the straight lines is played by geodesic lines similar to meridians, for example, and the fact that parallel meridians which are perpendicular to the equator intersect at the poles surprises no one. It is also not surprising that the sum of the angles in a triangle formed by two meridians and the equator is greater than two right angles.

Now let us imagine a curved space, not two-dimensional but three-dimensional. 'Imagine' here means the exact opposite of the literal meaning of the word. In this instance it means: 'let us reject the visual representation' (or let us rethink the meaning of visual images) and consider the curvature of space to mean the loss of Euclidean and the acquisition of non-Euclidean characteristics. Let us take still another step away from the usual visual image and 'imagine' a curved space-time, a curved four-dimensional continuum. In this instance world-lines will curve, that is, particles will deviate from their straight path and they will be absolutely accelerated. Generally speaking space-time will have different curvatures at different world-points.

Einstein identified *gravity* with the curvature of space-time. In the presence of heavy masses – centers of gravity – space-time deviates from the Euclidean. The measure of such deviation, which is the measure of the curvature of space-time, is the intensity of the gravitational field. This is the fundamental assertion of the theory of general relativity.

In addition to the curvature of space-time due to local fields, space possesses a general curvature just as the surface of the earth, with its local curves of hills, knolls, mountain peaks, possesses a general curvature: the surface of a sphere. From this point of view, independent of the local fields of the planets, stars and galaxies, a body, left to itself, will move in a curve and will return to its point of departure. Such a representation of a finite universe, such a closed model of the universe, was the origin of *relativistic cosmology*. The next step was the hypothesis of the continuous expansion of the universe, the dependence of its radius on time. This supposition of A. A. Fridman was confirmed by Hubble's observations of the receding galaxies.

Let us return to the world-lines, which form the framework of the universe, the objective *ratio* of the world.

Does this framework of four-dimensional lines possess physical being? What distinguishes the presence of a real particle at a given moment in a given place (the real physical event at the world-point) from its space-time localization? What distinguishes real movement from the four-dimensional geometric world-line representation depicting it?

These questions are a new modification of the basic problem of

Descartes's physics. Moreover, they modify the basic problem of classical rationalism if it is viewed from a contemporary perspective. For the present, we will limit ourselves to preliminary, cursory explanations.

Descartes identified space with matter; for him, bodies have no properties except the geometrical. Thus space gained several physical characteristics: its parts were impermeable, they moved relative to other parts, and were material bodies. Nevertheless, as far as space and matter are concerned, the locations of bodies and the bodies themselves represent one and the same thing. It is not possible for us to separate the body from the medium surrounding it, hence the movement of the bodies, their interaction, and their impermeability, represent concepts which in principle lack observable equivalents. The search for such equivalents, and the resolution of problems of being within matter as distinguished from [understanding them through] geometrical concepts, these are the major 'questioning' tendencies of classical rationalism.

And how will contemporary science resolve these problems?

At this point, a short digression is necessary. Quantum mechanics refuses to trace the world-line of a particle, to ascribe a definite space-time localization to it, without an experiment in which the particle interacts with another material body. A thin screen with a small hole can serve as such a body. When the particle passes through this pinhole, its spatial coordinates correspond with the known coordinates of the hole, and time is determined by means of a clock. Another body, then, allows the velocity, momentum and energy of the particle to be recorded. But the interaction has physical meaning only if it alters the behavior of these bodies, their space-time localization, their momentum and energy, i.e., alters their world-lines. The effect of interaction that alters world-lines is substantial and noticeable when we have before us not macroscopic bodies, but light-weight particles, such as electrons. Therefore, this effect was detected at the microscopic level within the framework of atomic physics, since atomic physics explained the structure and properties of the atom by the behavior of light-weight subatomic particles. In this instance how can the world-line of a particle be determined; how can the classical concepts of space-time localization, momentum and energy be applied to the study of the microcosm?

Quantum mechanics resolves this problem in the following manner. It introduces the concept of the macroscopic body of interaction. Such a body is considered to be classical. In relation to it, a kind of classical liberalism is manifested: the macroscopic body is freed from quantum specifications, its atomic structure is not observed, and the changes of the world-lines of the interacting particles comprising this body are ignored. When, for example, a screen with a pinhole is presented, then the absolute immobility of the screen during its interaction with the particle, and the possibility of a precise determination of the coordinates of the pinhole, are postulated. Such an (in principle) macroscopic interaction body (the 'classical device') allows us to determine the coordinates of the particle and the moment of its passage through the pinhole, but during this process the momentum of the particle changes. In its turn the device, which allows us to determine the momentum, changes the coordinates of the particle. As a result, it seems possible to determine one of the variables of the particle in terms of the complete indeterminacy of the other variable.

Quantum mechanics originates from the fact that the particle itself, without the macroscopic body of interaction, possesses neither a determinate space-time localization nor a determinate momentum and energy. It possesses only the probability of certain values of these variables. This probability is measured by *probability waves*; it is possible to determine this probability by using a certain wave equation. The experiment — the interaction of a particle with a macroscopic body by means of a classical device — allows us to go from probability waves to reliable values. But probability waves change in the presence of interaction with the device in such a way that precise determination of the particle's position changes its momentum in an uncontrolled manner and vice versa. The narrower the hole in the screen, that is, the more precisely defined the position of the particle at the moment of passage through the screen, the greater the diffraction of the probability wave, the more the particle deviates and the more uncertain its momentum. Just such a situation describes the indeterminate relation between the time when the experiment takes place and the energy of the particle: the more precisely determined the moment when the particle interacts with the device, the less precisely determined, the more indeterminate the energy of the particle. The reverse is also true: the more precisely determined the energy, the less precisely determined the time.

The uncertainty or indeterminate relations connecting the conjugate variables — position and momentum, time and energy — are deduced from the more general *complementarity principle*. The particle possesses corpuscular characteristics, which are inherent in the particle at each moment, i.e., position, momentum and energy. But the particle possesses wave properties as well. There is a frequency and amplitude of the oscillations for each point in space in which the particle moves, and for each moment of its motion. The wave and corpuscular properties of the particle are complementary. The particles lose their physical meaning if they lose their complementary polar opposite.

Niels Bohr wanted to extend the complementarity principle into other spheres beyond atomic physics. Apparently the complementarity principle can be extended further in time: into the past, becoming the fundamental re-evaluation of classical ideas; and into the future, to be modified by the study of the subnuclear world, the world of elementary particles.

How does the understanding of complementarity change when we pass into the subnuclear world?

Apparently in very small space-time regions (perhaps of the order of 10^{-13} cm and 10^{-24} sec, perhaps even smaller), the representation of the self-identical, continuously moving particles ceases to be elemental, stable and self-evident. Accordingly, the understanding of the world-lines — the continuous succession of world-points, the space-time localizations of the particle — becomes somewhat awkward in these regions, since they come from significantly more extensive regions. The quantum mechanics created in the 1920's had already proven that world-lines were somewhat blurred since, as mentioned previously, the space-time position of the localized particle, its spatial position and the time when it occupies that position, cannot be precisely determined if at the same time the momentum and energy of the particle are determined. Now, not only the determination of the world-line but its continuity as well is threatened. In ultramicroscopic regions the very being of the self-identical particle becomes problematic.

Let us assume that space-time is quantized, that the four-dimensional world-line of the particle is interrupted, in other words, that the particle in the minimal spatial interval (for example, it may be 10^{-13} cm) does not pass from one point to another, which differs infinitesimally from it (according to the coordinates), but instead is annihilated, transformed into a particle of a different type, which is in turn transformed into another particle of an elementary type. Such a regeneration takes place in the space of 10^{-13} cm and in 10^{-24} sec. Thus, within space-time, a time would exist inside of which space and time no longer appear as dynamical variables of the moving particle, and lose their physical being. We observe these elemental regeneration-displacements in macroscopic approximation, identify the particle arising as a result of the regeneration with an elementary particle, ascribe to it self-identity and continuous being, and trace its continuous trajectory. In order that this trajectory not equal zero during the course of some macroscopic time, i.e., in order that the particle move, asymmetry of the probability of the elementary displacements is necessary. If, in each instance, the particle has an equal chance of being displaced to any side during regeneration, then after a large number of similar displacements or chance wanderings, the particle will turn out to be near the point of origin. If, however, there is a tendency toward asymmetry in space, where the displacements to one side have a greater probability than those to the opposite side, then after a large number of elementary displacements, the particle will advance macroscopically for some distance. The macroscopic velocity of the particle is proportional to the degree of asymmetry, which we can identify with the momentum. The greater the asymmetry, the nearer the macroscopic trajectory and the ultramicroscopic trajectory consisting of elementary displacements are to each other. The velocity of the ultramicroscopic trajectory is always the same; it is equal to the elementary displacement made in an elementary interval of time (10^{-13} cm; 10^{-24} sec), i.e., the speed of light. Apparently, the velocity of the macroscopic trajectory cannot exceed this constant. The mass of the particle is the intensity of the symmetry of space, the symmetry of the probability of the elementary displacements. This symmetry can be explained by the influence of a homogeneous metagalaxy on the particles (Kuznetsov 1970b, pp. 413—416; 1966a, pp. 199—209).

The prediction that involved the results of atomic physics, the quantum theory created in 1925—27, subatomic physics and a more general theory [of elementary processes] can be obtained from the scheme cited above or from another scheme. Similar schemes illustrate two interrelated tendencies in nonclassical physics. One of them is the inevitable generalization and modification of the complementarity principle in predicting the future evolution of science. The other is the inevitable generalization of the complementarity principle when viewed from a historical perspective, as we re-evaluate the most fundamental and pervasive ideas from classical antiquity which have been modified but which have not disappeared from the spiritual history of mankind.

Let us take as an illustration the scheme of discrete space-time discussed above. It may seem that this is a question of the replacement of the mechanical picture of the world (mechanical in the most general sense: the building-blocks of the universe are the movements of self-identical bodies) by another *transmutational* picture (the elementary transmutations of particles serve as the building-blocks of the universe). In reality, the scheme of discrete space-time illustrates a more radical transformation of the picture of the world. The very concept of 'the building-blocks of the universe' — the elementary bodies and processes from which the world and its evolution are ultimately formed — becomes extremely problematic. What is transmutation? It is the transformation of a particle of one type into a particle of another type with another mass, charge, and life-time. But all these variables make no sense if there are no world-lines, no macroscopic world of continuously moving self-identical bodies. The mass of the particle and its charge are an indication of its possible world-line, of its behavior in one or another field. We cannot ascribe to elementary transmutations any physical sense other than the transformation of one possible macroscopic world-line into another. Without world-lines, both elementary displacements and concepts of minimal distance and minimal time, as well as concepts of symmetry and asymmetry, lose their meaning.

In their turn, world-lines are a physically meaningless concept without some ultramicroscopic processes which alter, limit or reject continuous macroscopic movement. Without this 'non-Cartesian' supplement, world-lines become a purely geometrical concept and lose their physical meaning. Obviously, complementarity in a very general sense — complementarity of concepts, mutually excluding each other and at the same time losing physical meaning without each other — now becomes the initial definition of being, without which it is difficult, or more likely, entirely impossible to say in which direction science is going with respect to its fundamental conceptions.

It seems that such a paradox — the existence of concepts which exclude each other and lose meaning without each other — represents the most general paradox of knowledge. It is, moreover, the fundamental paradox of being, which has assumed an explicit form in contemporary relativistic quantum physics, but has always existed to a greater or lesser degree, and has been reflected in scientific and philosophical thought.

Let us conceive the basic antinomy of time as the representation of space-time existence and being. All people in all ages have considered the past as no longer existing and the future as not yet existing. But the present — slipping away, zero-order in duration, on the verge between the past and the future — also exists in the passage of zero-order time, or, in other words, does not exist.

In essence, this is at the root of Zeno's paradoxes. Their meaning lies in the fact that in dividing time and space we try to approach the utmost instantaneous being, which already exists but has not yet disappeared, which has appeared from the past already-nonbeing, and has not yet entered the future not-yet-being. Without such being in the zero-order interval of time, the very passage of time loses real meaning. When the moving 'now' is carried away by the flow of time, this flow signifies something real. But if this 'now' is of zero-order duration, we will not find in it anything existing, and this means that the very flow of time ceases to be a real process of the change of something real. 'Now' proves to be a phantom, the arrow does not reach its mark, and Achilles does not overtake the tortoise.

Conceptions of discrete time, put forward as palliatives against the devastation and disappearance of being, have been advanced from very remote ages, thousands of years removed from us. But here Heraclitus's definition of being comes into play: what exists, moves. This definition contradicts the conclusion of Zeno and [his master] Parmenides: what exists is motionless. An atom of time is time when nothing occurs; there is nothing there but abstract duration: a known temporal interval passes. But in essence it does not pass; within its bounds are no processes, nothing that would divide up this elementary interval. Here time is not broken down into nothing, not reduced by Zeno's complete endless division to zero, but is stopped, what's more, compressed into zero.

In order for being to avoid this occurrence, a new representation of localized being comes to the fore, one that is not only Heraclitean but also Archimedean in origin. Precisely because movement never leads to rest, precisely because division is never complete and never arrives at the zero-order barrier, being is retained. This is a differential representation of being. It identifies being with movement in time, with process, and regards the 'now' as a kind of result of the past and as something containing the future within it. 'Now,' local being, is the frozen past and the potential future. It is characterized by statements about potential being derived from time, that is, compressed in an instant, and by the relations of potential processes to the expansion of time. But this also means that the 'now' acquires being while ascribing it to the future, to that future which excludes the 'now,' replacing the given existing 'now' with new, as yet, potential 'nows'.

This antinomy of time is connected with the antinomy of knowledge. The past is no longer influencing the sensory organs; the future is not yet exerting an influence. This is apparently the source of an absolutely local empiricism [of immediate sensed reality]; the sources of local impressions possess real being, but what passes beyond their limits is abstraction, devoid of being. But we may ask whether immediate, local impressions are perhaps the result of the dividing of nonlocalized complexity? Perhaps these nonlocalized complexities are not merely abstract names, but concrete processes?

These nonlocalized, enduring complexities may be grasped by reason. Reason is capable of seizing what is local, prolonging and identifying it, extending the 'now,' and finding that potential being which is the savior and hangman of actual being. Actual being is threatened with transformation into zero, into nothing, if it remains only local and immediate. Reason connects it with the past and the future, and saves the present by means of that which threatens it with destruction. This connection between the past and the future is the objective basis of being; it is the *ratio* of the Universe.

It would be desirable to strip away these purely logical pseudonyms of concrete philosophical and scientific controversies, and to trace these antinomies through the history of science and philosophy, beginning with the evolution of rationalism in the new forms which it adopted as a result of the influence of science. But, as we have seen before, it is impossible to avoid a nonclassical historical perspective. Nonclassical physics has not only made antinomies of the 'now' and potential being, of the individual and the general, the immediate and the prolonged, the concrete and the abstract, by means of fundamental problems without which it is impossible to proceed; it has not only connected their resolution with the perspectives of civilization as a whole, it has also significantly altered or elucidated their meaning.

Let us take the theory of relativity. It has demonstrated the physical emptiness of the absolute, timeless *present*, that momentary photograph of being which at one time seemed to be a representation of reality. Thus the theory of relativity connected epistemology with ontology in this problem: space taken in zero time, the absolute present, absolute simultaneity, do not hide from reason's knowledge of nature, indeed, they do not exist. The *ratio* of the world is not a spatial but a spatio-temporal scheme. It is not purely a spatial scheme of simultaneously existing masses distributed in space, connected with each other by momentary interactions taken in zero time. Rather, it is an arrangement of world-lines, a pattern of movements. Movements do not consist of the simple existence of particles at points in space, and world-lines do not consist only of world-points; they possess a determinate direction (a statement of the velocity of the particles) and curvature (a statement of their acceleration).

Immediate and localized being will be complete only if the particle is not only found at a world-point, i.e., possesses space-time localization, but also possesses these differential predicates: velocity, momentum, and further, acceleration, energy. From the clash of these predicates with space-time localization, from the fact that the interaction of the particle determining its location influences its differential predicates, momentum and energy, it follows that as the world-line becomes less determinate, it acquires physical content and becomes an experimentally recorded event, no longer only a four-dimensional geometrical concept, but also a physical one.

From the perspective of nonclassical science, we see that the genesis of the idea of representing being as contradiction proves to be the fundamental path of rationalism. Accordingly, rationalism is no longer perceived as an attempt at knowledge of the world by means of a logic of identifying. Logical deductive chains (of the type "concept A is in essence concept B, concept B is concept C, etc.,") are invariably accompanied by sensory [experiences]. Empiricism constantly introduces dissonances, contradictions, nonidentity into these chains. But empiricism itself takes part in the evolution of knowledge only by being interwoven into the logical chains.

Rationalism in the traditional sense, accompanied by and interwoven with empiricism, also existed in classical times — but it was hidden. In nonclassical physics it has become explicit. In this sense nonclassical physics unveils the mystery of classical rationalism. But the contemporary Oedipus, revealing the riddles of the classical Sphinx, reveals his own secrets as well, including new lines written on the time-table of his next steps. Therefore the conception of a *tabula rasa*, on which new lines must be written after the old have been eradicated, does not correspond perfectly to the contemporary style and aims of science. Such a conception, close to the representation of a 'history of errors,' stops with the most recently discovered truth. Sometimes it adopts the form of a caricature of a 'nonclassical' conceit, when any child who has read a contemporary textbook is thought to have an exhaustive knowledge of the world, unmatched by the conceptions that satisfied all those poor devils from Aristotle to Newton. The point is that Aristotle and Newton were not all that satisfied with the outlook of their own age. 'Nonclassical' nihilism may seem to be based on a form of dogmatized classical science, resting on its laurels. But this is not how it was; nonclassical nihilism included searching, *aporia* and dissatisfaction with itself. The science of antiquity also included these same factors. Probably the ancient idea of the isotropism of space, the idea of the antipodes not falling 'downwards,' will never be excelled in boldness and paradox. Likewise, in our time, the continuous passage toward ever new paradoxical assertions is characteristic of nonclassical science. Accordingly, the demonstration of the 'historical tradition of anti-traditionalism', passed by ancient science to classical science, and then by it on to nonclassical science, where anti-traditionalism has acquired an incomparably more explicit form, is typical of our nonclassical historical perspective.

4. THE STRUCTURE OF THE BOOK

The structure of this book corresponds to the sequential stages in the evolution of such a tradition.

The first chapters are devoted to the problem of 'being' in the evolution of classical rationalism. Long before the appearance of rationalism as a definite system of epistemological and ontological views, rationalist thought had already clashed with the antithesis between being as immobility and homogeneity, and being as mobility and heterogeneity. Such an antithesis first acquired a distinct character in the conflict between Heraclitus's conceptions and those of the Eleatics, which is examined in the first chapter. The resolution of this clash, far from conclusive, was the substance of the basic tendencies in ancient Greek thought, examined in Chapter 2. Among these tendencies, the atomism of Epicurus is analyzed in more detail and compared with contemporary science. Epicurus's idea that atoms spontaneously deviate from macroscopically determined paths is viewed as the basis of the pervasive conception which continues throughout the entire evolution of philosophical and physical thought. In this discussion, we return to the idea of local, individual 'rebellions' against the macroscopic deterministic regulation of being, to the filling-out of the rational framework of the universe with the substance of ultra-microscopic events. Chapter 3, the comparison of Epicurus with Einstein, includes not only physical, but also moral conceptions, especially in relation to the problem and fear of death, which Epicurus, Spinoza, and Einstein connected with the most general conceptions of the universe. Here the link between cosmic harmony and moral harmony, which is so typical of both classical rationalism and nonclassical science, becomes perceptible.

In Chapter 4, three thinkers of the Renaissance and the Baroque — Leonardo da Vinci, Bruno and Galileo — are analyzed. Here rationalism is connected first with the aesthetic program, the scientific-technical creative works and cosmological conceptions of Leonardo; then, in Bruno, with humanism moving beyond traditional limits, including Bruno's search for a new picture of the world; and finally, with the creation of such a picture by Galileo.

Classical rationalism in the true sense is presented in Chapters 5 and 6, devoted to Descartes and Spinoza. Spinoza's conceptions are considered as a higher stage in the evolution of rationalism, a transition from the rational explanation of the *behavior* of substance to the rational explanation of its being.

The last two chapters of Part I (Chapters 7 and 8) discuss the philosophical results of the two fundamental ideas of the end of the seventeenth century — the mathematical calculus of the infinitely small, and the [physical] idea of force (the dynamism of Newton) — which left their distinctive imprint on both the character of rationalism and the entire science and culture of the eighteenth century, the Age of Reason.

Part II, 'Reason and Movement', opens with the basic idea of Chapters 9 and 10, the differential criteria in the understanding of being. The purely localized statement, 'The particle is located at a given point,' does not yet mention the being of the particle. Alongside this statement is the fact that the particle possesses a velocity (in this particular instance, zero-order) and acceleration. In order for reason to be able to grasp such differential predicates, it must go beyond the limits of the 'here and now,' and represent the future to itself, that is, the potential existence of the particle (at a given moment it is only space and time — 'nonbeing'), and then return to the 'here and now.' In the differential calculus, the contraction of an increment into a point and the achievement of a maximum value of a derivative, corresponds to such a return to the 'here and now.' But such a return has an ontological sense. In order for the particle to have being, potential existence must supplement local existence; being cannot be realized without movement, without nonidentity; movement cannot be realized without that which moves, without a self-identical subject. Accordingly, common-sense judgment, grasping something that is still a static being, must move on to the constructs of reason and grasp the dynamics of being. In order to become the doctrine of the dynamics of being, rationalism must become the platform of reason; not reducible to common sense, reason must become a dialectic.

This part of the book will try to show the connection between classical science and the contemporary nonclassical equivalents of dialectical philosophy — the successors to classical rationalism. To this end (in Chapter 11) I shall first discuss momentum as it was understood by philosophy in the terms of seventeenth- and eighteenth-century classical science; then momentum as reflected in nineteenth-century science, revealing the hierarchy of increasingly complex forms of motion and the irreducibility of complex movements into simpler ones.

The theory of being in Hegelian philosophy connected 'being' with motion. Concepts of coming into being, [existentially] present being, of essence, existence and reality, Hegel's series of ever more concrete and physically substantial categories, represent the next step from the initial abstraction 'pure being,' equal to pure nothing. This series of categories reflects the genuine transition of historically developing reason toward a picture of the world that is increasingly rich in concrete definitions. This was the immanent tendency of classical rationalism. In the nineteenth century this tendency achieved a consistent form in dialectical materialism. It generalized the distinct 'nonclassical' current in classical science — the contradictions which were not resolved within the classical framework and threatened to break this framework apart. Within physical science, these future-oriented tendencies developed to the greatest degree in connection with the principles of least action, the principle of entropy and especially field theory.

Finally, in Chapter 12, the problem of alienation is analyzed in connection with the evolution of rationalism, and also with the content and predictions of the development and possible effects of nonclassical science. Our initial conception is one winnowed from nonclassical science, especially from the conception of a [physical] vacuum, for an understanding of the categories of 'nonbeing,' the Hegelian 'nothing,' the current physical idea of 'annihilation,' and finally alienation and its evolution. These categories are connected with that rupture of the macroscopic with individual existence which has been the initial problem for scientific and philosophical concepts since antiquity. The absolutization of this real division of science into macroscopic and microscopic aspects is evidenced by the absolute and abstract geometrization of science (which Einstein warned against in his 'autobiographical notes'); by the disappearance of physical content from scientific concepts and, on the other hand, by the disappearance of a macroscopic ratio for the universe; by the picture of its chaotic, maximally entropic state [that emerges from] the concept of thermal death; and finally by energetics, as a fictional account of the exclusively macroscopic nature of physical processes. The basic tendency of rationalism to draw close to, and to transform itself into science in its study of being consists in the connection between the macroscopic and microscopic aspects of being, in the demonstrations of their inseparability.

Such inseparability was demonstrated by classical macroscopic thermodynamics, which ceased to be a theory of being after it split off from kinetic theory and completely disregarded molecules, those physical 'negligibles' if one may use Dostoevsky's term, originally far removed from physics. Contemporary physics, in its search for an ultramicroscopic supplement to macroscopic world-lines, demonstrates this inseparability and tries to present it in simple forms. However, this is not reducible to examples, and the problem by no means consists in taking concrete scientific steps under some scheme or other. On the contrary, the entire history of science is a development of the thoroughly concrete dialectical correlation between the existence of the macroscopic *ratio*, embracing the entire universe, and the individual existence of autonomous ultramicroscopic objects.

5. EPISTEMOLOGICAL OPTIMISM

I shall conclude this Introduction with some remarks about the outcomes of a relativistic- and quantum-related historical perspective as applied to the evolution of rationalism and to an analysis of contemporary science. These outcomes lead to epistemological optimism. From the trends of nonclassical physics there stems a change — in the form of well-founded predictions — in its increasingly profound, general and fundamental principles. Insofar as it is possible to foresee its future, science will not enter an 'asymptotic' phase; instead, it will continue to take new tacks, to deviate from traditional paths. On the contrary, the amplitude of oscillations in the 'world-line' of science will increase. But this does not mean that every radical (increasingly radical!) change in science will cancel out its predecessor. Such changes will give to the truths already discovered a new, more profound and precise meaning, will define concretely the conditions of their applicability, and will deduce them from more general principles.

The increase in the amplitude of the oscillations of science, the transformation of its increasingly fundamental principles, will become (and already is becoming) a spontaneously moving force in the transformation of production and all of civilization. This is a transition from epistemological to social optimism. The repudiation of absolute space and absolute time, of the ether, of the constancy of mass, of the sum total of positive assertions in the theory of relativity, has led to the physics of the energy released during the splitting of atomic nuclei, to the use of this energy and to the atomic age. The rejection of classical continuity in the electromagnetic field, and of the classical determinacy of conjugate variables — i.e., the sum total of the conceptions of atomic physics growing out of the principles of quantum mechanics — has led to the possibility of the most radical transformation of the nature of labor. Future, post-atomic civilization will realize the potential of subatomic physics, the physics of elementary particles.

Our age is the dawn of nonclassical civilization. Perhaps at this point another image is appropriate — the image used by Alexei Tolstoy to give a title to his historical novel about the beginning of our revolution. He called it *A Bleak Morning*. Nonclassical science and nonclassical civilization will pass the shroud of clouds which prevent us for the time being from seeing the contours of a unified theory of elementary particles — the ideal of contemporary science which eludes physicists while at the same time luring them after it. There is also another cloud, much more threatening — the prospect of the destructive application of science threatens the very existence of civilization. Among the forces called upon to dispel these clouds is the rationalist spirit of contemporary science. It is strengthened by historical and scientific investigations which reveal in the past, present and future the irreversible movement of reason toward truth. We must not think that this irreversible movement has ended, that reason is [now] yielding its place to irrational feelings, that Logos will be replaced by Eros (as it seems to Herbert Marcuse). Reason itself is becoming much more supple; logical canons are altered depending upon their ontological content. Their physical content has been demonstrated by each fundamental change in classical science, and in nonclassical science this demonstration has become continuous. This has brought about the new dynamism in contemporary civilization. When more or less stable scientific conceptions receive increasingly complete and perfect embodiment, the continuing velocity of the progress of civilization is sustained. If these scientific conceptions are continuously changed in practice, progress will be continuously accelerated; it is precisely this dynamism that is typical of the modern age. If logical canons and the most general principles of science are changed, civilization acquires a still greater dynamism.

This characterization of our time, and these predictions for the coming decades, rest upon the establishment of the radical character of transformations in science; but this does not signify a break with the past; the new science cannot be written on a *tabula rasa*, for its radicalism signifies a revision of the increasingly remote historical sources of scientific and philosophical thought. Not only does such a revision not disparage the dynamic, future-oriented tendencies of the old science, but it also shows the heuristic force of that science which pervades the ages. And now we shall begin with ancient Greek science and philosophy.

NOTE

1 Lenin 1963 (*R*), pp. 325—332; 1961 (*E*), pp. 367—374.

PART I

1. THE HERACLITEAN-ELEATIC CLASH

1. HERACLITUS

The problem of being, throughout their history a problem of classical science and philosophy, and now a critical one in nonclassical science, consists in the transition from the concept 'is' as the copula between subject and predicate ('Bucephalus is a horse'), to the concept of 'is' as an absolute concept ('Bucephalus *is*!'), to a concept which is associated (we will see later to what extent) with the concepts of 'exists,' and 'exists in reality,' etc.

Is the concept of being as belonging to the objective world an abstract concept?

Ethnography and the study of language have accumulated numerous illustrations of the gradual rise, long before the beginning of science, of increasingly general, abstract concepts from those which are more concrete. The Laplanders have 20 names for ice and 41 for snow. The South American Bakairi [an indigenous Amazonian people] did not have a general concept of 'animal.' Other South American tribes had different names for different types of rain, but no general designation. But neither in language nor in thought have there ever been absolutely concrete concepts corresponding to single impressions. Similarly, absolutely general concepts are not connected with individual empirical impressions. These polar opposites have never existed without each other. The evolution of thinking has consisted in the movement between empirical and abstract categories.

At some stage in this process, it was necessary to be able, in principle, to subject abstract categories to empirical verification. Thus immediate or direct impressions ceased to be passive, having been channelled by already formed abstract conceptions. Historically, such actively channelled impressions grew within technology, specifically in *developing* technology, originating as they did not in tradition, but in the empirically observed actualizations of some general conceptions about the course of events in the external world. Later, this kind of actualization became independent, and cognitive problems became its goal; it became experiment. But the acceptance in principle of the connection between abstract schemata and empirical impressions was expressed much earlier. In this sense, the criterion of *external confirmation* lay at the origin of science (although of course in a very implicit, undeveloped, primordial form).

But at the same time, empirical impressions obviously became impossible without some kind of general categories or, at least, association. Since the time of Helmholtz, it has been widely known that empirical impressions are impossible without preliminary general conceptions, categories or associations. As an example let us take a well-known ethnographical report: Darwin's story about the aborigines who paid no attention to the great ship moored at their island, but stared in astonishment at the launches which they somehow associated with their canoes.

Such a link between separate empirical impressions and general conceptions existed in very ancient times. General conceptions, which to a certain extent have lost their link with the empirical, are typical for the later stage of the genesis of science. The connection did not disappear, but it became less explicit. At the same time science required that the results of empirical observations stem from more general conceptions, and that limited generalizations derive from more fundamental ones. Thus the criterion of inner perfection, in some primitive form or other, existed at the beginning of science.

From a combination of the initial forms of the two criteria — external confirmation and inner perfection — arose the problem of being, the concept of 'is' without indication of a predicate. Or rather, the predicate of objective existence appeared, and 'is' came to mean: There is something really existing, as distinguished from something illusory. Each predicate 'is' signifies the entry of a given individual into an ensemble of other such individuals possessing the same predicate. The absence of a predicate after 'is' signifies the entry of the individual into a very general ensemble, the ensemble of the objective world, of real being, of the Universe. This is not yet a resolution of the problem of being; it is only a program for its resolution. It is the program of science, the program of the *scientific* resolution of the problem of being — if we are dealing with the ensemble of the external world: something encompassing being as a whole, but differentiated, having separate elements, something capable of being the object of both differentiating experiments and empirical observation, in other words, something of not only the utmost abstraction, but the utmost concreteness as well. The problem of being, as a problem of the 'is' without indication of a predicate, means the problem of inclusion, not in an abstract set defined by the *identity* of the elements included in it, but in a concrete set defined not only by identity but also by nonidentity, individuality, by the opposition of the elements included in it.

This definition of a program for the scientific resolution of the problem of being seems modern. To some degree it is. In our Introduction, scientific ideas of the past were discussed as questions addressed to our time, as the questioning, programmatic side of science. The program always includes what does not yet exist but can follow from it. In science, the future flows from the heterogeneity of the past. Therefore modernity signifies not the identity of ideas of the past with contemporary ideas, but nonidentity, heterogeneity, the openness of the past.

The program for the scientific resolution of the problem of being originated from the unity and diversity of nature. Of what does this unity consist? What comprises the immutable substratum which remains the same through time and space? What is the subject of spatial distinctions in nature, and its changes in time? The first answers to these questions were given by Eastern civilizations and later, in a considerably more developed and distinct form, by the Ionian colonies of Greece. Thales considered water to be the prevailing, self-identical substance of the world; Anaximander — something more abstract: *ἄπειρον*, capable of movement and qualitative changes; Anaximenes — air; Heraclitus — fire.

The conception of Heraclitus differs essentially from the conceptions of Thales and Anaximenes. Fire, as the substance of the world, was by no means some new concrete matter, following water and air, as claimant to the title of 'substance' — the prevailing substratum of changes. For the Greeks, fire was not only, and not even so much, matter as the universal agent of the destruction of matter, the real and concrete bearer of the destruction of any given matter. Indeed, it was the bearer of transformation, of the destruction of every stable form of matter; it was considered matter enduring and self-identical, modifying but disappearing.

Heraclitus's conception was a very unexpected and radical turning-point in ancient thought with regard to the definition of being. The concept of being is diametrically opposed to the concept of nonbeing, destruction, change. The Eleatics saved being by rejecting nonbeing, by rejecting changes in time, and by rejecting movement and variability in space. Heraclitus expressed an idea no less paradoxical: change itself, active nonbeing itself, becomes being. There is nothing enduring, conserved, immortal save *mors immortalis* and the process of change itself, the violation of identity.

This is one of the chords of that overture — ancient thought as a whole — to the entire future development of science and philosophy. Heraclitus's conception is one chord which later became a powerful and resounding melody. It is like a rehearsal for the transition from the static harmony of being to dynamic harmony, an event of the seventeenth century. When Galileo declared motion [velocity] to be an immutable state and reduced all changes to acceleration, it was a repetition of the Heraclitean dynamic being, only more restricted: motion here was understood in the sense of Aristotelian localized motion, displacement. Galileo limited the problem to the behavior of substance; Heraclitus was referring to the very being of the world. Heraclitus's being is guaranteed by change, by the world-process, or rather, being is itself the world-process.

2. PARMENIDES

Compared to Heraclitus's conception, Parmenides's conception, for all its explicit paradox and obvious divergence from experience, seems simpler and less unexpected. Motion, and change in general, contradict the concept of being; therefore Parmenides ascribes to being immutability and homogeneity — absolute self-identity in time and space. Being is indivisible and immobile. As for apparent motion, with the introduction of the concept of substance, ancient Greek thought rejected naive faith in the truth of empirical data. Hegel cites the story of Diogenes of Sinope who, replying to Parmenides's denial of movement, began to walk. But this story ends with an unexpected twist: when Diogenes's student agreed with this argument, Diogenes began to beat him: in logical analysis it is impossible to be satisfied with reliance on the senses.[1]

In Eleatic philosophy, Greek thought confronted the *aporias* of movement, variability, and heterogeneity. But simultaneously it clashed with another *aporia* — the aporia of identity. The *aporias* of Zeno are the *aporias* of nonidentity. Zeno cited them in order to draw a negative conclusion: there is no motion. The *aporias* of identity are what were hidden in the *positive* assertions of the Eleatics.

One of the Eleatics' most fundamental positive assertions — that being is identity, homogeneity and immutability — has quite serious, real roots. Reason perceives in nature that which corresponds to its logical functions: the identification and inclusion of individual impressions and private conceptions in unifying sets of identified elements whose common factor is substance. If Heraclitus went from concrete matter to the very process of the transformation of one type of matter into another and made change itself into substance, Parmenides did the opposite. He viewed invariability, identity and homogeneity as substance itself. In fact, if only variability, identity and homogeneity exist, then the subject of these predicates disappears; there is nothing to which identity and invariability can be ascribed. If all moments are filled with the same content, if they are in this sense identical, then time is gathered into one non-extended moment. If space does not possess heterogeneous content, it is gathered into one non-extended point. Thus the subject of invariability, the subject of identity, disappears.

Parmenides's concept is a mirror image of Heraclitus's. An analogous danger — the annihilation of the subject of movement, and the suicide of movement itself — lay in wait for Heraclitus. If *everything* changes, then we approach change without that which changes, without the immutable which is the subject of change. If everything is conserved, then we approach an analogous suicide, devastation, annihilation: the subject of invariability, which must change and cannot exist without change, itself disappears.[2]

Thus the Heraclitean-Eleatic clash — the conflict between the two components of being — is a conflict between the invariant substratum and the changing predicates, a conflict which has permeated science and philosophy. The crux of the matter is conflict. Identity is the negation of difference; differences are the negation of identity. If one examines the philosophical schools of the past from their questioning, 'programming' side, the school of Heraclitus (if one existed) and the school of Parmenides expressed not only conflict but also the indivisibility of the components of being. Heraclitus's philosophy resounds with the future-oriented question about the invariant substratum of change, Parmenides's, with the question of nonidentity. In the case of nonidentity, which is denied by Parmenides's philosophy, the question does not arise from philosophy but is more likely a cry for help from being, which the philosophical schools burst apart by isolating its unisolatable poles.

3. DEMOCRITUS

Help came quickly; the answers to the unexpressed question were heard very soon. If one does not consider the negative answer of the Sophists — that authentic knowledge of being is impossible — the subsequent answers contained a new conception of identity and nonidentity, which we shall examine below. Let us simply point out now that the Heraclitean-Eleatic clash and the attempts to resolve it included epistemological problems closely linked to the conception of being. The question of the relationship between reason and being was by no means the subject matter of the new period inaugurated by Socrates, as is sometimes maintained. In essence, reason and being, epistemology and ontology, never radically parted from each other during the entire history of science and philosophy. The search for the self-identifying substratum of the universe stemmed from the identifying function of the understanding; the search for a world process, for heterogeneity in the world, stemmed from that which distinguishes reason from understanding, that is, from the individualizing function of reason. The dialogue between 'a' and 'the' entered epistemology and was projected onto ontology. Descartes's classical rationalism sought the objective harmony in nature which would correspond to the *clarity* of reason and found it in the scheme of moving parts in a homogeneous material that is indistinguishable from space. Descartes sought the precision characteristic of reason, while trying to individualize the elements of being. But in antiquity there already existed a series of attempts, other than the Heraclitean-Eleatic confrontation, to find the objective equivalent of reason, the objective *ratio* of the world. Among them were the atomism of Democritus, the philosophy of Plato, and the system of Aristotle. These schools may be seen as fundamental attempts at a way out of the clash between identity and heterogeneity and, in addition, as fundamental means for bringing reason and being together, ontologizing rationalism. The fate of these attempts varied: the proposed routes divided into narrower paths, many of which ended abruptly as they turned out to be blind alleys. But the basic trends remained.

Democritus's atomism replaced the absolute identity of substance with another, more flexible identity. For the Eleatics, movement, change and heterogeneity were entirely illusory; the question was one of the immutable substratum of absent, zero-order changes, of *trivial self-identity*. In essence Leucippus and Democritus clearly expressed the concept of *nontrivial self-identity*: the subject possesses various predicates which exclude each other but, nevertheless, it remains identical to itself. At different moments an atom has different positions in space, but it is still the same atom. If one unites position and moment into one concept of a world-point, i.e., a space-time localization, then the motion of the atom represents a continuous series of non-coinciding space-time localizations, [or] varying world-points. If we consider only space, then the motion of the atom is a continuous series of non-coinciding positions. At each given moment the atom possesses one real position and an infinite set of possible positions which comprise space, i.e., the Democritean 'nonbeing.' The world is broken up into 'being' and 'nonbeing.' Self-identical atoms, possessing substantial, immutable predicates, which are conserved during a change in spatial position, form the Democritean 'being.' The changing real and possible positions of atoms are Democritean 'nonbeing,' empty space.

Which are the immutable predicates of being? Which predicates are the ones that permit the identification of the atom, its recognition at another, different point, which guarantees its self-identity? It is possible to identify an atom located at one point with the atom appearing at another point if the shift from one point to another is simply determined and regular. The genesis of atomism coincided with the genesis of the conviction that it was possible to determine motion, that there existed some law or other which connected one spatial position of an atom with another and allowed the prediction of the atom's subsequent movement. The subsequent development of science was, to a considerable degree, a series of answers to the question of the character of such laws. Each answer is expressed in terms of the retention of some predicate. When an atom is in motion, it changes its position, that is, its distance from other atoms changes, time changes, its world-point (its location in time and space) changes. But at each point its velocity, i.e., the ratio of the possible space [distance] through which the atom must pass to the possible time interval, is constant. If the velocity changes, another law, the law of constant acceleration, is in effect. If acceleration changes, the mass of the atom is conserved, [constant], etc. These constant quantities determine some possible *world-line* of the particle, the aggregate of its world-points, its space-time localizations. The particle is distinguished from its space-time localization because it possesses velocity, energy, mass, charge. . . . But all these predicates may be determined by the possible world-lines of the particle and they lose their meaning without the space-time continuum, without 'nonbeing.' On the other hand, we have seen that 'nonbeing' also represents an aggregate of possible, changing predicates of being; space enters the physical picture of the world as an aggregate of the possible positions of the atom.

At this point it is necessary to pause and to postpone for a while further analysis of 'being' as it is distinguished from and related to 'nonbeing.' As it is, we have skipped far ahead and have almost ascribed to Democritus and Epicurus concepts, including those of world-points and world-lines, which appeared more than two thousand years later. It must be said that in such a *rapprochement* there is not a trace of anachronistic modernization. The historical being of an idea is determined by its possible development just as the physical being of a particle is comprised of its possible world-line. In this case the issue is not the *rapprochement* of the positive content of ancient atomism with the positive content of the ideas of modern science, but is rather that of a question echoing over two thousand years, until an answer, also incomplete, re-addressed the question to posterity . . .

In this sense, ancient atomism was the prototype of rationalism — an undifferentiated prototype where the motifs, which were later differentiated, while retaining their genetic and logical connection, merged and crossed over each other. The world of moving atoms is accessible to reason; reason identifies empirical impressions, groups them into classes, and introduces general concepts within which individual differences are lost. Comprehension of the world is impossible without such identification to introduce order into the chaos of immediate impressions. But when reason identifies and orders it does not go beyond the bounds of what in the nineteenth century was called common sense. Reason in the true sense, when its function is not reducible to common sense, takes individual differences into account, which fracture and deform old identities and differentiations, modify old concepts, and render rationalism inseparable from its sensual accompaniment. This complex structure of a real rationalism which grasps reality is already found in a basic form in ancient atomism — in the form of an atom which moves, but which remains identical to itself.

Let us look more closely at the idea of space as an aggregate of the possible positions of atoms. Plato in the *Sophist* says that "each thing is what it is and is not what it is not." Such a formula, in the words of Bergson, signifies the plurality of the elements of being (". . . il n'y a pas . . . l'être . . . mais . . . êtres" (cf. Hyppolite 1953, p. 144)).[3] This formula really includes the negation of other elements of being during the process of determining the given, and accordingly, the negation of the given object during the process of pointing out other objects. As Spinoza says, *determinatio est negatio*. But there is an even more subtle thought, which is connected with the non-trivial self-identity of the atom. The spatial determination of a given object (in this instance, the atom) includes its limitations, its reference to those points in space where the atom is not *now*. But this is not a reference to an absolute nothing, but to some nothing which can become *something*, that is, the predicate of the atom. The interpretation of empty space as unoccupied 'nows' but possibly capable of being filled with spatial predicates derives from the concept of movement. A spatial point (or rather, a spatial volume) is the *spatial* determination of the atom 'here' without the temporal determination of 'now.' The atom is *here*, but it is not here *now*.

The idea of space as a heterogeneity of possible positions of an atom allows us to examine nonbeing as something real, as an element in the picture of the world, as something guaranteeing the reality of movement, change, heterogeneity in nature.

In this way, thought becomes content; rationalism (in its early, embryonic form) becomes ontological. Today when rationalism has taken the form of nonclassical science, it is easy for us to maintain that to *think* about an object means to *repeat* some definition, and not to think about an *object* means not to *repeat* its definition. We now know that thought rich in content is not a tautology, that the influence of the object of thought on the sensory organs can contradict the repeated definition, can be paradoxical, can provoke a 'flight from paradox,' change, a modification of the repeated [accepted] definition. But all this was found in classical science as well, which knew the prototypes of Einstein's 'flight from paradox'; but for classical science the intervention of heterogeneity and nonidentity seemed less a physical statement than a philosophical one. The intervention was in fact philosophical, and as such it proceeded from ancient thought, which included heterogeneity, nonidentity and, ultimately, nonbeing in its picture of the world.

The question of whether nonbeing could be included in the definition of an atom as a possible spatial predicate later received a number of positive answers. One such answer was Faraday's concept of the field: the possible behavior of the particle at a point where it was absent was viewed as something actual, as the state of some medium already there. Another such answer was the principle of relativity; the position of a body is determined not simply by distance, but also by spatial distance as a projection of a world-line, that is, its possible movement. Still another such answer, even more complex and paradoxical, far removed from the question raised, was the quantum theory: here, the possible presence of the particle becomes a *probable* presence, with a determinate meaning for the 'probability of presence' for each point. The interpretation of Democritean *nonbeing* as possible being reveals these historical and logical connections.

The differentiation of being and nonbeing in Democritean atomism refers to the transformation of *being* from the copula ('Bucephalus *is* a horse') into being without indication of a predicate ('Bucephalus *is*!'). The absence of a specific predicate means that the subject is granted entry into the most general physical set, the objective world. This set is the most abstract and, at the same time, the most concrete; it has an infinite wealth of determinations. Nonbeing, the void, allows us to differentiate the separate elements of the set and to grasp the richness of its determinations.

4. PLATO

Let us now turn to another attempt to resolve the Heraclitean-Eleatic confrontation.

Plato, like Parmenides, unites the concept of being with the concept of invariance of duration, immutability. If one takes a number of concrete sensations, individual images and individual objects, and transforms them into a general concept, what is general, what is characteristic of all the individual elements of the group comes to the fore. We transform a concrete group into a general idea. This unified, self-identical idea, is embodied in individual images. In such a self-identical idea, Plato sees the abiding, the truly existing. He sees being in it. The conception of a self-identical and therefore real, actually existing idea is one of Plato's most fundamental notions.

But Plato, unlike Parmenides, does not consider nonidentical, transient, individual images as simple illusions. No, the images exist, but their existence is on a lower level; they are devoid of substantial being; they are only embodiments of an idea. This hierarchy of reality, this concept of the embodiment of abiding substance in concrete images, makes substance active and creative. Later, the Platonic real world of abiding ideas was transformed into a world of concrete, but non-extended, spirits which are not intelligible to the senses. The neo-Platonists replaced the creative activity of lasting Platonic ideas with the emanation and degradation of the spirit, with the descent of the *Logos* from God to the world. They moved from the philosophy of general concepts to religious demonology.

But for Plato himself, the point of departure for philosophical constructs is the world, the fundamental confrontation of the identical and lasting on the one hand, and the nonidentical and transient on the other. Plato replaced the ascent from the real, mutable and heterogeneous world fabric to abstract, immobile concepts, with the reverse process – the embodiment of these concepts in concrete images. We are dealing primarily with geometrical concepts. Here, at the sources of scientific thought, we encounter a differentiation remarkably like the confrontations within modern science. This is a differentiation between the geometrical and the physical image. Geometry identifies diverse physical objects and transforms them into geometrical, homogeneous spatial concepts. Contemporary science ascribes being to physical objects and seeks predicates that distinguish them from nonbeing, from geometry, that distinguish the particle from the world-point. This is the tradition of Democritus. Plato does not stop at ascribing being to geometrical images but carries the ascription of reality further, to still more abstract, nonspatial objects. Later this line hardens, loses its ancient flexibility and is infused into the mysticism of the Neoplatonists.

5. ARISTOTLE

The transition from the philosophy of Plato to the philosophy of Aristotle is a logical one. Aristotle started with problems which could not be resolved within the conceptual framework of ideas as the only reality. He extended the limits of being, bringing it under the influence of the internal logic of philosophical thought. But even the purely historical sources of the transition are now presented comparatively clearly. Aristotle seemed extremely sensual to the Platonists; and so he was, in comparison with Plato. They both were true Hellenes and their work was to a considerable extent a work of art. But in the style of his work Plato is more like a sculptor – he comprehends the world in its forms; while Aristotle, more like a painter, comprehends the world in its colors.

For Aristotle's generation, regardless of how little it differed from Plato's generation, the world was represented considerably more colorfully and dynamically. Plato's generation lived during a comparatively stable period when resources were only being accumulated for the conquest of the universe. This conquest began during Aristotle's life, and was accompanied by widespread, organized studies of nature and man. Not only the staff of Alexander the Great and his followers, but all of Greece as well received new information and learned new facts illustrating the heterogeneity and creative power of nature. While the earlier period is characterized by a very powerful intellectual drive directed toward the logical ordering of the world, the age of Alexander the Great is characterized by an even more powerful drive to embrace the flow of new information that stimulated a sensual interest in the world and an understanding of its many colors. This sensual and evolutionary tendency, recording the evolution of life and its forms, is typical of the new period.

During this period, the edifice of Aristotle's philosophy appeared: it was startling in its grasp of the entire universe and even more startling in its vital connection with experience, in its sensual perception of nature. Initially completely nondogmatic, this doctrine was later the most dogmatized. A powerful current of living sensualism forces its way through it: living, but by no means turned away from the logical comprehension of world harmony.

However, even Plato, whose drive toward the logical ordering of the universe arrived at the conception of general ideas as the real framework of being, did not really break with the perception of the concrete, the multicolored, the individual, so typical of Greek genius. "For a genius of an artist in a race of artists," writes Brunschvicg, "no special effort is needed to succeed in bringing nature to life and deifying it; this genius and this race see directly the soul and godhead in nature."[4] But what was temperament for Plato became doctrine for Aristotle. Aristotle was to a considerable degree a sensualist, but his sensualism embraced *everything*, the entire Universe. He eliminated from the picture of the world whatever could not in principle be the object of sensory perception (infinite and empty space); but everything that can be understood by the senses is represented as a coherent whole, a single system. It is fundamentally static; the universe is founded on the stationary framework of the center of the Universe, its boundaries, and the natural locations of bodies. But whatever the general plan of the universe, it remained comprehensible to the senses, and represented the *ratio* of the world, its objective reason and its logic. "Plato," says Gilson, "was at once a philosopher and an artist; he spoke concretely about abstract concepts. Aristotle, at least insofar as we know him, was not an artist, but rather a philosopher and a scholar; he spoke abstractly about concrete things."[5]

Perhaps it would be more precise to say that for Aristotle, abstract analysis and concrete analysis of being demonstrate their connection more distinctly than for Plato. This connection, uniting rationalism and science, reflects the objective link which is abstractly and concretely grasped in nature. For the medieval Aristotelians, this vital, flexible connection became rigid, devoid of nuances and shades of meaning. The picture of the world, as it appeared through direct sensory perception, was identified with reality. "The Aristotelian physics of the scholastics was entirely based upon the following hypothesis: the child's universe is the real Universe."[6]

To a certain extent, this idea comes from Aristotle himself. But for Aristotle the child's perception of reality had a different meaning than for the scholastics. In ancient philosophy, the perception of the world through the eyes of a child meant not just naiveté but a freshness of thinking that takes in all of nature at first glance, a perception that has already become a conception but has not yet ceased to be a glance. The infantile quality of medieval Aristotelianism is the naiveté of childhood become the tradition of old age.

That conception, which is still a glance, that idea which has not yet lost its link to a direct, spontaneous vision of the world, is Aristotle's conception of form and matter. For Plato, form, order, identity and geometry are opposed to matter, and the world is transformed from chaos into cosmos, in a repudiation of itself, its [spatial] extension and its ability to act upon the sensory organs. For Aristotle, the world becomes an ordered, formed world as a result of the evolution of matter, of the fact that force, *δύναμις*[7] ([capable of existing or acting, potentiality, hence] a propensity to achieve a certain form), is characteristic of matter. Matter contains potential form. Here at last we encounter the concept of potential being which plays such a major, although for the most part implicit, role in classical philosophy and nonclassical science. Matter exists because it exists dynamically [virtually] (*δύναμειον*) that is, the *possibility* of form, structure, a hierarchy of structure, and development is defined in it.

It should be emphasized that for Aristotle, matter contains only the possibility of form. In particular, in this sense of the Aristotelian concept of force (*δύναμις*), form is opposed to matter as reality: in this way the meaning of energy (*ἐνέργεια* [actuality]) is attributed to form. Both concepts referred to here are extremely fluid and fluctuating, reflecting the characteristic plasticity of Greek thought. But in them, as in other recent concepts, are found the beginnings of a very deep and pervasive tendency which acquired an explicit physical form in non-classical science. Let us turn to this contemporary form.

Let us take a certain volume of matter (in the contemporary sense), for example, a certain volume of a gas. In itself it contains only possibilities of temperature differentials, of macroscopic structure. All this is created by external forces, by the activity of an enveloping system, of a Whole. In quantum mechanics the very presence of particles at given points is determined by a macroscopic device (at the expense of the indeterminability of other variables).

For Plato, the macroscopic structure of being was hidden under a distinctly non-physical concept of forms existing outside of matter. Aristotle seeks a link between the Whole and the localized — that is, he is searching. For it is precisely in searching, in uncertainty, in the unfixed character of such categories as *δύναμις* and *ἐνέργεια* that we find that which links Aristotle's conception with the future.

The historical evaluation of a philosophical and scientific idea includes not only a *rapprochement* (including the *rapprochement* of question and answer), but also differentiations, a statement of inadequacy, the absence of necessary concepts and information in the ancient prototype of the modern conception. Here, as everywhere, *determinatio est negatio*. In the Arisfotelian conception of form and matter, the microscopic mechanism of shaping and forming is lacking. Here lies the reason for its plasticity, and moreover, for the conception of an expedient influence of form on matter, of entelechy (*ἐντελέχια* [full, complete reality]). Aristotle (like all ancient scientists) does not know about the mechanism of the interaction of the Whole with its local elements. He does not know that form is coded. This neologism is appropriate here because the question is the *absence* of this modern concept. Aristotle does not know about the internal mechanism leading to the transformation of the chaos of individual microscopic (or better, ultramicroscopic) objects and processes within the ordered world of macroscopic forms. The search for such a mechanism and its subsequent discovery are in fact the fundamental goals of our classical science, distinguishing it in a fundamental way from Aristotelianism. Atomism was the most important ancient source of the subsequent search for the microscopic mechanism of that developmental process which, for Aristotle, was connected with concepts of form as 'energy' and of matter as 'force.'

It must be pointed out that Aristotle's ἐνέργεια was not as close to the modern concept of energy as it was to the entirely new concept of *negative entropy*, or to the even newer concept of *information*. Negative entropy is the negative magnitude of entropy, that is, chaos, disorder, the absence of macroscopic transitions, processes and objects. Negative entropy is the measure of the macroscopic ordering of being, the measure of the transformation of chaos into cosmos. For Aristotle the idea of evolution is the idea of increasing negative entropy, the structure in the world which man comprehends in increasingly complex and sublime forms. And precisely because Aristotle lacked a conception of the mechanism of evolution, he discovered the road to the concept of purpose as the moving force of evolution and he travelled, although at times unsurely, along this route.

Evolution itself ceases when we are dealing with the plan of the Universe as a whole. The motionless scheme of natural locations defines the movement of bodies only under limited conditions. Movement for Aristotle was always from something to something. In contrast to the classical differential concept of movement — from point to point and from moment to moment — Aristotle knows the initial form of the concept of continuous movement, which existed in ancient Greece and became best known in the illustration of the contradictions between the continuous and the discontinuous. But it was easy for Aristotle to avoid the *aporia* of Zeno because he essentially lacked [an understanding of] both the continuous and the discrete mechanisms that maintain movement. Aristotle speaks of continuous matter and its continuous movement, but these are negative definitions; they signify only that Aristotle's physics does not examine the microstructure of matter and movement — at a moment and at a point, nothing takes place; differential concepts entered science much later.

What does Aristotle's conception mean when contrasted with the viewpoints of Plato and the Pre-Socratics? What does it mean for the problem of identity and non-identity — the basic problem at the root of the Heraclitean-Eleatic confrontation? And what does it mean for the problem of matter and space to which it is connected?

Infinite empty space is identity which goes beyond the limits of being. Thought that identifies considers the possible location of a body as real, as a location of the body that exists in the absence of the body. From here there are two paths: reality as ascribed to thought itself and only to it (Plato), or space considered as a reality independent of thought and also independent of bodies. Aristotle chooses a third path: space is real, but it is inseparable from bodies, it is filled and finite.

NOTES

[1] See Lenin 1963 (*R*), p. 230; 1961 (*E*), pp. 256—257.

[2] ". . . si tout est identique, rien n'existe" [If everything is identical, then nothing exists], writes Émile Meyerson (Meyerson 1921, **II**, p. 347). We shall soon return to Meyerson's well-known views on the concepts of identity and nonidentity and to their link with the evolution of rationalism.

[3] For an analysis of Plato's thesis and Bergson's commentaries, see J. Hyppolite 1953, pp. 139—146.

[4] Brunschvicg 1949, p. 50: "Pour un génie d'artiste, chez un peuple d'artistes, il n'y a pas un effort spécial à faire pour aboutir à l'animation, à la divinisation de la nature; ce génie et ce peuple voient directement l'âme et la divinité dans la nature."

[5] Gilson 1948, p. 46: "Platon était à la fois un philosophe et un artiste: il parlait concrètement de notions abstraites. Aristote, tel du moins que nous le connaissons, n'était pas un artiste, mais plutôt un philosophe et un savant: il parlait abstraitement de choses concrètes.

[6] Gilson 1951, p. 170: ". . . la physique aristotélicienne de la scolastique repose tout entière sur cette hypothèse que l'univers de l'enfant est l'univers réel . . ."

[7] [In Western philosophical parlance, *δύναμις* is almost always translated as 'potentiality', *ἐνέργεια* as 'actuality', and *ἐντελέχια* as 'entelechy'. In this and following passages, the author translates these terms variously into Russian, and his variety has been retained in the English. — Eds.]

2. PARADOXES OF BEING

1. PARADOXES OF IDENTITY

Between the 6th and 4th centuries B.C., Greek philosophy arrived at a series of paradoxes, the best known of which are the paradoxes of Zeno mentioned above ("the arrow does not reach its mark . . ." etc.), and the paradoxes of Epimenides ("all Cretans are liars . . .") which will be discussed below. To what do these paradoxes attest — to the paradoxicality of the construction of knowledge or to the paradoxicality of being?

This is not only a fundamental question in the history of ancient philosophy. It is also the beginning of a pervasive and pivotal problem in the entire history of philosophical and scientific thought, namely an epistemological problem: the problem of common sense ordering the world, stating contradictions, and moving on to the new ordering given by reason. It is, furthermore, the beginning of an ontological problem, the problem of paradoxicality and contradictoriness in the objective world. If we are faced with paradoxes of being, if statements of the paradoxes of knowledge possess ontological value, if epistemology is inseparable from ontology, then the evolution of paradoxes becomes the most important subject matter in the history of science and philosophy.

The initial paradox of being has already been examined in the introductory chapter: the past is no longer, the future is not yet, the present is the boundary, of zero-order duration, between the past and future, the flow of time transforms the not-yet-existing into the no-longer-existing, and real existence, the non-extended moment when such a transformation takes place, is illusory as well.

Aristotle states this thesis in a very precise form: in comparison, similar views expressed earlier now seem unclear and admitting of different interpretations. But for Aristotle, as well as for those who preceded and followed him (e.g., St. Augustine), the initial paradox of being by no means led to the denial of being. It was a paradox because it contradicted the unquestionable reality of being, that is, it did not lead to the denial of being, but to the statement of its contradictoriness, to the statement of an objective, ontological paradox. The words 'did not lead' and 'led' are somewhat ambiguous here: the first ('did not lead') refers to the nature of the views of Parmenides, Aristotle, St. Augustine, *et al.*, while the second ('led') characterizes the evolution of philosophical thought as a whole. The philosophers of antiquity overcame the difficulty in different ways. Basically two possibilities were already known in the 6th and 5th centuries B.C. They have already been mentioned: the Heraclitean-Eleatic clash demonstrated such possibilities. Parmenides avoided the problem of the zero-order duration of the 'here and now' by identifying the localized being of 'now' with the past and future, with 'earlier' and 'later,' that is, by rejecting the very concept of time in its pervasively destructive and changing function, which is hostile to being. He also rejected the concept of space as heterogeneity, denying spatial distinctions in the world, and identified 'here' with 'in another place.' On the other hand, Heraclitus identified the very change now taking place with the past and future. In essence — in the form of a qualitative guess — he did roughly what seventeenth-century science did: for the self-identical natural *locations* in the physics of Aristotle he substituted the self-identical *velocities* of the mechanics of Galileo and Descartes as the rational scheme of the universe.

Such an identification saves the 'here and now' from the spectre of nonbeing, from the threat of the no longer existing past, the not yet existing future and the present of zero-order duration. The present enters the past and future; in thus moving beyond its limits, in becoming externalized, in identifying with the past and the future and in denying change, it acquires real, enduring being.

But the danger still exists. The paradox of nonidentity, change and heterogeneity is replaced by the paradox of identity. In his many historical and scientific works Émile Meyerson wrote that in rendering those states identical, which occur at different times in the world, time is compressed into a moment of zero duration. If, in a series of consecutively occurring moments, moment t_1 is identical to moment t_2 ("The universe without events," as Ivan Karamazov was told by the devil who visited him),[1] then nothing happens in the world and time itself stops. The same thing happens with space: the absence of spatial distinctions compresses space into a non-extended point.

But 'events,' nonidentity, the distinction between moments t_1 and t_2 or between points x_1 and x_2, lose all meaning without the concept of identity. This was already discussed in the example of transmutation of particles: the concept of nonidentity loses meaning without the concept of the world-line of the self-identical particle, because to differentiate a particle of one type from a particle of another means to differentiate one potential world-line from another.

While nonclassical science does so practically continuously, classical science did not resolve the paradoxes of identity at its radical turning point either by revealing variation, heterogeneity, nonidentity, or by arriving at new schemes of identity (from the Aristotelian identity of locations, to the Galilean identity of velocities, the Newtonian identity of accelerations, the identity of masses, and then the identity of rest masses and so forth). Already in ancient science the inevitable clash of identity and nonidentity led to new paradoxes.

2. PARADOXES OF INFINITY

As we have seen, one of the things emerging from the Heraclitean-Eleatic clash was the idea of motion in its simplest form (as it seemed then and even later, right up to quantum mechanics): the form of mechanical displacement, Aristotelian *φορά*, local movement. The subject of movement is self-identical in all its predicates except for its space-time location. Thus at this point the externalization of the 'here and now' rests upon the identity of predicates other than space-time localization. This identity guarantees the self-identity of the moving body. Zeno wanted to transfer the change in space-time predicates, the motions of a body, into the sphere of what is sensually comprehensible, but in essence illusory. Zeno's paradoxes achieve this end.

Zeno's paradoxes are paradoxes of infinity, the infinite division of the path of the perpetually moving body. The paradox of identity, the clash between identity and nonidentity, was resolved (by no means conclusively, however) by the concept of nontrivial self-identity: the subject of motion changes its localization but remains identical to itself. This self-identity is guaranteed by the continuity of change in localization, continuity of motion, the impossibility of reaching the end by a division of the path of the body into ever smaller segments. Therefore concepts expressing the static aspect of being, the trivial physical self-identity of the object, the identity of the localization of the object (the same point, the same moment) lead to the impossibility of understanding motion, of expressing it in *concepts*. Zeno's paradoxes prove nothing else. However strange it may sound, Zeno's paradoxes are, in essence, arguments *against* Parmenides' conception of being. The negation of movement, the negation of temporal and spatial heterogeneity, mean that space is compressed into a non-extended point and time into a non-extended moment, that the 'here and now' is self-identical, and that the Parmenidean element of being in its purest form does not exist.

Zeno's first paradox, the paradox of measurement, demonstrates the impossibility of forming an extended quantity from the nonextended 'here and now.' Other *aporias* essentially demonstrate the same fundamental fact: the elements, from which movement is formed, are by no means nonextended in an absolute sense; to some degree they reflect the infinite quantity encompassing them. Classical mechanics had already given this conception very precise form. At point x, the moving particle possesses determined coordinates, a determined position. But it also possesses *velocity* and possibly *acceleration* as well — predicates that express the ratio of the increment of space traveled to the increment of time, or the increment of velocity to the increment of time. This is more clearly explicit in the nonclassical situation already mentioned: the particle at point x possesses charge and mass, but these predicates signify one or another world-line in given fields and express at each 'here and now' the structure of the fields, and, in general, the structure of the set encompassing the infinite number of these 'here-and-nows'.

From approximately the middle of our century, the paradoxes of infinity have become the fundamental physical paradoxes of modern science. The application of relativistic relations to ultramicroscopic regions, the calculation of the interaction of a particle with a vacuum, the appearance of the completely paradoxical, infinite values of energy and change, give contemporary physical thought its characteristically predictive style. It has already been mentioned that the prescribed methods for excluding infinities are proposed 'on credit' in the expectation of a theory possessing what Einstein called inner perfection — a theory flowing from the most general assumptions possible. We also mentioned that such a theory would perhaps connect the events of the 'here and now' in a localized space-time region on the order of, say, 10^{-13} cm and 10^{-24} sec, with the megaworld, and in some sense would even identify the particles with the megaworld. This style of physical thinking also appears in our book as the origin of an historico-philosophical review of the past. In contemporary science, the age-old problems of being, knowledge, as well as the problem of infinity acquire unprecedented precision. In any case Aristotelianism and the evaluation of Zeno's paradoxes are associated in the consciousness of contemporary scholars with their own current, immediate preoccupation. Let us return to Aristotelianism and the evaluation of Zeno's paradoxes, limiting ourselves, of course, to brief remarks.

Zeno did not deny the sensual certainty of movement. Therefore, the paradoxes of infinity became not so much proof of the unreality of movement as authentic paradoxes — clashes of incompatible assertions that contradicted each other, but were nevertheless unambiguously deduced from each other. There arose a clash between the sensual and logical certainty of movement, neither of which could be denied by ancient Greek thought. In analyzing Zeno's third paradox (Achilles does not overtake the tortoise), Hegel notes in passing that "the Ancients loved to clothe difficulties in sensuous representations" (Hegel 1974, **1**, p. 272). This casual remark is very important. The sensual perception of the world, so characteristic of the folk artist, suggested the idea of determining [the nature of] motion. The concrete, sensory image of Achilles and the tortoise reminds us of the conditional character of Zeno's denial. But the thinking of the folk philosopher could not be limited to a purely empirical discrediting of paradoxes. In order to be understood, they needed a rational interpretation expressed in concepts.

Hegel quotes Aristotle as saying: "the quick will overtake the slow body if the limits to be traversed be granted it" (Hegel 1974, **1**, pp. 272—273). What is this limit?

It is the limit of the abstract opposition between separate 'here-and-nows' and their infinite, continuous series. A body is located at a given point and then leaves it. "But movement means to be in this place and not to be in it, and thus to be in both alike" (Hegel 1974, **1**, pp. 273—274).

A nonclassical historical perspective not only renders the connection between the interpretation of Zeno's paradoxes and the conception of true infinity more comprehensible, but makes it more physically tangible as well. The Hegelian concept [of a true infinity] destroys the absolute barrier between the finite and the infinite; it allows us to examine the finite element of an infinite set as something which brings infinite correlations into being; infinity defines the relation of the finite element to others, defines its fate, its behavior. In its turn, the integral structure of the finite object is transformed into a law determining the being of the infinitely large number of its infinitely tiny elements. Hegel considers the paradoxes of Zeno a demonstration of the contradictions in a *bad infinity*, the unceasing accumulation of always newer elements of being (Hegel 1974, **1**, p. 268). Let us note that the Hegelian critique of Zeno's paradoxes demonstrates their connection with the *identity* of the elements expressed in an infinite set. Hegel says that the successive divisions of the segment into ever smaller parts can never reach the point itself, because the point possesses being as something nonidentical, individual; its being is being-for-itself. On the contrary, "self-identity or continuity is absolute cohesion, the destruction of all difference, of all negation, of being-for-self" (Hegel 1974, **1**, p. 268).

I shall return to this conception below and show convincingly that it explains not only the paradoxes of infinity in the strict sense, but also other logical paradoxes of antiquity that are still with us today although expressed in a new form. Right now I want to concentrate just on Hegel's concept of true infinity, its embodiment in classical, and even more so in nonclassical science and how it modifies the contradiction between potential and actual infinity. Let us take Riemann's conception, which is nonclassical in its subsequent physical application but entirely classical historically: he states that the curvature at a given point is the *local* definition of infinite space; if the curvature is equal to zero and the metric is Euclidean, then space is infinite; if the curvature is not equal to zero, then space may turn out to be finite; infinity is the metrical correlation, as distinguished from *limitlessness* — the possibility of more and more additions of new segments to the given.

Such a local infinity exists at a given moment, but it has nothing in common with actual infinity in the traditional sense, that is, infinity that is calculated. The point is that here we are dealing not with calculation but with the local correlation which is defined in an infinite sphere or, more precisely, in an infinite set of infinitely small 'here-and-nows.'

It stands to reason that such a radical modification of the concept of infinity allows Hegel not only to overcome Kant's antinomies, but also to demonstrate their logical kinship to Zeno's paradoxes (Hegel 1974, **1**, p. 277).

We need not dwell on this and can turn our attention briefly to the subsequent evolution of the paradoxes of infinity, limiting ourselves as previously to only a few of its stages.

True infinity is infinite being; it is the Universe, infinite in space and time, the real four-dimensional world. True infinity informs each of its elements; it determines the infinite being of the finite and thus itself acquires local, finite being. This was how the infinite Universe appeared to science when the system of differential physical laws connected local events in infinitely small 'here-and-nows' with macroscopic processes. In a note on Nägeli's lecture, 'The Limits of Natural-Scientific Knowledge,' Engels says that the laws of natural science have a universal character and that, therefore, knowledge of finite objects is knowledge of infinity.[2] Such a precise natural-scientific interpretation of what Hegel called 'true infinity' becomes especially pellucid when the question of differential laws arises, when the laws determine events in infinitely small 'here-and-nows.' These events consist of the uniform repetition of one and the same results under repeated conditions: for example, acceleration is repeated while the mass and intensity of the field of force are unchanged (of course this repetition is limited by the framework of classical mechanics; the theory of relativity includes supplementary conditions). But the repetition becomes senseless without change: it is identity without nonidentity, the maintenance of velocity without a change in position, uniform acceleration without a change in velocity. Nonclassical science, as already mentioned, complicates the relation of the infinite to the finite; it introduces the conception of the limited interaction of the infinitely large with the infinitely small, the cosmos with the microcosm, the universal law with the 'here and now,' and as we will see, this alters the conception of the structure of being, of inclusive systems and their included elements.

The differential conception of motion, the analysis of events in the infinitely small 'here and now,' represented a transition from the Aristotelian picture in which the *ratio* of the world consisted of a static arrangement of natural locations, to the picture of continuous movement, examined from point to point and from moment to moment. Zeno's paradoxes became the fundamental paradoxes of being; they acquired physical meaning. But the analysis of the problem of infinity at this point proceeded not from the conceptions of the 6th—4th centuries B.C., but from a later conception of infinity, that of Archimedes. For Galileo, the conception of infinity had still not derived an adequate mathematical form from the analysis of the infinitely small, but it did come quite close to it. For him the logically paradoxical and sensually incomprehensible, the calculated and the incalculable quantity, no longer served as the image of infinity; instead he used any curve where the direction changes at each point — something sensually comprehensible in principle, capable of being investigated in local experiments, and logically correct.

Galileo also generated the conception of infinite sets of equal power, a thesis which became the basis for the constructs of Georg Cantor two and a half centuries later. They were paradoxical constructs: in order to evaluate the power of infinite sets, the one-to-one correspondence of the elements of one set with the elements of another set, we arrive at the concept of transfinite numbers which remain equal to themselves during the addition or subtraction of another number that is not equal to zero: $\alpha \pm \beta = \alpha$ (the whole is equal to a part), or during multiplication by a number greater than one, $2\alpha = \alpha$. But this is the paradox of being: the comparison of the elements of infinite sets is the mathematical equivalent of local correlations (e.g., between force and acceleration), the physical prototypes of infinity, the laws of natural science, true infinity — that which makes the Universe not chaos, but cosmos.

When we arrive at such a conception of infinity, if the old paradoxes concerning a calculated infinity do not disappear, at least they exit to the wings. New paradoxes take the stage. We are dealing not so much with the paradoxical assertions $\alpha \pm \beta = \alpha$ and $2\alpha = \alpha$ as with other paradoxes which have become fundamental as a result of the appearance of set theory, and which have acquired physical meaning in nonclassical science.

As far as the paradoxes $\alpha \pm \beta = \alpha$ and $2\alpha = \alpha$ are concerned, they, like Zeno's paradoxes, demonstrate one very general law of knowledge. In the first paragraph of this section, we mentioned the transitions from one kind of identity to another, for example, from identity of locations to identity of velocities and, further, from identity of velocities to identity of accelerations. In each such transition, the mutability, the nonidentity, the motion which common sense reduced to identity, to immutability, to a law, to rest, is reborn like a phoenix, obliging us to move on to a *new* identity, to a *new* law. And thus infinity, having obtained its embodiment in the finite, again appears on the stage and demonstrates its paradoxical essence, which cannot be reduced to the finite. In accordance with traditional terminology, [recognizing this irreducibility] is the function of reason as distinguished from common sense.

Let us take the first historical antecedent of the conception of transfinite numbers. Galileo listed a series of whole positive numbers and their squares. These numbers can be placed in a one-to-one correspondence, contradicting the thesis that the whole is greater than any of its parts.

What are the physical prototypes of this paradox?

Zeno's paradoxes acquired an ontological meaning and a fundamental significance for science when science passed from a static design of the universe to an analysis of motion. We examine uniform motion from point to point and from moment to moment. If the law of motion is established in the form of a law of inertia, it is no longer necessary for us to divide motion into an infinite set of infinitely small segments.[3] A statement about motion, about change in location, is transformed into a *statement of the immutability* of motion itself, of the immutability of velocity. It is now possible to perform operations with entities of finite magnitudes. But then it turned out that motion takes place by means of acceleration. And so once again we find ourselves dividing space and time into infinitely small intervals. Once more an infinity irreducible to the finite comes into being, and demonstrates this

flicts in philosophy, those pervasive conflicts in its history which took on very different, sometimes extremely irrational, forms. One example is the controversy between the nominalists and the realists, which expressed the profoundly conflicting, contradictory aspects of knowledge. These aspects are also connected with a group of equally pervasive paradoxes, beginning with 'All Cretans are liars . . .' But before we turn to these paradoxes, we must remark briefly on the contemporary physical paradoxes of this same logical type.

A distinguishing feature of real physical sets is that their analysis cannot entirely ignore the interaction of the elements that come into the set. We can examine their abstract reflection in a set of situations (falling bodies, their impact, the passage of electricity through a liquid, the influence of one chemical substance on another, etc.) that are subject to one and the same laws. But in physics (in an obvious form in nonclassical physics), the existence of a set is not reduced to subordination to general laws. In the history of physics, a conception of the interaction between the elements in a set — their nonlinear connection as well as the distinction between the real set and the abstract set — always went before, continually clearing the path. The intensity of the interaction lies at the base of the structure of the world. Newton had already portrayed the hierarchy of increasingly inclusive systems which have their origin in the diminishing intensity of interaction: each system becomes an individual system when its internal interactive forces are essentially more intense than the external forces acting upon it.

Is it possible to regard an entire system as one pole of an interaction? In other words, does the system itself enter itself as one of the components, one of the elements determining its structure, character, behavior? Such an assumption led classical physics into difficult paradoxes, one of which is the paradox of gravity: if the Universe is infinite, then its gravitational field, acting on each heavy body, will be infinite, and the local, finite fields will not move this body from its place. Nonclassical physics made it possible to conceptualize the reciprocal interaction between the all-inclusive system and its local element. It paved the way for the conception of inclusion-in-a-system as a physical process, an experiment which alters both the included element and the system including it. It stressed the important question of the *intensiveness of inclusion, of the degree and limits* of the subordination of local events to general laws.

What is the relation of these conceptions to the paradoxes of set theory, which are connected with the inclusion of an element in a set and, particularly, with the inclusion of the set itself in itself as one of its elements?

These paradoxes are logical. But at the same time they demonstrate the inevitability of ontological content. Interaction is not a problem here; but on the other hand, there is a problem of reinterpretation, a change in meaning, a change in what may be called the *intensiveness of deduction* related to the inclusion of an element in a set, corresponding to the concept of the intensiveness of physical inclusion introduced above.

The first of the paradoxes that may be called paradoxes of inclusion, or paradoxes of structure, is the paradox of Epimenides, or the 'liar's paradox'. The Cretan announces that 'All Cretans are liars.' If this statement is true, it negates its contents, and turns out to be a lie; but if it is false, then it supports its content and is therefore true. The statement ascribed to Eubulides, 'The statement I have made is false,' follows this pattern as do a number of analogous paradoxical situations.

In systematically rendering logic and mathematics physical, in imparting physical content to them and modifying their foundations, can contemporary science shed additional light on such paradoxes? Let us wait before we answer this question. First we will touch upon the paradoxes of set theory, which appeared at the end of the nineteenth and the beginning of the twentieth century. Let us just mention Russell's paradox about the set of all sets which are not elements of themselves. Will this set be included in itself? If it is included, it destroys the rule of selection and cannot be included in itself. If it is not included, it turns out to be one of the sets which is not an element of itself, that is, it is included in itself. Each of the two possible answers leads to the other, which in turn excludes it.

Let us note that Russell's paradox and the other paradoxes concerning the inclusion of a set within itself, are logically connected with Zeno's paradox, and thus support Bolzano's comment in the beginning of his *Paradoxes of the Infinite*: if not all, then the majority of the paradoxes in science flow from the concept of infinity.

Now let us turn to what is common to and immediately striking about both Zeno's and Russell's paradoxes. In both, the paradoxical situation arises when an assertion is made relating to assertions about the entry of the elements under consideration into a set. Every statement of each Cretan enters the set of false statements. Does the set include the statement 'All Cretans are liars' as it relates not to each individual local utterance of the Cretans but to their set?

The point is that this statement is ambiguous. It has two aspects: first, it is all-inclusive in its objective and is related to the set; but, on the other hand, it is an isolated, local statement. This statement does not enter the set on equal terms with the others because its content, which relates to the set, denies the formulating, forming, defining property of the set, and thus denies its inclusion in the set.

Let us turn our attention to another question connected with a positive, rather than a negative, way out of the paradox; not with a reassuring return to a noncontradictory, nonparadoxical situation, but with an intensification of the paradoxical situation. We will assume that the paradox of Epimenides, as well as its latest formulations, are paradoxes of being, contradictions of being, which take the form of paradoxes in a system of concepts that are not sufficiently flexible to deal with these contradictions.

Why do the ambiguity of the statements of Epimenides and the ambiguity of the inclusion in itself of Russell's set of sets not included in themselves seem paradoxical to us? They seem paradoxical as long as the presumption of the existence of each separate element of the set is retained. In the paradox of Epimenides, each Cretan has veto rights in the resolution of the question of the included property, 'liars', in the set 'all Cretans.' This is an oligarchic set. Each person who enters it has, by his entrance, already influenced the determination of the composition of the set, but his influence may negate that property which allowed us to include the person entering the set.

What is common to both Zeno's paradoxes and the Epimenides-Russell situation?

We are interested here in the common ground between the paradoxes of infinity and the paradoxes of inclusion, which becomes apparent when these are viewed ontologically, especially and precisely from a nonclassical, historical perspective. Let us take Zeno's *aporia* of measurement. The including set is an extended finite magnitude comprised of an infinite number of elements; if these elements are extended, their sum will be of an infinite magnitude; if they possess zero-order extension, their sum will be just as nonextended. We shall examine, as an inclusive magnitude, a finite world-line composed of an infinite number of world-points. If the world-points are really nonextended points, the world-line will be found to be nonextended; if they are extended, then their sum will be infinite. Zeno's *aporia* in this instance has such a form.

Let us see how the magnitude of the world-line appears in a non-classical historical perspective. The world-point belongs to the world-line; it is included in this set. But isn't there a change in the form of the world-line as a result of the inclusion? Isn't there a connection between its increment, the inclusion of a new element, and the variation in the line? And, on the other hand, isn't there a connection between the invariant direction of the world-line at a given world-point (in other words, a determinate velocity of the particle) and the oscillations of the world-point, the indeterminant significance of the inclusion, the indeterminant increment of the world-line, the uncertainty of its co-ordinates, its position? The quantum object either imprecisely fulfills the obligations placed upon it by its inclusion in the set, or it oscillates with respect to the inclusion itself. We arrive at the correlation of uncertainties as we record position and velocity, which are conjugate variables.

What does this mean for the Epimenides-Russell paradox? The quantum object will resemble Epimenides's Cretan, if the Cretan 'oscillates.' Must he limit, in a local way, the precision of the macroscopic rule 'All Cretans are liars,' allowing local variation of the law; or, alternately, must he limit the precision of the increment to the number of Cretans, condition his belonging to the lying population of Crete, and lean toward uncertainty in his coordinates?

Such a quantum way out of the paradox of Epimenides makes possible a retrospective reinterpretation of Zeno's *aporia* of measurement. The dilemma — does an element of the world-line possess extension or not — comes to an ambiguous resolution. The world-point possesses both here-and-now being and beyond-the-here-and-now being. Probability waves pass over the world-point, and a particle is located at it. The world-point is the space-time localization of a *physical* object. If we ascribe to this object 'here-and-now being' and compress its position into a nonextended point, the world-line is not compressed into a nonextended sum of such localizations, and thus makes its way out of Zeno's conflict. This way out — 'the denial of inclusion' in the case of 'subordination to a law', and the reverse — illustrates the 'denial of inclusion' in quantum mechanics in the historical evolution, or rather, modification, of Zeno's *aporia*. It is a unique historical and philosophical 'uncertainty principle'!

In quantum mechanics, the world-line of the particle is represented as blurred. If we view this world-line as a set of space-time localizations, these localizations cannot become precise for conjugate variables. The reason for this lies in the entry of the set — the world-line — into its local element and thus into its own self. As was already mentioned, the derivatives of place and time, which enter the expression for momentum and energy of the particle, are the intervention of the world-line itself in itself, the compression of the interval extended in space and time into a point, the appearance of infinity, the rapprochement of the paradox of Zeno with that of Epimenides. But as soon as we allow such an intervention, it blurs the world-line and makes space-time localization uncertain. In striving for precision and rigor in the description of the quantum mechanical situation arising out of the Heisenberg uncertainty relations, there is no need to present these observations in detail. Even with such a brief hint, the connection between quantum mechanics and the paradoxes of infinity, and the paradoxes of inclusion in a set, becomes clear. Now, the classical reduction is possible: the representation of the classical integral law, for example, the principle of least action, as the entry of the set itself into itself.

Does this mean that 'oats grow according to Hegel,' that concrete processes, including even the paradoxical, are subject to the most abstract principles?

No, the paradoxicality of principles — standing out in bold relief in the example of quantum mechanics — ultimately means that concrete local mechanisms cannot be reduced to macroscopic laws. The element in the set is not only subordinate to the set, but it influences the set as well. This is true in the history of philosophy and science. A pervasive general idea enters the history of thought as a set of concrete conceptions which, in entering the set and revealing their genealogy, are not only modified themselves but also modify the set. Similarly, the formula $\alpha \varepsilon A$ indicates a change in both α and A. ε is the symbol of the operator.

4. PARADOXES OF FREEDOM

What has been said of the paradoxes of infinity and structure allows us to view differently not only the Kantian antinomies of the infinity of the universe, but also the contradiction of transcendental freedom and 'physiocracy' (Kant's term designating absolute necessity in nature).

Let us consider the classical atom, not the atom of classical chemistry and classical physics, which always had the potential of being broken into subatoms, but the atom of classical antiquity, that is, something absolutely indivisible. This indivisible atom entered into nonclassical physics in the form of the elementary particle, which, in spite of its designation, is the most complex image in modern science. The elementarity of the classical atom was expressed in its complete autonomy; given the disappearance of the rest of the entire world, the atom would remain, in a manner similar to the illusory 'Unique' of Max Stirner.[4] The classical atom was dependent in its behavior, [insofar as] it collided with other atoms, but its own being was autonomous. There was nothing paradoxical in its being. But true being did not exist. What distinguishes Democritean 'being,' i.e., atoms, from Democritean 'nonbeing,' i.e., space? Space occupied by the atom is occupied by it here and now, because it serves as the source of sensation. Space not occupied by the atom now is empty space – and it is real insofar as it may potentially be occupied by the atom. But is actual being real, without the potentiality?

Not one property of the real atom, which distinguishes it from a purely geometrical representation, has meaning without 'nonbeing,' without other atoms, without the interaction between them and the given atom. This is also true for geometrical properties – not *purely* geometrical, but geometrical properties possessing a physical meaning. The dimensions of the atom are an invariant of motion, which loses meaning without reference bodies. Both the position and form of the atom lose meaning without these bodies. Impermeability and mass lose physical meaning without interaction with other bodies.

Later, the interactions of bodies and the existence of 'nonbeing' were united into the concept of a field of force. The atom acquires charge, and this concept has meaning if there are other atoms with similar charge. The atom acquires properties to the extent that it is included in a set: its externalization, its subordination to some general laws that give physical meaning to the dynamical variables of the atom, to its mass, to its charge, to everything that distinguishes 'being' from 'nonbeing,' are its properties to the extent that the atom is included within the real structure of the world.

But such an inclusion makes the being of the atom paradoxical. It is time to clarify further the concept of paradox. Mutually exclusive properties are paradoxical if they belong to one subject. It is precisely this which guarantees the subject's authentic being, one that is not equal to 'nothing'. This is precisely what occurs with the inclusion of 'nonbeing' in 'being', when the atom acquires properties which are simultaneously incompatible and compatible. Such ambiguity lies at the bottom of the paradoxes of infinity and the paradoxes of structure. The 'here and now' includes the 'beyond here' and the 'beyond now.' The element included in the set is included in it and, at the same time, opposed to it. By analogy, the atom does not exist if it does not possess properties separating it from space, from the surrounding world; but these properties are the result of several factors: the interaction with the surrounding world, the world's structure, the macroscopic laws of being, and the subjugation of the behavior of the atom to macroscopic laws. But in its *complete* subjugation to external laws, the atom's individuality disappears in the paradoxical combination of its separation from the surrounding world and its inclusion in it. These poles of the paradox are interchangeable. Epicurus understood that the freedom of the atoms, its *clinamen* [spontaneous swerve], is a necessary component of being. (There will be further discussion of this in the next chapter.) *Clinamen* loses its meaning if there is no macroscopic structure of the world, and no possibility of introducing macroscopic systems of reference. But the macroscopic structure of the world also lacks physical meaning if the elements in the set do not possess spontaneous properties.

How will the conflict be resolved between freedom, spontaneous existence, individuality, exclusion from the set, the one point, on the one hand, and subordination to macroscopic laws, the necessity of entry into a real set, [spatial] extension on the other? Such a conflict is resolved by the transition to new, and likewise, paradoxical, forms. As classical science eliminated the paradoxes of the Aristotelian static scheme of being, so the quantum-relativistic picture of the world eliminated the paradoxes of classical science. The paradoxical independence from macroscopic law signifies not indeterminism, but the transition to a new form of determinism: from the Aristotelian scheme of natural locations and natural movements to the Cartesian inertial picture to Newtonian dynamic determinism: the positions of the bodies depend on forces, the forces depend on the position of the bodies, a concept that found its expression in the ability of Laplace's 'superior reason', which, knowing the coordinates and velocities of all particles, could foresee the future universe. Then there is a transition to the statistical determinism of thermodynamics, to relativistic and quantum causality, and, in the future, to the knowledge of increasingly complex ultra-relativistic causality in ultramicroscopic spheres.

The paradoxes of a given form of determinism, the conflict of freedom and necessity in its given form, is a transition to new conflicts. Thus the epistemological meaning of the paradoxes is revealed in their history.

5. PARADOXES AND THE HISTORY OF PHILOSOPHY AND SCIENCE

In science, at least in classical science, transitions from one prevailing scheme of identity and ordering of the world, from one type of determinism to another scheme and another type, occur in the form of scientific revolutions, which are separated by several periods of comparatively stable conceptualizations. Hence the possibility of static sections of developing science existing simultaneously [leads to] its classification [while] the transition from one prevailing scheme of the universe to another takes time to be constructed, and this lies at the basis of the periodization of science. The classification of science reflects the heterogeneity of the world picture, periodization, its fundamental dynamism, which is related to the most general principles of knowledge.

In *Dialectics of Nature*, Engels spoke of the classification of science, that is, a hierarchy of scientific disciplines as a series of transitions from one form of motion to another more complex form. Each of the disciplines could be characterized by its field invariant, those stable principles which remain unchanged in the transition from one problem to another within a given discipline. This would be analogous to Felix Klein's *Erlanger Programm*[5] if, in this program, the transition from metric to affine geometry, and so on, flowed from the contradictions of metric geometry, etc., as is exactly the case in the classification of science. In the new, higher and more complex form of motion, a conflict arises between the laws that are reduced to the most general (in Engel's terminology, to the subsidiary forms of movement; see Engels 1954, p. 328) and those not reducible to them (the major form of motion). Such a conflict leads to paradoxes which are removed by the transition to a successive, even more complex form of motion. It would probably be possible to trace such an evolution in the history of the separate disciplines; or rather to some extent, this has already been done.

Some kind of conventional *Erlanger Programm* is possible in periodization. There exist invariants common to all of science, unaltered principles transferred from one branch to another. In classical science all branches are inseparable from these principles insofar as they are irreducible to simple mechanical images of the moving body. When the paradoxes of science force the rejection of such invariants, a new period in the development of science as a whole arrives.

Invariability, the identity of scientific concepts and laws, testifies to the unity, homogeneity and continuity in the structure and evolution of science. Their realization is a function of common sense in science. Destruction and change in the invariants, as a result of unavoidable and intensifying paradoxes, express the heterogeneity in the structure and evolution of science. The realization of heterogeneity is a function of reason going beyond the framework of common-sense thought.

6. PARADOXES OF SCIENCE AND SCIENTIFIC PREDICTION

In nonclassical science, a new, previously nonexisting relation was created between the classification of the sciences and the periodization of knowledge. The classification of science changes practically continuously. New problems are constantly arising, the word 'new' signifying here not only that they did not exist earlier, but something more profound as well: a change in the style of scientific thought in the given branch, in its relation to other branches, in its location within the classification of knowledge. But this is not all. Some discoveries lead to fundamental paradoxes. Changes in classification, structure, the specific gravity of problems, become inseparable from periodization, from change in fundamental principles characterizing a major period in the development of knowledge.

In our time the hierarchy of increasingly inclusive systems has radically altered. The snake has bitten its own tail. The movement of knowledge into the depths of matter has collided with its expansion, with its envelopment of increasingly greater space-time spheres. Astrophysical problems have come close to the problems of elementary particle theory and have sometimes merged with them: The processes occurring in the metagalaxy during the course of billions of years are explained by processes in the subatomic world occurring in a billionth of a second. The atomic and cosmic epoch has not only a scientific and technical, but also an epistemological attribute. At the junction of astrophysics and elementary particle theory, new conceptualizations arise, altering the character of the connection between macroscopic and microscopic objects, the character and meaning of the inclusion of the smaller systems (beginning with the smallest presently known) in the greater (including the greatest presently known). The paradoxes of inclusion, infinity, and of quantum, relativistic, and ultrarelativistic causality, have been embodied in concrete physical paradoxes.

The removal of paradoxes, which is the essential function of science, is noticeably outstripped by their accumulation. Generally speaking, the paradoxes of science are not only (and sometimes not so much) paradoxes of the new, but also paradoxes of the old: they remain paradoxes until a new theory appears from which the former paradoxical facts and deductions follow in a simple manner. The present paradoxes of quantum electrodynamics as well as the paradoxes of cosmology and astrophysics will remain paradoxes, as long as there is no noncontradictory general theory.

The origin of new paradoxes, the appearance of new paradoxical effects, is difficult to foresee. However, the already existing paradoxes are points of origin for predictions. A paradox does not indicate the routes to the resolution of the problem, but it does allow the problem to be formulated. Everyone remembers the problems formulated by Hilbert at the turn of the century and their heuristic role. At present, the fundamental paradoxes of knowledge have revealed their ontological basis to an unprecedented extent; they have been embodied in physical paradoxes and, furthermore, in physical problems.

It is possible to list hundreds of articles and dozens of books in which there are analyses of the paradox of the infinite energy of the particle and its interaction with a vacuum. The analyses include predictions and indications of ways for an experimental (e.g., by accelerators) and theoretical elimination of the paradox. But in reality, this paradox is the modern physical embodiment of the traditional, all-pervasive paradox. In an era when science in its paradoxes embodies the most general paradoxes of being known to us, a comprehensive analysis of paradoxes, including logical, epistemological and historical components, will have an essential heuristic effect.

NOTES

[1] [F. Dostoyevsky, *The Brothers Karamazov*, Pt. 4, Bk. 11, Chap. 9: 'The Devil. Ivan's Nightmare' — Eds.]

[2] [Karl Wilhelm Nägeli (1817—1891), a German botanist, delivered this lecture (in German, 'Die Schranken der naturwissenschaftlichen Erkenntnis') to the Munich Congress of German Natural Scientists and Physicians on 20 September 1877 (cf. Engels 1954, pp. 431—432, n. 156). In his note, Engels wrote:

> . . . all real, exhaustive knowledge consists solely in raising the individual thing in thought from individuality into particularity and from this into universality, in seeking and establishing the infinite in the finite, the eternal in the transitory. The form of universality, however, is the form of self-completeness, hence of infinity; it is the comprehension of the many finites in the infinite (Engels 1954, p. 310) — Eds.]

[3] This was first pointed out by Galileo in *Discourses and Mathematical Demonstrations Concerning Two New Sciences* . . . (Galileo 1974, pp. 148—149).

[4] [Max Stirner is the pseudonym of Johann Kaspar Schmidt (1806—1856). 'Unique' is the literal translation of the first part of the title of his book, *Der einzige und sein eigenthum*. For full information see Bibliography under 'Schmidt' — Eds.]

[5] [Felix Klein (1849—1925) sought to "characterize the various geometries not just on the basis of nonmetric and metric properties and the distinctions among the metrics, but from the broader standpoint of what these geometries and other geometries, which had already appeared on the scene, sought to accomplish. He gave this characterization in a speech of 1872, 'Vergleichende Betrachtungen über neuere geometrische Forschungen' (A Comparative Review of Recent Researches in Geometry), on the occasion of his admission to the faculty of the University of Erlangen, and the views expressed in it are known as the *Erlanger Programm*" (Kline 1972, p. 917; see *ibid*., pp. 917—921 for fuller discussion, particularly p. 919, which gives a schematic chart of Klein's classification of the principal geometries) — Eds.]

3. EINSTEIN AND EPICURUS

1. ATOMS AND SPACE

In 1923 a new edition of Lucretius's poem, *De rerum natura*, was published in Germany. A foreword by Einstein was included in the second volume, which contains a German translation of the poem (Lucretius 1923—24). In it Einstein wrote: "For anyone who is not completely submerged in the spirit of our age, who feels instead like a spectator as the world goes past him, especially, from time to time, vis-à-vis the intellectual attitudes of his contemporaries — on him will Lucretius's poem work its magic" (Lucretius 1923—24, **2**, p. VIa).[1]

Why, in fact, is it necessary not to submerge oneself in the spirit of the modern age and to look at it from the sidelines? Apparently, Einstein considers all this an intellectual advantage, perhaps even a characteristic demand of the modern age itself. And so it is. As we mentioned in our Introduction, nonclassical science requires an analysis of the contemporary situation in science from the point of view of its dynamics: that is, there must be some prediction, a conjecture as to the future of contemporary conceptions, as well as some excursions into the past, some re-evaluation from an historical perspective. The theory of relativity could not have emerged without a re-evaluation of conceptions that were deeply rooted in the past and seemed independent of time, unchangeable. In searching for what he called the "inner perfection" of the theory (its natural deduction from the most general assumptions), Einstein returned to the controversies of classical physics, which he transformed and generalized so that they acquired new meaning. At the same time, Einstein saw the open character of the theory of relativity itself and sought more general solutions, which had still not acquired "external confirmation," that is, they could not be compared with the results of observation.

The more nonclassical science moves (and *looks*) forward, the more it finds itself in the midst of its own retrospective self-scrutiny. Now, almost half a century after Einstein wrote his foreword to the new edition of Lucretius's poem, it has become obvious that the most vital conflicts of modern theoretical physics are connected with problems that have pervaded the entire history of science; and that surmounting the most acute and serious difficulties in elementary particle theory demands a radical revision of the fundamental conceptions of space, time and movement, of their continuity or discreteness, and of the relation between Democritean 'being' (in particles) and Democritean 'nonbeing' (in space).

The obvious transitional, 'expectant' character of modern conceptions, the difficult *aporias*, such as infinite energy and charge in the calculation of particle-vacuum interactions, as well as the solutions and experimental methods advanced 'on credit' in the expectation of a non-contradictory general concept to substantiate them — all these demand from the modern age a scholarly "refusal to submerge oneself" in previous achievements, a detachment, a position outside the framework of established results, and an evaluation of their historical mobility and flexibility. This refusal to be subjugated to the present and the ability to observe with detachment is an integral aspect of the truly modern style of scientific thought. So today, even more than in 1923 when Einstein wrote his foreword, the flexibility and dynamism of ancient thought is attractive, with its initial naive view of the world, a view which created the prototypes of the most heterogeneous physical ideas, including those which have yet to be embodied in simple forms.

In particular, there are distinctions between bodies that can act on the senses, and empty space, which cannot do so and can be accepted only as an interval between bodies — an accumulation of the distances between them. Such a physical distinction is relativist in its tendencies: space is viewed only as an accumulation of distances which lose their meaning when there are no bodies other than the given one, when there are no reference bodies.

This relativist tendency demonstrates that the comprehension of the world begins with bodies, with reality comprehended empirically, by means of the senses. In Lucretius's systematic exposition of the philosophy of Epicurus in *De rerum natura*, this is one of the most fundamental ideas. Even the first definition of a body as distinguished from space — its ability to act on the senses — introduces an empirical characterization as the setting for a rationalist comprehension of the world. Let us take the first book of Lucretius's poem, specifically the lines which describe the action of bodies not immediately visible. These lines should substantiate the corporeality of invisible atoms. Having given examples of wind, sound, warmth, etc., Lucretius concludes:

> Yet all these must of necessity consist of bodily
> structure, since they can act upon our senses.
> For nothing can touch or be touched, except body
> (Lucretius 1975, *1*, lines 302–304).[2]

However, the passage is not about effects on the senses, but rather about the theoretical possibility of such effects. Between these positions there is a gap separating the idea of corporeality from extreme phenomenology. "Yet all these must of necessity consist of bodily structure, since they can act upon our senses." The *theoretical* ability to act on the senses means that we can speak of corporeal objects which we do not observe. This possibility is the point of departure for speculation. It creates a world-picture that is broader than the sum of isolated direct representations. Speculation seems to reject the sensory basis of knowledge; it deprives the isolated representation of its individuality as it identifies it with other isolated representations and, at the same time, cannot disregard the individual representations, nor become a comprehension of reality without them. In their turn, the isolated representations, and even isolated impressions, are impossible without the general representations, although, in denying identity, they break down general representations into discrete elements.

Epicurus already had an elementary epistemological scheme for the formation of concepts. Isolated representations (*φαντασας*) are grouped according to their similarities into general representations or concepts (*προληψεις*). Starting with the 'Letter to Herodotus,' in which Epicurus states his epistemological position, and throughout the entire history of philosophy and science, thought that identifies or generalizes and individualized empirical impressions cannot be freed of each other. In his very important *Autobiographical Notes*, Einstein speaks of the transition from separate perceptions to imagined pictures and of the continuous series which have come about in this way, where each link calls forth the next by association. Such series cannot even be formed without some resemblance to concrete images. But, beyond that, thought identifies the images belonging to various series and then organizes and groups these series; the identified images become concepts, instruments for organizing associated series.

> What, precisely, is 'thinking'? When, at the reception of sense-impressions, memory-pictures emerge, this is not yet 'thinking'. And when such pictures form series, each member of which calls forth another, this too is not yet 'thinking'. When, however, a certain picture turns up in many such series, then — precisely through such return — it becomes an ordering element for such series, in that it connects series which in them-

selves are unconnected. Such an element becomes an instrument, a concept. I think that the transition from free association or 'dreaming' to thinking is characterized by the more or less dominating rôle which the 'concept' plays in it (Einstein in Schilpp 1949, p. 7).

For Einstein this epistemological conception was a working element of his world-view, which led to the research that culminated in the theory of relativity. By no means was Einstein's theory the logical outgrowth of experimental discoveries based on unrealized epistemological tenets; rather, it was based on a precise rationalist conception, one moreover not opposed to sensualism and the empirical sources of knowledge. The two criteria under discussion for the selection of a scientific theory — *inner perfection* (natural, logical deduction from the most general premises possible) and *external confirmation* (experimental confirmation of the theory) — both demand that the concept have physical content: the initial concepts that form part of logico-mathematical deduction must lead to experimentally verifiable results, even if only in principle, or only in the form of an intuitive guess. Let us consider the principle of relativity in its most general form, embracing not only Einstein's theory, but also the classical principle of relativity, and the most elementary sources of relativism. The principle of relativity, if it claims to be a fundamental principle of science, demands that the world-picture include spatial positions and movements of bodies, and further that these positions and movements relate not to space itself but to other bodies, to something corporeal, capable of acting on the senses. On the other hand, rationalism demands that physical processes be changes in distances, objects that are incomprehensible to the senses. But these are distances between objects that are comprehensible by means of the senses and, thus, relativity is a sensual category. The point is that logical deduction and the empirical, the rational and sensual components of knowledge, are complementary; they exclude each other and, simultaneously, they lose real meaning without each other.

It is possible to demonstrate that for Epicurus and Lucretius the relativist tendency was connected with such complementarity. For Aristotle, the relativist tendency consisted primarily in the definition of rest and motion in terms of contiguous bodies. In addition to that definition, Aristotle had another definition which was also essentially relativist: rest and motion are considered in terms of a moving body. In 'peripatetic' cosmology, the earth, the natural location for heavy bodies, served as a reference body for the sublunar world. The movements of heavy bodies are relative to an object, the earth, which is perceived by the senses. But in addition to sensorily comprehensible properties, the earth possesses still another quality: it is the center of the universe. Here the sensualist tendency is combined with another: the geometrical structure of the universe, its center and boundaries, which are rationalist concepts, come into play. They cannot be eliminated from the sensualist picture, from the matter which fills space. Bodies that can be comprehended by the senses and places that can be comprehended by reason alone, coincide. For Aristotle, the schemes of natural locations, the center of the Universe, and its boundaries, would have formed a framework over which absolute space was stretched; for him, the systems of coordinates of the earth would have been a privileged system. But space as such, empty space, is absent from Aristotle's system. The concepts of the absolute and the relative, of the rationally comprehensible and the empirically comprehensible, are given in peripatetic philosophy in their initial, undifferentiated form.

These concepts are also interwoven in the philosophy of the atomists, but with a difference. Here there is empty space. But it is always the totality of the distances between bodies, and without these it has no meaning. Empty space, Democritean 'nonbeing,' is, as previously stated, a negative predicate of a body, a place which is not occupied by a body, a place where there is no body at a given moment. But the negative predicate has meaning in the presence of a positive one: empty space is the potential location of a body, a location which can be, was, or will be occupied by a body and thus can be attributed to a body as its own predicate. But such an attribution is distinguished from real possession by the impossibility of its affecting the senses, by the absence of the sensory, empirical component of knowledge. This is the origin of purely speculative constructs which claim independence. Motion serves as the real basis for including nonbeing as something real, i.e., as incomplete being, lacking the empirical component. It permits us to view nonbeing — space — as deprived of a body, or as a place which the body can occupy, as a potential location, a potential predicate of the body. Such an equating of the potential location with the real one allows us to view space as distance, to identify locations that are at equal distances from a point, an axis, etc., to introduce systems of reference, metrics and, in general, to create geometrical and logical constructs.

The idea of space as a totality of potential predicates of bodies is a relativist idea. It deprives empty space of physical meaning if there are no sensorily perceived bodies in it and, accordingly, no connections between the spatial location of a body and its distance from other sensorily perceived bodies (or bodies theoretically allowing sensory perception!). Spatial distance must enter the world-picture as the potential trajectory of motion.

Similar ideas could probably be found even in Democritus, more likely in Epicurus, and we can see them with our own eyes, as it were, in Lucretius. One reservation is appropriate here. We can "see them with our own eyes" from a modern point of view, that is, from a modern, relativist, historical perspective as we examine Epicurus and his contemporaries through Einstein and *his* contemporaries. But historical vision is similar to simple physiological vision, which, according to Helmholtz, is impossible without apperception. 'Pure description,' without some system of coordinates, some point of reference, is as impossible in historical studies as in all other spheres. We must remember that an historical perspective has as its object not so much concrete and positive answers as questions; contemporary science, as it takes shape, considers only what is really addressed to it.

Lucretius writes about empty space as a necessary condition of motion. One property of corporeality, which distinguishes bodies from the void, namely, their ability to act on the senses, is connected with dynamic properties, specifically with impermeability. Therefore the idea of bodies in motion, that is, of their corporeality realized in motion, is inseparable from a representation of the void.

> Yet everything is not held close and packed everywhere
> in one solid mass, for there is void in things: which
> knowledge will be useful to you in many matters, and will
> not allow you to wander in doubt and always to be at a loss
> as regards the universe and to distrust my words. Therefore there is
> intangible space, void, emptiness. But if there were none,
> things could not in any way move; for that which is the
> province of the body, to prevent and to obstruct, would
> at all times be present to all things; therefore nothing
> would be able to move forward, since nothing
> could begin to give place. But as it is, we discern
> before our eyes, throughout seas and lands and the
> heights of heaven, many things moving in many ways and
> various manners, which, if there were no void, would not
> so much lack altogether their restless motion, as never
> would have been in any way produced at all, since matter
> would have been everywhere quiescent packed in one solid mass
> (Lucretius 1975, *1*, lines 329—345).[3]

Empty space is a condition not only of the movement of bodies but of their individualization as well. This problem of individualization, a fundamental problem for the mechanical picture of the world, applies only to bodies without fundamental qualitative distinctions. If a body does not have qualitative properties, but only geometrical ones, if it is not distinguished [in a qualitative way] from its localization, then can we speak of its existence? what distinguishes the body from the surrounding medium? This question, which Descartes was unable to answer, was resolved by the atomists through reference to the void. For them, the atom is surrounded by space, which lacks impenetrability, and, accordingly, does not act on the senses. Therefore the void limits the body. As we have seen, the void itself enters the world-picture in its turn because it is the totality of the distances between bodies and is limited by them. Hence we have the conception of the void and the bodies alternating [cf. below] which allows them to be determinate and is therefore infinite.

> Furthermore, nature withholds the sum of existing
> things from providing a limit for itself, because
> she compels body to be bounded by void and that
> again which is void to be bounded by body
> (Lucretius 1975, *1*, lines 1008—1011).[4]

This deduction of the *infinity* of the Universe from the spatial *limitation* of bodies and the void throws light on the epistemological nature of the Epicurean view of the world.

The universe of Aristotle was finite, and this corresponded to the peripatetic combination of the sensory and the rational comprehension of the world. These are indivisible, and speculative thought could not manipulate objects that were, in principle, inaccessible to sense perception. The spatially infinite world would have been such an object. For Epicurus the sensory and the rational comprehension of the world were no longer fused, but complementary. Speculation, operating on the possible predicates of the body, the places for its possible dislocation, determined the boundaries of the body itself in its actual being, its boundaries at the given moment. The body in its actual being limits space and attaches reality to it. Here we find that which is complementary, mutually exclusive, and confers reality on one another: it is a going out beyond the boundaries of a body, a determining of its localization and its individuality (*determinatio est negatio*), and a limiting of such an extension — a reminder to us that bodies are boundaries for the manipulation of purely spatial definitions, distances, geometric images, empty space. Objective thought, thought about being, reflects the similarity of a given individual object to others and replaces individual objects with general concepts in which their dissimilarity is suppressed. But each step in this speculative direction demands a limitation, namely the empirical record of nonidentical, individual, actually existing bodies. Such an operation cannot stop. Cessation would be an incorrect concept without physical meaning precisely because, from Epicurus's point of view, there are such things as bodies that are not limited by the void and a void that is not limited by bodies.

The idea of the homogeneity of space flows from the representation of the infinity of the Universe, in the sense that in space there is neither a center nor boundaries, nothing to which a privileged system of reference could be referred. But Epicurus presents an idea missing from both Democritus and Aristotle – the anisotropy of space. Atoms move downwards and the concepts of 'up' and 'down' have absolute meaning. Here Epicurus's positivist views are retrogressive in comparison with those of his immediate predecessors. But what does the 'questioning' side of the idea of the anisotropy of space consist of?

It consists of an idea of a cosmic law to which the motions of atoms are subject, a law which has lost its connection with the *static* scheme of the Cosmos. For Democritus there was no such scheme, but there was also no law to which the cosmos as a whole was subject. Atoms move in all directions, and do not possess as a background any regulated or ordered movement common to the cosmos. For Aristotle the cosmos is ordered by a *static* system of natural locations. Epicurus depicts the *kinetic* harmony of the world: straight-line movements in one and the same direction are characteristic of atoms, stemming from the general cosmic law.

Epicurus's imaginative conception of kinetic harmony, in ordering being and levelling individual variations, clashes with two opposing questions. One of these questions refers to the physical: from cosmically ordered parallel movements, how is the observable world formed, a world in which atoms are grouped into macroscopic bodies and precisely because of this, are able to act upon the senses, thus acquiring physical being? The second question refers to man: will man keep his freedom, i.e., his human 'being' if nature is subject to a single cosmic law determining the behavior of the atoms comprising that nature?

2. *CLINAMEN*[5]

The answer to these questions is connected with the Epicurean conception of the microcosm — or more precisely, the ultramicrocosm, which we cannot penetrate by direct observation. Is the absolute power of the cosmic law of the parallel paths of atoms, and of macroscopic laws as a whole, preserved in the ultramicroscopic world? There is a still more radical quandary: are the continuous motion and continuous existence of atoms preserved here?

Let us return to the problem of infinity, no longer the infinitely extended Universe, but rather infinitely small segments of space. The idea of infinity derives from Epicurus's idea of the alternating [in the sense that void is bounded by body, and body by void] boundaries of bodies and the void, which result from the unlimited addition of finite elements. But from this idea we can derive another conception of infinity that results from the unlimited division of space into ever smaller parts. Epicurus, following Democritus, views space as a totality of potential positions belonging to the moving atom, that is, as a predicative heterogeneity, a series of diverse spatial predicates of one and the same self-identical subject. If we can make the differences between these predicates as small as we wish, then we obtain a continuous predicate series — infinity as a result of unlimited division. If the differences between these predicates (in this instance, distances) can be as large as we wish, then we obtain infinity as a result of expansion. The Greek atomists and Lucretius held the latter concept. But what about the former, i.e., the representation of the continuity of space and movement?

To this complex question there is probably no one complete answer covering both Democritus and Epicurus and his students. We are dealing with an historico-philosophical and historico-physical problem based primarily upon the defective, fragmentary quality of the extant works of Democritus and Epicurus. A purely logical analysis of the problem is another matter. Does the idea of discrete or continuous space and motion follow from Epicurus's epistemological position?

A purely rational representation of the world, to the extent that it is possible, does not conflict with the limit of the division of space into ever smaller parts. Such a limit would interfere with the empirical in the process of speculative thinking and would, moreover, destroy its basic direction. In ascribing various predicates to a subject, we must be certain that we are considering the same self-identical subject. Its identification is the foundation of the rational understanding of motion. It was stated above that the preservation of the atom's predicates [as they relate to the atom as a physical substance] guarantees its self-identity. But here another guarantee is required. The possibility, in principle, of observing an atom from point to point and from moment to moment guarantees that, in a given instance, the prince is not replaced by Tom Canty, as occurred in Mark Twain's story, *The Prince and the Pauper*. But does not such continuity of motion destroy the physical being of the atom, its differentiation from its spatial localization, the differentiation of its motion from its trajectory, the differentiation of 'being' from 'nonbeing'?

Let us answer at once: if the continuity of motion is necessary for speculative thinking, which uses the potential positions of the self-identical atom, then a localized event, which cannot be reduced to the presence of the atom on its continuous trajectory, is necessary for the actual existence of the atom at a given moment. We shall see how this clash between continuous and discontinuous motion was realized by Epicurus and Lucretius.

These two attributes of motion refer to two worlds: to the macrocosm which acts directly on the senses, and to the microcosm — to discrete objects and events which cannot become objects of direct observation. Such a limitation upon observation begins with the atoms themselves, which, while smaller than any observed part of matter, do possess finite dimensions. This distinction ignores the behavior of atoms and, to judge by certain indirect evidence of the views of the Epicureans, the very existence of the atoms.

The behavior of atoms in macroscopic spheres that are accessible to observation differs from their behavior in the microsphere. In macrospheres, atoms move along continuous trajectories which are fully determined by cosmic law — by the downward motion and collisions of atoms. In the microspheres the atoms are subject to spontaneous deviations (*clinamen*). The idea of *clinamen* has double roots: the physical, and what may be called the moral, criteria or roots. Let us begin with the latter.

For Epicurus, the basic idea behind all human creative work is man's liberation from the power of religion, from the fear of death, from everything that paralyzes his personality and robs it of its truly human 'being'. Also included among those forces inimical to man is the absolute determinism of nature which excludes human freedom. In a letter to Menoeceus Epicurus says:

> For, indeed, it were better to follow the myths
> about the gods than to become a slave to the destiny
> of the natural philosophers: for the former suggests
> a hope of placating the gods by worship, whereas the
> latter involves a necessity which knows no placation
> (Epicurus 1926/79, § 134, p. 91, lines 1—5).[6]

The scientific ideas of Lucretius are less subordinated to the idea of moral harmony and freedom as the basis of human existence than are those of Epicurus. His artistic temperament leads the Roman poet to independent descriptions of nature with no concealed moral nuances. However, even for Lucretius, *clinamen* is connected with the defense of human freedom against scientific determinism. The destruction of macroscopic determinism in microscopic spheres serves as the guarantee of freedom.

> For it is weight that prevents all things
> from being caused through blows by a sort of external
> force; but what keeps the mind itself from having
> necessity within it in all actions, and from being as it
> were mastered and forced to endure and to suffer, is
> the minute swerving of the first-beginnings at no
> fixed place and at no fixed time
> (Lucretius 1975, *2*, lines 288—293).[7]

The moral motifs are combined with the specifically physical motifs. The strictly vertical paths of atoms rule out their collision with one another and the formation of macroscopic bodies in that way. Lucretius also begins the exposition of the concept of *clinamen* with this same consideration:

> One further point in this matter I desire you to understand: that while the first bodies are being carried downwards by their own weight in a straight line through the void, at times quite uncertain and uncertain places, they swerve a little from their course, just so much as you might call a change

> of motion. For if they were not apt to incline, all would fall downwards like raindrops through the profound void, no collision would take place and no blow would be caused amongst the first-beginnings: thus nature would never have produced anything.
>
> (Lucretius 1975, *2*, lines 216—224).[8]

3. INDIVIDUAL BEING

In the history of philosophy, the idea of *clinamen* given in the lines above was not viewed as a fundamental principle that could acquire increasingly new concrete forms when new phenomena, unknown to antiquity, were analysed. Marx was the only thinker to see a deeper clash of being in *clinamen*. Marx's youthful doctoral dissertation, 'On the Difference between the Democritean and Epicurean Philosophy of Nature'[9] is a work of the early left-Hegelian period of Marx's intellectual development; but even here, the lion can be recognized by his claws. Marx sees in *clinamen* Hegelian categories which were not yet turned right side up. But here Marx (as distinguished from the left-Hegelians, the Hegelians [proper] and all the philosophers who analyzed ancient atomism) proceeds from the questioning, future-oriented side of the views of Democritus and Epicurus, from the *aporias* and contradictions, from what is demanded for the resolution of ever newer empirical data and logical constructs. Thus Marx reveals what ancient atomism has in common with modern science, because, as we have already mentioned, they are united not by the similarity of their positive results, nor by the general levels of a static system where the old and the new are placed one after another, but by the link between old questions and new answers.

Included in such old questions is the question of individual being and its alienation. This question unites the ideas of Epicurus not only with [our] contemporary science, but also with its social effect. For Marx himself, this question turned out to be remarkably important for the ideological evolution that he underwent in the 1840's.

What fundamental questions did Marx see in the Epicurean conception of *clinamen*?

The straight line described by the falling body is not a totality of bodies, but a totality of places, or, in the simplest case, of points. Moreover, the points which make up the straight line are deprived of independence; the straight line absorbs them and they disappear within this line. Nothing happens at the points, which would distinguish the body. "Every body," writes Marx, "insofar as we are concerned with the motion of falling, is therefore nothing but a moving point, and indeed a point without independence, which in a certain mode of being – the straight line which it describes – surrenders its individuality [*Einzelheit*]" (Marx 1975–, **I**, p. 48).

Later, Marx introduces Aristotle's observation concerning the behavior of Pythagorean constructs: the motion of a line forms a surface; the motion of a point forms a line. In such an instance the motion of a monad also forms a line. If so, if the motion of a monad or an atom consists only in the transfer from one point to another, then the atom, as it is distinguished from space, no longer exists. "The consequence of this for the monads as well as for the atoms would therefore be – since they are in constant motion – that neither monads nor atoms exist, but rather disappear in the straight line . . ." (Marx 1975–, **I**, p. 48).

Marx says that the swerve of the atom from the straight line is not a peculiar, individual, chance attribute, but the expression of a law pervading all of Epicurean philosophy. In Marx's opinion, "Lucretius therefore is correct when he maintains that the declination breaks the *fati foedera*, and, since he applies this immediately to consciousness, it can be said of the atom that the declination is that something in its breast that can fight back and resist" (Marx 1975–, **I**, p. 49). Here Marx is referring to the verses of Lucretius which describe man's resistance to external force:

> . . . yet there is in our breast something strong
> enough to fight against it and to resist?
> (Lucretius 1975, 2, lines 279–280).[10]

The rebellion against *fati foedera* [bonds of fate] within the breast of the atom, which is analogous to the ability to struggle that is concentrated in the breast of man, is not only a metaphor. For Epicurus, this rebellion shatters absolute scientific fatalism, and liberates man. Therefore, the generalization of the 'rebellion' is by no means an arbitrary *rapprochement* of phenomena which are different in nature. Moreover, in physics itself, the problem of individual being is a real problem. It pervades not only Epicurean philosophy but also modern science, and allows us to see the limitations of the Epicurean concept of 'rebellion.'

For Epicurus, individual being has a defensive character. "The purpose of action is to be found therefore in abstracting, swerving away from pain and confusion, in ataraxy" (Marx 1975–, **I**, p. 51).[11] The modern conception of individual being is different, dynamic, active, in the area of not only the moral ideal (which we will touch on later in connection with Epicurean hedonism), but also the modern ideal of physical explanation. Modern science by no means views elementary particles as only passive entities, ignored by the deterministic macroscopic laws of the universe, in contrast to classical thermodynamics, which had in fact relegated such a role to particles. Modern science no longer confines itself to the statement of a 'local rebellion' by the particle, the local uncertainty of its dynamic variables. The local rebellion becomes the basis for a chain reaction; it alters the movement of macroscopic processes too. Furthermore, it is only when the local rebellion alters the macroscopic world-line (relatively macroscopic, as we are dealing with a particle track, the trace on the photographic plate) that the local event becomes physically significant.

Here we find a fundamental property of science: it always gives more than it is asked for. Whatever the applied, technical or moral problems presented to it, science responds by giving them a more general twist. Flight from *ad hoc* explanations, aspiration toward inner perfection, toward the overall solution which Poincaré called 'elegance,' is an integral characteristic of science. In solving applied problems, science always attempts to follow the example of Hercules's sixth labor — diverting the flow of the river Alpheus in order to fulfill Augeas's demand to cleanse the stables. For Epicurus, as well as for Lucretius, the driving force in creative work was the striving to free man's consciousness from the fear of death and the gods. However, Epicurus's thought tended toward more general physical conceptions which, while they could not yet be embodied in simple and reliable results, could still find a moral equivalent, not in the passive evasion of pain, but in the active transformation of life. For Lucretius such an expansion of physical and moral conclusions was deepened by a poetic love for nature and an active social temperament. We shall now examine the philosophical conclusions of Epicurus and Lucretius which went far beyond the boundaries of the scientific and moral ideals of antiquity, and turned out to be closest to the scientific and moral ideals of Einstein.

4. ISOTACHY

As previously discussed, in a comparison of Epicurus's physical ideas with modern ones, the key is the questioning, predictive, future-oriented aspect of ancient atomism. But modern science also experiences its own peculiar form of castling. This can be seen in the example of the *clinamen*. In 1915–16, V. I. Lenin summarized Hegel's lectures on the history of philosophy and marked Hegel's attack on Epicurean conceptions with a note, "And electrons?"[12] In those days, the relatively established views held that electrons did not possess spontaneous deviations that limited the strict macroscopic determinism of their movements, the simple dependence of their movements on fields of force. The physics of that time had not yet achieved any such simple results. But Lenin's treatment of physics in his philosophical works by no means led to the qualification of simple results. Lenin was interested in the tendencies of science, its perspectives, its dynamics. If we now re-read the [scientific] literature of the beginning of the century, we will encounter a number of hypothetical views which, as a whole, do not present a simple picture of the states of the electron, but on the other hand do characterize the state of physics, that is, the great amplitude of its oscillations, the considerable confusion concerning possible subsequent directions, and the threat of a revision of extremely well-entrenched conceptions. Many surprises were expected from the electron, and it must be said that it justified such expectations. In particular, the electron removed the odium of paradox from spontaneous movements, thereby destroying the total power of simple and precise determination in the microworld, and, perhaps even more important, replaced this odium with the recognition of the heuristic value of paradox in science.

At the present moment, the role of the electron as the transformer of the historical perspective has not come to an end. Rather, it has been absorbed within the collective role of the entire group of elementary particles. But here the idea of *clinamen* becomes inevitably interwoven with another Epicurean idea, that of *isotachy* — the single uniform velocity of atomic motion.

This idea is expounded in Epicurus's 'Letter to Herodotus' in connection with the differentiation between continuous time which is accessible to the senses and discrete time which is accessible only to thought. One point must be noted in passing, which is essential not only for Epicurus, but also for the history of rationalism in its connection with science, and for the problem of being and knowledge as a whole. Epicurus is considered the harbinger of the empirical and sensualist trend, distinguishing him from Democritus. This is a just distinction: the legend about Democritus (who blinded himself so that sensory, visual impressions would not interfere with the activity of reason) is as typical of the philosopher from Abdera as the frequent remarks about the truth of sensory impressions are of Epicurus.

But even here, at the sources and prototypes of all the subsequent trends of philosophy, the indivisibility of the sensual and the rational comprehension of being is apparent in principle. When Democritus deprived the atom of sensually comprehensible, qualitative properties and distinctions, he apparently became involved in an area of difficult *aporias* in a colorless, world picture. On the other hand, for Epicurus, sensory perception and its rational comprehension relate to different worlds: the senses perceive what can be immediately identified as the macroscopic world; reason comprehends the microscopic world. For Epicurus both are valid. Democritus went from the true world of atoms to the illusory, multicolored world of sensually comprehensible bodies. Epicurus shows the way from the true, objective world of atoms to the true, objective, macroscopic world of observable bodies. This epistemological principle is distinguished from the old absolute opposition — empirical *vs.* rational comprehension; it transfers the problem to the objective world, makes it ontological, sees in reality itself something that can be comprehended by both reason and the senses. But these worlds cannot be torn away from each other. Macroscopic movements are real because *they* are formed by microscopic processes accessible to reason; the processes are real because the motion that is accessible to the senses is formed *from them.*

Accordingly, the rationally comprehensible world where atoms possess an equal and invariant velocity plays a part in the physics of Epicurus. This is the primary constant of the microworld — the continuous, microscopic, very great but finite velocity of the atoms.

Epicurus can define it only subjectively and qualitatively. He calls the velocity of the atoms the *velocity of thought*, or incomprehensible velocity (referring to sensory perception and the concretely imagined form). Atoms move in the void at this velocity until they collide with other atoms, after which, although they change their direction, they do not change their absolute velocity (Epicurus 1926/79, §§61–62, pp. 36–39). The collisions of atoms lead to the idea that a macroscopic velocity of a body consisting of atoms is variable (". . . for the motion of the whole body," says Epicurus, "will be the outward expression of its internal collisions . . . (Epicurus 1926/79, §62, p. 39).[13]

Some obscure lines of Epicurus's 'Letter to Herodotus' enable us to interpret this idea as follows. The elementary free paths of the atoms (called '*kinema*') give a variable macroscopic velocity to the body owing to the variable asymmetry of these paths. If these paths were uniformly symmetric, the body would remain macroscopically motionless; if they were uniformly asymmetric, if all the atoms consistently ran their elementary courses in one and the same direction, the body would move in that direction with a constant 'velocity of thought,' that is, with a maximum speed equal to the limiting velocity of the *kinema*. In reality, the atoms collide and change direction, and the real macroscopic movement of the body as well as, approximately, the macroscopic movement of each atom for "the period of time accessible to sensory perception," is very small in comparison with the limiting velocity, and depends upon the prevalence of random displacements in one direction or another.

In the second book of his poem, Lucretius explains a special case of macroscopic motion: the body itself at rest during a period of equilibrium among the random displacements comprising it and the atoms in motion:

> One point in these matters need cause no wonder, why, though all the first-beginnings of things are in motion, the sum total seems nevertheless to abide in supreme quietude, except for anything that may show movement with its own body. For the nature of the first things lies all hidden far beneath our senses; therefore, since you cannot get so far as to see the things themselves, they must necessarily steal their motions too from your sight, especially when things that we can perceive do yet often conceal their motions if they be withdrawn at a great distance. For often on a hill, cropping the rich pasture, woolly sheep go creeping whither the herbage all gemmed with fresh dew tempts and invites each,

> and full-fed the lambs play and butt heads in fun; all which things are seen by us blurred together in the distance, as a kind of whiteness at rest on a green hill. Besides, when great legions cover the outspread plains in their manœvres, evoking war in mimicry, and the sheen rises to the sky and all the country around flashes back the brilliancy of bronze, and beneath, the ground quakes, resounding with the mighty tramp of men's feet, and the mountains, stricken by the clamour, throw back the sounds to the stars of heaven, and horsemen gallop around and suddenly course through the midst of the plains, shaking them with their mighty rush, yet there is a place on the high mountains, from which they seem to stand still, and to be a brightness at rest upon a plain (Lucretius 1975, *2*, lines 308–332).[14]

Does motion exist?, or more precisely, does the Aristotelian *φορά*, 'localized movement,' displacement in ultramicroscopic intervals of space and time, in "moments comprehensible only by thought", exist? Epicurus writes of such motion. However, did not he and his students have suspicions as to the mechanical nature of microprocesses? The combination of the subject's self-identity and the changes in its predicates by means of which the atomists wanted to resolve the Heraclitean-Eleatic confrontation, was based on the continuous existence and motion of the atom. But when the behavior of the atom was separated from the sensorily perceived macroscopic world, the certainty that the atom repeats on a small scale the acts that are characteristic of macroscopic bodies could be shaken. Or rather, the certainty that the atom repeats the *mechanics* of the visible world, the motion of self-identical bodies, could be shaken. Other macroscopic processes, which Aristotle united in the concepts of qualitative and substantial motion, could also claim the role of model of the microworld. We are concerned here with uncertainty, hesitations, searching, and not with a kind of carefully elaborated nonmechanistic conception of the behavior of the atom; and this (together with the disappearance of many ancient works) explains the indefinite answer that the history of science and philosophy will give to the question of the existence of such nonmechanistic conceptions. But the uncertainty of the answer relates to the time, the representations and the dissemination of nonmechanistic conceptions. That they did exist is beyond doubt.

Atomism explained both qualitative changes and the origin and disappearance of bodies in terms of the motion of atoms. In a nonmechanistic atomism, the atom ceases to be a self-identical substratum of motion. The processes of destruction and creation, as elementary processes from which continuous motion is formed, pervade the atomic picture of the world.

At the beginning of the second century A.D., Alexander of Aphrodisias wrote of the Epicureans: "In maintaining that space and motion and time consist of indivisible particles, they maintain that the moving body moves over the entire expanse of space which consists of indivisible parts, and in each of the indivisible parts included in space there is no motion, but only the result of motion."[15]

"There is no motion, but only the result of motion." This phrase can signify the annihilation of the atom in a given cell of discrete space and time, and its regeneration in a neighboring cell. This result coincides with the result of motion. But, bearing in mind the Aristotelian concepts of coming-to-be (*γένησις*) and passing-away (*φθορά*), how can we explain these processes, which are located even deeper than the continuous motion of atoms from which the dynamics of the universe are formed? It is difficult to imagine that such questions did not occur to the Epicureans. But it is easy to imagine that Epicurus and his students reconciled themselves to the lack of an answer. Science goes further than is demanded of it, but it stops in its tracks if it does not receive new practical and moral impulses. For Epicurus, the moral impulse of philosophy was its striving to free people from the fear of death and the gods. Later, as Epicurean ideas developed and changed, they provided new answers connected with new, more radical, active and widespread stimuli. In the history of philosophy and science, the modification of ideas is not reduced to a logical deduction which guarantees the pervasive character of the problem and the self-identity of the object of historical analysis. The modification of ideas follows from new experimental evidence which makes the demonstrated self-identity non-trivial. But this is the least of it. A change takes place in the stimuli which move science. And precisely because of this, the history of philosophy and science is a part of general history, whose central component is the evolution of mankind's moral ideals.

5. A MODERN EPILOGUE TO EPICUREAN PHYSICS

The demarcation of the macro- and microworld was one of the critical stages in the history of knowledge primarily because such a demarcation demonstrated the indivisibility of rational knowledge and empirical observation, and, at the same time, the relative autonomy of these approaches, both of which were subsequently demonstrated in ever more precise and concrete forms. However, these demonstrations seem insufficiently precise in comparison with those of modern nonclassical science, which has made indivisibility and relative autonomy a component of knowledge by virtue of their physical connection.

In 1932, in his paper, 'Remarks on the Change in the Formulation of Problems in Theoretical Physics,' Einstein wrote:

> At the present time it is well known that science cannot grow on the basis of immediate experience alone and that in the construction of science we turn to free invention whose utility may be verified *a posteriori* by means of experiment. These factors may have escaped the attention of previous generations to whom it seemed that theoretical creations developed out of experience by induction, without the creative effect of the free formation of concepts. The more primitive the state of science, the easier it is for the scientist to live in the illusion that he is a pure empiricist. Even in the nineteenth century many people believed that Newton's fundamental principle — "hypotheses non fingo" — must serve as the basis of every sound natural science.
>
> Recently the alteration of the theoretical system has led to the fact that the recognition of the speculative character of science has become a universal property (Einstein 1932, p. 363).[16]

Ancient Greek philosophy sensed the immediate connection of its positive constructs with a similar declaration. Ancient atomism was a direct and conscious transition to the sphere into which observation cannot penetrate. For Epicurus it included the representation of the impossibility, in principle, of seeing ultramicroscopic movements of atoms. At the same time, for both Epicurus and Democritus the fundamental problem was the empirical proof of the atomic structure of matter. Epicurus resolved it by the hypothetical transition from *clinamen* to motions subordinate to macroscopic laws, and from *kinema* to macroscopically observable continuous motions. For Lucretius the atomists' real or illusory empirical proofs formed the basic content of his poem.

As a result of the powerful and fruitful intervention of pure speculation ('pure,' naturally, in the conventional sense, i.e., subject to empirical verification), the representation of nature included the concept of empty space. Space was viewed as the possible locations of bodies, separate from them but necessary for their motion and, consequently, for their influence on the sense organs. The concept of empty space never displaced the picture of a filled-in continuum. It must just be stressed that in Newton's system, this concept played a fundamental role. Space was no longer only a totality of the possible locations of a moving body; it played the role of distance across which forces acted. After the Aristotelians and Cartesians, this new filling of space had a dynamic meaning. The forces acting between bodies were united in the concept of the field as acting between bodies and a medium that for a long time had claimed the role of an all-pervasive body that filled all the space between ordinary bodies. The theory of relativity put a stop to these claims, and the ether disappeared from the world-picture. The field could not serve as a reference body: after having been given the role of all-pervasive substance, the field could not simply be included in Democritean 'being' along with atoms and the bodies they comprise. After the appearance of the theory of relativity, the Democritean antithesis changed. Within the framework of both ancient atomism and classical science, one side of this antithesis — 'nonbeing', the void — was the domain of *ratio*, and was subject to rational comprehension. The other side — 'being', the atoms — was the source of sensory feelings, and acted upon the sense organs. The unification of sensation with speculation occurred when spatial properties, positions, and distances were ascribed to atoms; the rationally comprehensible scheme of the universe was reduced to these properties of the atoms.

In relativity physics, distances lost their purely rational, geometrical character. Distance is the distance traversed by a body, the spatial component of a four-dimensional space-time process occurring in time — *motion* as something comprehended by the senses and not only through speculation. A certain materialization or reification of the rationally comprehensible 'nonbeing' took place. But it became *new* 'being', and was not simply equated with atoms as in the mechanical theories of the ether.

The discussion above refers to the positive content of the theory of relativity and to ancient atomism as a whole. As an epilogue to Epicurus's specific ideas, it is necessary to touch upon the unresolved, future-oriented, 'predictive' problems of the theory of relativity.

Einstein said that macroscopic statements of the theory of relativity should be deducible from the microstructure of bodies.[17] This is a new step which makes the indivisibility, complementarity and synthesis of Democritean 'being' and 'nonbeing' apparent. In the theory of relativity, distances are straight edges [rigid rods] that measure the positions of particles, i.e., their distances from reference bodies, and the distances from one position of a particle to another — the spatial component of the segment of the world-line. Einstein made these distances depend on material reference bodies, a dependence, however, which had to be made concrete by picturing the effect of the atomic structure of the straight edges on their length. Thus what we are dealing with is the deduction of the properties of the macroworld from those of the microworld.

We encounter this problem in Epicurus and his students. *Clinamen* is a characteristic of the microworld; it is impossible macroscopically to foretell and record the spontaneous deviation of an atom. Nevertheless, it is precisely because of *clinamen* that macroscopic bodies are formed. *Kinemas* are elementary swerves of atoms with one and the same macroscopically unattainable velocity; this is also a characteristic of the microworld. From *kinema* the macroscopic movements of bodies with variable velocity are formed. Both *clinamen* and *kinemas* belong to history. These concepts raise a question that concerns the here-and-now as well as predictions of the future: how is the macroscopically ordered world formed from the chaos of microprocesses?

The theory of relativity can answer this question only on the basis of concepts introduced by quantum mechanics, primarily the concept of the probability of the elementary processes of being.

To a considerable extent quantum mechanics makes concrete the mutual penetration of 'being' and 'nonbeing'. In other words, in a more precise way, it reveals *being* without quotes, in which mutually exclusive definitions are united. Quantum mechanics regards the behavior of the atom (the modern atom, i.e., the microparticle) as something subordinate in the general case only to the laws of probability. For each point in space, at each moment of time, there is a certain probability magnitude for the location of the particle. The localization of the particle is the point of the greatest probability of its presence. Space (Democritean 'nonbeing') is a mixture of the probable localizations of the particle (Democritean 'being'). Within the framework of ancient atomism, 'nonbeing' is real because it consists of the possible locations of atoms. For Democritus and Epicurus 'being' is the actual particle, 'nonbeing', the potential one. But today this idea has not only been transformed from a qualitative contrast into a quantitative one, nor is it only a transition from potential to probable presence (with a precise quantitative characterization of the probability). Here the particle — 'being' — is inseparable from the waves of probability which fill space — 'nonbeing'.

When we compare the positive content of ancient science with the positive content of modern science, we are within the bounds of a static generalization of science. However, when we compare the unresolved problems of the past with the unresolved problems of the present, we find ourselves within the framework of dynamic generalization, i.e., within the boundaries of proper historical analysis: historical transformation is not only expressed in the difference between contrasted conceptions but also permeates each of them. In the case at hand, the most important conceptions in ancient atomism are *clinamen* and *kinemas*, and in contemporary science, the conception of the vacuum.

The problem of the vacuum is one of the least developed problems in modern physics (in the sense of having a unified noncontradictory theory). It raises serious difficulties and far-reaching hopes, both relating to the mechanism of the transition from the chaos of virtual particles and processes to an ordered world, where there are (in comparison with 10^{-24} sec) long-lived particles in motion, and the world-lines of such particles exist. This is already something new, something entirely different from the ancient controversy between 'being' and 'nonbeing'. Ancient atomism saw in atoms something *empirically* comprehensible and saw in the void the sum total of *speculatively* comprehensible geometrical forms. But the nonclassical 'vacuum,' in replacing the classical 'void', turns out to be the source of those predicates of the particle which endow it with empirical comprehensibility, and the concept of a world-line, with physical meaning. Vacuum processes give mass and charge to the particle (unfortunately, the calculation of this leads to infinite values which are difficult to eliminate).

Was there even a hint in antiquity of the problem of the transition from ultramicroscopic chaos to the macroscopically ordered world? The concepts of *clinamen* and *kinema* were such hints. The quotation from Alexander of Aphrodisias demonstrates that ancient thought perhaps connected ultramicroscopic processes, in principle inaccessible to empirical comprehension, with the absence of internal motion ("there is only the result of motion") and with the discreteness of space and time.

But these tendencies were not realized, and the hints remained only hints. Apparently this is not only the result of the lack of many ancient works. In his poem, Lucretius touches upon the concepts of *clinamen* and *kinema* only in passing. Why was so little attention paid to a problem which seems so interesting now? This question, going beyond the framework of ancient atomism, is connected with the general philosophical and moral ideals of Epicurus and Lucretius.

6. THE PROBLEM OF DEATH AND THE FEAR OF DEATH FOR EPICURUS AND FOR EINSTEIN

As soon as we pass from the question "How is the world constructed?" to the question "How does our representation of the construction of the world change?", we immediately build a bridge from studies of nature to studies of man. It stands to reason that the objective logic of facts and generalizations is the basis of the history of science; in this sense the history of science is inseparable from science itself and functions with the immanent forces of its development. But the history of science is not reducible to the analysis of immanent forces. The law of gravity follows from Kepler's laws, but this link explains neither the tempo of the transition to a gravitational representation, nor the concrete forms of the transition. In such problems it is already impossible to ignore the ideals, interests and psychological peculiarities that are specific to the age. Nonclassical science — with the inborn tendency of both the theory of relativity and quantum mechanics to review the possibility of comprehending truth by means of expounding the truth, with its constant attraction to epistemological concepts for the formulation of ontological problems, with its inclusion of theoretical knowability within the definition of being — is in this sense a very 'human' science.

Therefore Einstein's moral and generally humanitarian interests, his attitude toward specifically human problems, is not only a biographical trait but also an historical one, characteristic of the new science, not in the sense of the unification in one person of scientific and moral ideals, but rather in their very interweaving and even their synthesis. Still more characteristic is the unification of scientific and moral authority; the scholar becomes the conscience of the epoch and combines within himself a great intellect and a great heart.

Was such a combination a realization of Epicurus's ideal: man, in understanding the structure of the world, is freed from the fear of death and the gods and acquires a serene, harmonious joy of being?

Again, this is not only a biographical question, nor even only an historical and philosophical one: It is inseparable from the more general question of the influence of modern nonclassical science on the relation of man to his fate, on the realization of moral harmony, and on man's emotions. Here, within the comparison of the emotional effect of science on Epicurus and Einstein, we shall limit ourselves to the question that was of paramount importance for Epicurus, the question of death and the fear of death.

Let us introduce the well-known exhortation against the fear of death, one which has endured through the centuries — the lines from the letter of Epicurus to Menoeceus:

Become accustomed to the belief that death is nothing to us. For all good and evil consists in sensation, but death is deprivation of sensation. And therefore a right understanding that death is nothing to us makes the mortality of life enjoyable, not because it adds to it an infinite span of time, but because it takes away the craving for immortality. . . . So death, the most terrifying of ills, is nothing to us, since so long as we exist, death is not with us; but when death comes, then we do not exist (Epicurus 1926/79, §§124—125, p. 85).[18]

Probably the Greek thinkers chatting in Epicurus's garden sensed the logical irrefutability of this thought. Probably Epicurus himself did not fear nonbeing. Before his death he wrote to Idomeneus: "On this truly happy day of my life, as I am at the point of death, I write this to you" (Epicurus 1926/79, §30, p. 127).[19] He goes on to describe the agonizing pains which are the evidence of the approach of death, and then he says: "But against all this is the joy in my heart at the recollection of my conversations with you" (Epicurus 1926/79, §30, p. 127, 129).[20]

However, this logical overcoming of the fear of death was not the prevailing mood of the ancient world, even among the intellectual elite. Apparently another attitude was widespread — a reconciled, twilight sadness, the quiet regret for life which is passing that pervades the *Odyssey* [as, for example in Book V when we first meet Odysseus: "Life with its sweetness was ebbing away in the tears he shed for his lost home."].

For Einstein, the Epicurean formula "death is nothing to us" became so constant an attitude that he did not even return to the formula itself in his thought. We possess a certain amount of evidence on this point, and all that we know of Einstein speaks of its absolute reliability.

Leopold Infeld recalled the words of Einstein: "Life is an exciting show. I enjoy it. It is wonderful. But if I knew that I should have to die in three hours it would impress me very little. I should think how best to use the last three hours, then quietly order my papers and lie peacefully down" (Infeld 1941, p. 294, cited in Kuznetsov 1965, p. 360).

In 1916, during a serious illness that threatened his life, Einstein spoke of death so calmly that Hedwig Born (wife of Max Born) asked him whether he was not afraid of death. "No," he replied, "I feel myself so much a part of everything living, that I am not in the least concerned with the beginning or the termination of the concrete existence of any person in this eternal flow" (Seelig 1956, p. 36, cited in Kuznetsov 1965, p. 359).

Beyond its purely personal characteristics, such an attitude toward death expressed Einstein's own particular *Weltanschauung*, which was related to his scientific achievement, the style of his scientific thinking and the character of nonclassical science. As we come to this idea, we recall that Spinoza, whose ideas had a powerful influence on Einstein's *Weltanschauung*, said, "A free man thinks of death least of all things, and his wisdom is a meditation of life, not of death" (Spinoza (1982), *Ethics*, Pt. IV, Prop. 67, p. 193).[21] The idea of death and the fear of death are not only logically devalued (as in Epicurus), but, psychologically speaking, they are pushed out of the consciousness as well, by thoughts of life, by the sense of the continuity of that eternal flow of life which Einstein mentioned to Hedwig Born.

Why did Spinoza ascribe such a rejection to the *free man*? And what is the connection between nonclassical science and the relation of man to the eternal flow of life?

The infinity of such a flow acts on the individual consciousness and on each localized state of the individual consciousness because this is the 'true infinity,' in Hegel's sense: it is reflected in each of its elements, and is not reducible to simple repetition and the accumulation of finite elements. This reflection of the infinite in the finite, the whole in the local, was what Feuerbach had in mind when he said: "Every minute one drinks to the dregs the chalice of immortality which is filled again like the goblet of Oberon."

The sense of infinity and of merging with nature does not allow man to drain this chalice of immortality. The words of Schiller — "Are you afraid of death? Do you want to live for ever and ever? Live in the Whole! It abides when you are long gone."[22] — relate not to nature as an entity, not to a cosmic process, but to the process of infinite human life, infinite knowledge, and the infinite transformation of nature and the conditions of man's life. Human freedom, as something defined in positive terms, as something not reducible to a negative definition or to an independence from the whole, lies precisely in the factual possibility of individual, local influence on this dynamic, fluid, living whole;

freedom includes both the dependence of local events on the whole, and the dependence of the whole on local variations.

Nonclassical science has rendered concrete and, to some extent, altered the Hegelian concept of 'true infinity', which in classical science corresponded to the dependence of local events on a universal law. The function that serves as the physical equivalent of true infinity is defined in terms of an infinite plurality and in terms of the law that uniformly acts at each point and determines the local events, which thus become a reflection of the law, entire and infinite. The world-line — the endless number of world points — is an image of infinity, where the presence of the particle at each point, each local event, is precisely determined by a differential law. In nonclassical science, the world-line is not only blurred, but can also be altered as a whole under the influence of local variation.

This is certainly not a semantic matrix for the laws of human behavior and for the relation of the individual to the whole, to nature, to its knowledge, to the historical process. Nonclassical science possesses a specific effect in its practical embodiment: the fluid and changeable character of its laws provides a high degree of dynamism for all areas of production and culture, and, moreover, the integral changes in these areas can be the result of a 'chain reaction' begun by an individual and local creative act. In freeing man from the oppression and inorganic power of elemental social forces, nonclassical science helps to abolish the alienation of the personality and to fill the consciousness with the problems of immortal life, forcing from consciousness the fear of death, which was already logically discredited by Epicurus.

Still, the problem of death has not disappeared from people's consciousness. It is not completely expunged by the logically irrefutable words of Epicurus, the *amor intellectualis* of Spinoza, or the transformation of man's thought and action that resulted from the principles of nonclassical science and the realization of its potential. All these are stages in the discreditation and disappearance of the *fear* of death. But the thought of death and the sense of its irreversibility is retained. They are connected with the most general problems and ways of being and knowledge. What we are dealing with here is, in a very subtle sense, a result of the flow of time, which carries off everything and retains nothing. This sense permeates ancient culture, and if it is almost elusive in the *Iliad* where it is muffled by the sounds of battle, then, as was already mentioned, in the *Odyssey*, the reconciled lyrical sorrow for what is irretrievably passing away is the constant mood in the story and, even more, the undercurrent of the narration. It is not fear, not the agonizing image of impending nonbeing, but calm sorrow. This melancholy, twilight sorrow for that which is irretrievably passing away and for inevitable oblivion are exactly the same today. The grief is for each impression, no matter how fleeting, which will be buried in nonbeing. The image of Lethe was always imbued with grief for the concrete, the local, the individual attached to the given moment, with grief for the 'now', that, giving way to the new, is plunged into the waters of Lethe.

P. Muratov, in *Images of Italy* (1924), writes about a painting by Giovanni Bellini, 'Souls in Purgatory', a strange composition, which provokes an unaccountable sense of some estrangement from being. He considered the picture a representation of the waters of Lethe.[23] In general, the culture of the Renaissance was well acquainted with sorrow for what is irretrievably passing away — the passing day and even the passing moment, the constant passing of man, the passing life. Dante, in the *Purgatorio*, writes about the religious pilgrim who hears the distant bell for vespers:

> It was now the hour that turns back the
> longing of seafaring folk and melts their
> heart the day they have bidden sweet
> friends farewell, and that pierces the new
> pilgrim with love if he hears from afar a bell
> that seems to mourn the dying day . . .
> (Dante (1970—75) II: *Purgatorio* VIII, lines 1—6, p. 77)[24]

Sometimes it is not twilight, but morning sorrow: the new day erases the impression of the old; this is the image of the passing of the concrete and local into nonbeing. This is morning as described at the beginning of the ninth *canto* of the *Purgatorio*,

> At the hour near morning
> when the swallow begins her sad lays, perhaps
> in memory of her former woes, and when our
> mind, more a pilgrim from the flesh and less
> captive to thoughts, is in its visions almost divine
> (Dante (1970—75) II: *Purgatorio* IX, lines 14—18, p. 87).[25]

Sorrow for what has passed accompanies thoughts about the future.

It is felt by reason, which is "a pilgrim from the flesh," liberated from the old and penetrating the new, a reason both wise and prophetic. Odysseus's sad note of farewell to that which is passing is inseparable from the predictive component of thought.

Is it possible and necessary for wise and prophetic reason, in its optimistic and predictive drive toward the future, to free itself from sorrow at both parting with what is passing and foreseeing the death of the individual and the local? As far as we can judge the past and future of reason, such an attitude toward the local and individual is alien to it. For without the local and individual, being disappears. And it disappears without replacement, without the flow of time carrying off each 'now' into nonbeing; but it [also] disappears if the 'now' is *only* a momentary and nonessential splash of infinite being, and the individual only a nonessential embodiment of the immortal whole. Sorrow for the passing moment, for the day, for the period, and above all for the passing individual life of a man, is the emotional side of the acknowledgement of the value and uniqueness of individual and local being. It is the emotional side of the *ontological* value of individual being.

Later, when our discussion turns to the conceptions of Spinoza, it will be necessary to return to the problem of the fear of death and to the pure sorrow at the frailty of individual being that remains even while fear is overcome.

7. THE GODS OF EPICURUS AND THE 'GOD' OF EINSTEIN

The philosophy of Epicurus as a whole, and at first glance, if I may put it this way, does not require filling the consciousness with the problems of *change* in the representations of the world and of the changes in the world. This philosophy reflects the characteristic features of a slowly developing, quasi-static civilization. But, like any philosophy, Epicureanism also reflects those tendencies of an age that are claimed by the future and it is precisely these tendencies which allow us, at second glance, to see in Epicurus's views the prototype of the philosophical ideas of later ages, to compare them with the latest conceptions, and to discover not a simple similarity and difference, but the evolution of an idea.

From this point of view, let us look at the gods of Epicurus who lived in the space between worlds, absolutely indifferent to the affairs of the world, inert because of their perfection. People laughed at this idea. "And yet," writes Marx

these gods are no fiction of Epicurus. They did exist. *They are the plastic gods of Greek art. Cicero*, the *Roman*, rightly scoffs at them, but *Plutarch*, the *Greek*, has forgotten the whole Greek outlook when he claims that although this doctrine of the gods does away with fear and superstition, it produces no joy or favour in the gods, but instead bestows on us that relation to them that we have to the Hyrcanian fish,[26] from which we expect neither harm nor advantage. Theoretical calm is one of the chief characteristics of the Greek gods. As *Aristotle* says, "What is best has no need of action, for it is its own end" (Marx 1975–, **I**, p. 51)

In reality, the static quality of the ideals — which, because of that static quality, become no longer ideals but canons — is characteristic of all of Greek culture: of its art with its perfect, plastic[27] canons, of its stable codes of morality, and of [its conception of] the static harmony of the universe which achieved its most stable embodiment historically in the Aristotelian scheme of the fixed center of the universe, a universe with limits, naturally-occurring spaces and constant, perfect motion.

The conception of happiness for Epicurus lies within the framework of a static scheme. According to him, happiness consists in the absence of chagrin, including that of unsatisfied needs. Therefore happiness is guaranteed by limiting needs. This negative hedonism, this purely negative definition of happiness in terms of the absence of woe, means the limitation of needs by the traditional, the known, the unchanged.

The greatest sum of precisely such negative happiness is the ethical criterion. Einstein, having read Solovine's book on Epicurus, wrote to Solovine:

That he is basically right in his ethics can hardly be doubted. On the other hand it still seems to me that he has not exhausted the subject because the values he feels are positive are to some extent incommensurable and cannot be added or subtracted directly. If we assume, for example, that we were convinced that the sum of happiness for ants is greater than that for people, would it then be right from the ethical point of view for people to yield their place to ants?[28]

In principle elementary and traditional values can probably be made commensurate and added together. But the value connected with a change in tradition and with the re-evaluation of what is considered happiness are not commensurate with them, and it is precisely these values [related to change] that distinguish people from ants in the example. If we limit ourselves to measurable values, then the sum of people's happiness may be compared with the sum of ants' happiness, and the larger sum may be considered the criterion of the higher ethical level. If we equate happiness with the degree of satisfaction of traditional needs, then the sum of happiness for ants may seem much greater than that for people. The historical transformation of the concept of happiness is characteristic of man; the concept contains values that are, in principle, incomparable and are, collectively, not comparable with elementary concepts. They guarantee the transformation of life, the velocity and acceleration of this process, and include ever newer values, in particular, new knowledge of being as it conforms to laws.

For Epicurus, this knowledge is included in the sum of values, insofar as it frees man from sorrowful submission to the gods and from what is just as sorrowful and groundless, the fear of death. But the matter doesn't rest here: When the mortality of the soul is proven and man is liberated from the fear of death and the tortures of the afterlife, when the [self]-determinism of being is proven and man is liberated from the fear of the gods, when *clinamens* liberate man from enslavement to the "destiny of the natural philosophers," from fatalism, the stimuli of knowledge disappear. [See Epicurus above, pp. 77–78.]

In this respect Epicurus stands in sharp contrast to the tragic figure of Democritus who constantly sensed the contradictions and *aporia* of the atomistic world picture, who drove himself toward empirical knowledge, and in the end, according to legend, killed himself.[29] Democritus constantly found collisions and contradictions in nature. Epicurus, on the other hand, never encountered anything anywhere that could surprise him or cause him to doubt the truth of what reason and observation communicated to him.

But here the "second glance" comes into play, the dynamic and dialectical tendency of Epicurean philosophy. For Epicurus, reason and observation were two agents of knowledge, possessing equal rights. Both reason and observation give man an adequate representation of reality, but observation (Epicurean *φύσεως θεωρία*) uncovers the macroscopic world while speculation, reflection on nature (Epicurean *φυσιολογία*) reveals the microcosm. The Epicurean canon seems to immortalize this demarcation. But at the same time Epicurus takes one step toward the problem where the isolation of observation and reflection is broken because the isolation of the macroworld and the microworld disappears. This is the problem of the origin of the macroworld, the formation of observable bodies from microparticles that are comprehensible only to thought, and what is needed for this is the *clinamen* of the observed movements arising out of the *kinema* that is comprehensible only to thought. But this is only one step. It suffices for the negative goal, for the elimination of the fear of the gods. Epicurus goes no further.

Lucretius does. For him the representation of *kinema* and *clinamen* was not conceptually adequate and here he did go no further than Epicurus. But for Lucretius the encyclopedia of nature was an open book, wherein it is not only stated but also demonstrated that the sensorily comprehensible [macro]world is formed out of the rationally comprehensible microworld. Reason intervenes in the sensorily comprehensible [macro]world and sensory images intervene in the [rationally comprehensible] microworld.

Such intervention is, to a certain degree, connected with the embodiment of the artistic genius of Greece – "with the plastic gods of Greek art" on new, no longer Greek, but Roman soil. Epicurus ascribed to the gods a purely aesthetic value: They are excellent, perfect. This perfection excludes an active role in the universe and the search for the *rerum natura* leads to the gods' exclusion from the universe. But the artistic genius of Greece as embodied in these mythological images imparted to Lucretius's godless world picture its specific connection between the rationally comprehensible *rerum natura* and sensually known objects. In Ia. M. Borovskii's article, 'The Poetics of Proof in Lucretius' (1945), there is a very subtle analysis of the epistemological function, the aesthetic features and the poetic fabric of *De rerum natura.* Lucretius unites into one concept an examination [discrimination] and a judgment (*species ratioque*) and, as Borovskii says,

> The essential trait of his tenor of thought and poetic diction is reflected here. Both are distinguished by tremendous concreteness and an intimate connection with visual representation. 'Examination' for Lucretius is not a metaphor but an entirely adequate characteristic of his consistent relation to the subject. He sees before him those things and processes about which he speaks and obliges his listener to see them, not only in the narrative and descriptive parts of the poem but also where the direct goal is the scientific proof of the explanation given to the phenomena of nature by the atomistic system (Borovskii 1945, **II**, p. 202).

Here it is necessary to point out one circumstance which is most essential to the prehistory of rationalism and to the history of philosophy as a whole. Lucretius unites in the term *species ratioque* two concepts, *naturae species* and *naturae ratio*, which do not entirely coincide with the Epicurean equivalents. *Naturae species*, says F. A. Petrovskii in the commentary to his translation of Lucretius's poem (1945), translates the Epicurean *φύσεως*, the only difference being that *species* indicates the result of known action, that is, "external appearance, image," revealed by means of contemplation, and *θεωρία*, the act of contemplation itself. *Naturae ratio* is the Epicurean *φυσιολογία* but again with the same difference in point of view: *naturae ratio* itself is "the internal law, the principle at the basis of the phenomena of nature discovered by means of the contemplation of nature," while *φυσιολογία* is really "judgment, meditation about nature" (Petrovskii 1945, **II**, p. 324).

Thus Epicurus's purely epistemological concept acquires for Lucretius an ontological meaning, namely, the idea of the objective *ratio* of the world, which was indeed present in Einstein's, but in an unclear form. For Lucretius it acquired a more precise character and, most importantly, was no longer reduced to the status of a canon: it unites *naturae species* and *naturae ratio*, the sensorily comprehensible image and the rationally comprehensible essence of being. The search for the objective *ratio* of the world was no longer entirely subordinated to the moral and psychological task of freeing the human psyche from illusory fears, the fear of the gods and tortures in an afterlife that constitutes the fundamental reason for all crimes and all human sorrows. The search becomes an autonomous need for man; it arouses in him an emotional élan, and a sense of the beauty and grandeur of the search itself, independent of its moral effect. At the beginning of the third book of his poem, Lucretius, turning to Epicurus, says:

> Thereupon from all these things a sort of
> divine delight gets hold upon me and a
> shuddering, because nature thus by your power
> has been so manifestly laid open and uncovered
> in every part
>
> (Lucretius 1975, *3*, lines 28—30).[30]

Here it is no longer a negative and static task that arouses "a sort of divine delight and a shuddering" but the rational comprehensibility of nature ("nature . . . so manifestly . . . uncovered"). This is the psychological turning point from the inert "plastic gods of Greek art", to the new gods who are just as metaphorical as the gods of Epicurus.

Or they are even more metaphorical. Epicurus eliminated the gods from the Universe and their only reality turned out to be the plastic images in poetry, painting and sculpture. Now, the halo of beauty and perfection is transferred to rationally comprehensible nature, and the term 'God' comes to be a metaphorical synonym for nature in an essentially atheistic world-view.

Such is the God of Spinoza – *deus sive natura*. This is an irrational pseudonym for cosmic *ratio*, which is by no means a static scheme of immutable objects and movements. For Spinoza nature is rationally explicable not only in terms of its geography, motion, interaction or, in general, of the *behavior* of the bodies that comprise it, but also in terms of the very existence of these bodies. If statements about motion consist of the subject ("*what* moves?") and the predicate ("*moves*!"), then Spinoza explains in a rational, causal manner not only the predicate, but also the subject. For such an explanation, Spinoza introduces a concept of being which is its own reason (*causa sui*). This is a remote antecedent of the 'self-action' of substance that figures in several modern physical conceptions. It goes far beyond the framework of classical science, which, not having attained the paradoxical *causa sui* and the causal explanation of the very being of the Universe, sought explanations for the modal processes of the behavior of bodies. For Spinoza nature is not only created nature (*natura naturata*) but creating nature (*natura naturans*). Certainly this is no longer a static system; there are no *a priori*, and therefore, no fixed definitions.

Now let us turn to the 'God' of Einstein. The quotation marks are justified here: Einstein used the term 'God', for the most part, as an ironic pseudonym for world order, causal order transforming chaos into cosmos. The term 'religion' has a psychological and moral meaning for Einstein and signifies both the sense of world harmony and the purpose of existence. Accordingly, 'God' signifies being's material conformity with laws, its rational, causal conditionality which is material in nature. The 'God' of Einstein has nothing in common with a personal god. "I cannot conceive," wrote Einstein, "of a God who rewards and punishes his creatures, or has a will of the type of which we are conscious in ourselves. That an individual should survive his physical death is also beyond my comprehension, nor do I wish it otherwise; such notions are for the fears or absurd egoism of feeble souls" (Einstein 1934, p. 242).

Here the question of the terminological concessions to religion is a biographical one. The history of philosophy and science raises a question about the character of that cosmic harmony, which, for Einstein, is hidden under this (let us repeat, for the most part ironic) pseudonym. Let us limit ourselves to two aspects of this cosmic harmony: its paradoxical nature, and the precisely simple, dynamic meaning or, contrariwise, the probable meaning, of the laws of being.

On one of the stone fireplaces at Princeton University there is carved a quotation of Einstein: "Raffiniert ist der Herrgott aber boshaft ist er nicht" ("Subtle is the Lord, but malicious He is not").[31] The laws of being are paradoxical but comprehensible, and their paradoxical nature, so precisely demonstrated by nonclassical science, was revealed before and at each major turning point within the framework of classical science. At present, the paradoxical turning points of science have become the almost constant stuff of its evolution, and the great dynamism that relativity and quantum physics transmit to contemporary culture is, in fact, dependent upon them. The paradoxical quality of science and the dynamism of man's being are now directly linked. During the time of Epicurus, the static quality of man's being and the static quality of cosmic harmony as portrayed by ancient science corresponded to the static moral ideal and to the image of the gods, who did not interfere in the fate of the universe. For the modern scholar, Spinoza's God, who coincides with an infinitely complex, truly refined nature, is closer, posing unexpected riddles to man and then revealing to him increasingly unexpected answers.

Let us discuss first the dynamically simple or static character of the fundamental laws of being, or, to use Einstein's expression, whether God plays dice. "You believe in the God who plays dice, and I in complete law and order in a world which objectively exists . . ." wrote Einstein to Max Born in 1944 (Einstein 1971, p. 149).[32]

The God who does not play dice does not entrust to chance the determination of each event, but determines it himself in a simple and precise way — such a god is compatible with classical statistical physics. Behind the macroscopic laws of thermodynamics that govern only the probable behavior of the particles stand the precise, dynamic laws of mechanics, determining, in a simple way, the fate of each particle. In quantum mechanics, elementary events — the presence of the particles at world points — are determined by their probability.

But are these events really elementary? Does the concept of the uncertainty of the particle's behavior mean anything without its polar opposite — the strictly determined world-line of the particle? Quantum mechanics says no. The uncertainty of the particle's behavior is a physical concept without content if there are no classical concepts of the particle's impulse and position, and no image of the particle corresponding to these classical concepts, i.e., an image of the particle moving continuously along a simply determined trajectory. The most paradoxical aspect of quantum mechanics is not its rejection of classical concepts, but the inseparability of such a rejection from classical presuppositions, from the existence of the rejected image. Moreover, the concept of uncertainty, in its turn, provides physical meaning for these classical images and concepts. The particle, whose properties are all reducible to localization and to a change of localization, does not have physical being and its world-line remains a purely geometric concept. Quantum uncertainty blurs the world-line, but this is precisely what gives the world-line its physical being.

This complementarity of mutually exclusive concepts is most clearly apparent in relativistic quantum physics. Here are concepts recalling the Epicureans' views as expounded by Alexander of Aphrodisias: motion is the result of interrupted transmutations. Such transmutations provide the world-line with physical being, but are themselves an empty concept without the image of the continuous world-line.

This is something entirely new as compared with Epicurus's ideas. For the Greek thinker, ultramicroscopic destruction limited macroscopic conformity to law and protected nature and man from fatalism and enslavement to the implacable "destiny of the natural philosophers." In modern science, the representation of ultramicroscopic events does not so much limit as impart physical content to the power [Epicurean 'destiny'] of macroscopic laws. For Epicurus, the gods had nothing to do with this power. The 'gods' of modern science are pseudonyms for the causal harmony of being, for lawfulness and order in nature. They play dice with all events, determining their probability or directly pointing them out with a simple image. It turns out that the god of Einstein, in not playing dice, and the god of Heisenberg, Bohr and Born, in controlling the probabilities of events, represent complementary concepts. Macroscopic causality in general relativity really does govern the world precisely because it grows out of the ultramicroscopic events which are subordinate to ultrarelativistic causality.

This particularly paradoxical and contradictory situation in physics stands in contrast to the soothing tendency of Epicurus to eliminate contradiction. But it continues and renders concrete the dialectical tendency in Greek thought, which searches for contradictions and is oriented toward the future. What is the relation of the contemporary situation to the problem that most interested Epicurus, namely that of human happiness?

Here we must mention still another, by now entirely metaphorical god. Rousseau, in his *Discourse on the Arts and Sciences*, recalled a legend that was transmitted from Egypt to Greece, in which it is said: "... some god, who was an enemy to the repose of mankind, was the inventor of the sciences" (Rousseau n.d., p. 140).[33] This was true even in antiquity, truer still in the seventeenth century when differential laws were at the basis of science and the static harmony of the universe was replaced by a dynamic one, and is now completely obvious. To all appearances, the static hedonism of Epicurus – happiness as a negative category, as the absence of pain, based on the limitation of needs and desires – is now impossible. But happiness as something positive can take place only in the presence of the acceleration of the growth of satisfied desires and the acceleration of the degree of their satisfaction. Thomas More in *Utopia* asked whether people would not get used to the continuous satisfaction of their desires and whether, as a result, the sense of happiness would not disappear, and here the restless deity of science provides an answer. Science changes not only the productive potentials of man – the level of the practical embodiment of ideal physical schemata – but also the schemata themselves, thereby giving progress not only continuous velocity, but acceleration as well. However, science is hostile to human tranquility in more than its applied results. Knowledge itself and, moreover, the very change in representations of nature become a necessity for man, and the satisfaction of this need (not only for knowledge, but also for the tempo of its growth and modification) becomes one of the sources of human happiness.

It is precisely this tempo and, what is of particular importance, the continuity of the modification of fundamental representations of the world, the explicit character of such modification which is inherent in nonclassical science, that serve as the basis for contemporary epistemological optimism.

NOTES

[1] "Auf jeden, der nicht ganz im Geiste unserer Zeit aufgeht, sondern seiner Mitwelt und speziell der geistigen Einstellung der Zeitgenossen gegenüber sich gelegentlich als Zuschauer fühlt, wird das Werk von Lukrez seinen Zauber ausüben" (Einstein in Lucretius 1923–24, **2**, p. VIa).

[2] quae tamen omnia corporea constare necessest
natura, quoniam sensus inpellere possunt;
tangere enim et tangi, nisi corpus, nulla potest res.
(Lucretius 1975, *1*, lines 302—304).

[3] Nec tamen undique corporea stipata tenentur
omnia natura; namque est in rebus inane.
quod tibi cognosse in multis erit utile rebus
nec sinet errantem dubitare et quaerere semper
de summa rerum et nostris diffidere dictis.
quapropter locus est intactus inane vacansque.
quod si non esset, nulla ratione moveri
res possent; namque officium quod corporis exstat,
officere atque obstare, id in omni tempore adesset
omnibus; haud igitur quicquam procedere posset,
principium quoniam cedendi nulla daret res.
at nunc per maria ac terras sublimaque caeli
multa modis multis varia ratione moveri
cernimus ante oculos, quae, si non esset inane,
non tam sollicito motu privata carerent —
quam genita omnino nulla ratione fuissent,
undique materies quoniam stipata quiesset.
(Lucretius 1975, *1*, lines 329—345).

[4] Ipsa modum porro sibi rerum summa parare
ne possit, natura tenet, quae corpus inani
et quod inane autem est finiri corpore cogit,
ut sic alternis infinita omnia reddat
(Lucretius 1975, *1*, lines 1008—1011).

[5] *Clinamen* = a turning aside, swerving. Cf. Lucretius, *2*, 292, and below, pp. 77—78.

[6] "*ἐπεὶ κρεῖττον ἦν τῷ περὶ θεῶν μύθῳ κατακολουθεῖν ἢ τῇ τῶν φυσικῶν εἱμαρμένῃ δουλεύειν· ὁ μὲν γὰρ ἐλπίδα παραιτήσεως ὑπογράφει θεῶν διὰ τιμῆς, ἡ δὲ ἀπαραίτητον ἔχει τὴν ἀνάγκην*" ('Epistula ad Menoeceum', in Epicurus 1926/79, § 134, p. 90, lines 1—4).

[7] pondus enim prohibet ne plagis omnia fiant
externa quasi vi; sed ne mens ipsa necessum
intestinum habeat cunctis in rebus agendis
et devicta quasi cogatur ferre patique,
id facit exiguum clinamen principiorum
nec regione loci certa nec tempore certo.
(Lucretius 1975, *2*, lines 288—293).

[8] Illud in his quoque te rebus cognoscere avemus,
corpora cum deorsum rectum per inane feruntur
ponderibus propriis, incerto tempore ferme
incertisque locis spatio depellere paulum,
tantum quod momen mutatum dicere possis.
quod nisi declinare solerent, omnia deorsum,
imbris uti guttae, caderent per inane profundum,
nec foret offensus natus nec plaga creata

principiis: ita nil umquam natura creasset.
(Lucretius 1975, *2*, lines 216—224).

[9] ["This work, completed in 1840—41, before Marx was 23, was not published in full even in German until 1927, when it appeared in [(Marx 1927—35, **I**, i, 1, pp. 1—144) of which pages 83—144 consist of preparatory materials] . . . An English translation (replacing an inferior earlier one) has recently been published in [(Marx 1975—, **I**, pp. 25—107)]. Cyril Bailey, reviewing the original publication in [his 1928] was greatly impressed with its scholarship and its originality: he found it 'of real interest to a modern student of Epicureanism' and ended by saying that such a student would find in it 'some illuminating ideas'" (de Ste. Croix 1981, p. 23) — Eds.]

[10] tamen esse in pectore nostro
quiddam quod contra pugnare obstareque possit?
(Lucretius 1975, *2*, lines 279—280).

[11] ["Ataraxy — in ancient Greek ethics [*ἀταραξία*] — tranquillity. In Epicurean ethics — the ideal of life; state of the sage who has attained inner freedom through knowledge of nature and deliverance from fear of death" (Editor's note in Marx 1975—, **I**, p. 736, n. 21) — Eds.]

[12] [Castling is a defensive chess move (to protect the King from future attack) that also serves to advance the game. Here Kuznetsov seems to be using castling as a metaphor to reiterate his theme of modern science's backward-looking, historical perspective and the capacity of this perspective to inform present and future scientific problems: in this case it would seem to apply to Lenin's modern defense, from an historical perspective and against Hegel's attack, of Epicurus's atomism. Epicurus's words about the atom "are the guess-work of genius and *signposts for science* . . ." (italics in text). Just below this statement Lenin mentions the curvilinear motion ascribed to atoms by Epicurus and in a marginal note directly opposite is the question, "and electrons?", alluding to the possibility that perhaps they too, like the atoms, possess 'spontaneous deviation' (Lenin 1963 (*R*), pp. 266—267; 1961 (*E*), p. 294). — Eds.]

[13] "*ἀντικοτῇ γὰρ ὅμοιον ἔσται* . . ." (Epicurus 1926/79, §62, p. 38).

[14] Illud in his rebus non est mirabile quare,
omnia cum rerum primordia sint in motu,
summa tamen summa videatur stare quiete,
praeterquam siquid proprio dat corpore motus.
omnis enim longe nostris ab sensibus infra
primorum natura iacet; quapropter, ubi ipsa
cernere iam nequeas, motus quoque surpere debent,
praesertim cum, quae possimus cernere, celent
saepe tamen motus spatio diducta locorum.
nam saepe in colli tondentes pabula laeta
lanigerae reptant pecudes quo quamque vocantes
invitant herbae gemmantes rore recenti,
et satiati agni ludunt blandeque coruscant;
omnia quae nobis longe confusa videntur
et velut in viridi candor consistere colli.
praeterea magnae legiones cum loca cursu
camporum complent belli simulacra cientes,
fulgor ubi ad caelum se tollit totaque circum

aere renidescit tellus subterque virum vi
excitur pedibus sonitus clamoreque montes
icti reiectant voces ad sidera mundi
et circumvolitant equites mediosque repente
tramittunt valido quatientes impete campos —
et tamen est quidam locus altis montibus unde
stare videntur et in campis consistere fulgor.

(Lucretius 1975, *2*, lines 308—332)

[15] [Alexander's (fl. 198—209 A.D.) commentary on Aristotle's *Physics*, on which this passage is based, is not extant, and we have been unable to locate it in any other of Alexander's works. A later commentator, Themistius (*c.* 317—388 A.D.; and Simplicius later still) made an almost identical statement:

> But our clever friend Epicurus is not ashamed to use a remedy more severe than the disease — and this in spite of Aristotle's demonstration of the viciousness of the argument. The moving object, he says, *moves* over the whole distance, but over each of the indivisible units of which the whole is composed it does not move but *has moved.*
>
> [e] *ἀλλ' ὁ σοφώτατος ἡμῖν 'Επίκουρος οὐκ αἰσχύνεται χρῆσθαι φαρμάκῳ τῆς νόσου χαλεπωτέρῳ καὶ ταῦτα 'Αριστοτέλους τὴν μοχθηρίαν τοῦ λόγου προεπιδείξαντος. ἐπὶ μὲν γὰρ τῆς ὅλης τῆς* ΑΒΓ *κινεῖται, φησί, τὸ κινούμενον, ἐφ' ἑκάστου δὲ τῶν ἀμερῶν, ἐξ ὧν σύγκειται, οὐ κινεῖται, ἀλλὰ κεκίνηται.* (Themistius *Physics* 184.9 = Usener 1963, §278 — in Furley 1967, pp. 113—114).

As Furley notes, "Aristotle showed that an indivisible body could not move except on the supposition that time, motion, and extension are themselves all composed of indivisible units; and he claimed to have demonstrated the impossibility of this. Epicurus accepted the conclusion that *all* these things must be composed of indivisibles if any are, but asserted that this is in fact the case" (Furley 1967, p. 114).

Kuznetsov's sentence, "There is no motion, but only the result of motion" corresponds to "does not move but has moved," and "by saying not that the atom 'moves' (*κινεῖται*) over the minimum of distance in the minimum of time, but that it 'has been moved' (*κεκίνηται*): the action is instantaneous and complete as soon as begun" (Bailey 1928a, p. 316; see also footnote 1 on the same page. For the similar passage in Simplicius, see Usener 1963, end of §278; or Simplicius 196—, p. 934, lines 25—30) — Eds.]

[16] "Wir wissen heute, dass aus der nächsten Empirie allein Wissenschaft nicht erwachsen kann, dass wir bei dem Bau der Wissenschaft auf die freie Erfindung angewiesen sind, die erst a posteriori an der Erfahrung auf ihre Brauchbarkeit geprüft werden kann. Diese Tatsachen konnten früheren Generationen entgehen, denen die theoretischen Schöpfungen auf induktivem Wege aus der Empirie ohne das schöpferische Wirken einer freien Begriffsbildung herauszuwachsen schienen. Je primitiver der Stand der Wissenschaft ist, desto leichter kann der Forscher in der Illusion leben, dass er reiner Empiriker sei. Noch im 19 Jahrhundert glaubten viele, dass Newton's Grundsatz "hypotheses non fingo" jeder gesunden Naturwissenschaft zugrunde gelegt werden müsse.

In der letzten Zeit hat der Wandel der theoretischen Systeme dazu gefuhrt, dass die Erkenntnis vom spekulativen Character der Wissenschaft Allgemeingut geworden ist" (Einstein 1932, p. 363).

17 Einstein in Schilpp 1949, pp. 59, 61.
18 "*Συνέθιζε δὲ ἐν τῷ νομίζειν μηδὲν πρὸς ἡμᾶς εἶναι τὸν θάνατον· ἐπεὶ πᾶν ἀγαθὸν καὶ κακὸν ἐν αἰσθήσει· στέρησις δέ ἐστιν αἰσθήσεως ὁ θάνατος. ὅθεν γνῶσις ὀρθὴ τοῦ μηθὲν εἶναι πρὸς ἡμᾶς τὸν θάνατον ἀπολαυστὸν ποιεῖ τὸ τῆς ζωῆς θνητόν, οὐκ ἄπειρον προστιθεῖσα χρόνον, ἀλλὰ τὸν τῆς ἀθανασίας ἀφελομένη πόθον. . . . τὸ φρικωδέστατον οὖν τῶν κακῶν ὁ θάνατος οὐθὲν πρὸς ἡμᾶς, ἐπειδήπερ ὅταν μὲν ἡμεῖς ὦμεν, ὁ θάνατος οὐ πάρεστιν*" (Epicurus 1926/79, §§124—125, p. 84).
19 "*Τὴν μακαρίαν ἄγοντες καὶ ἅμα τελευτῶντες ἡμέραν τοῦ βίου ἐγράφομεν ὑμῖν ταυτί*" (Epicurus 1926/79, §30, p. 126).
20 "*. . . ἀντιπαρετάττετο δὲ πᾶσι τούτοις τὸ κατὰ ψυχὴν χαῖρον ἐπὶ τῇ τῶν γεγονότων ἡμῖν διαλογισμῶν μνήμῃ*" (Epicurus 1926/79, §30, p. 128).

This quotation as well as the one in n. 18 are both cited verbatim in Diogenes Laertius's 'Life of Epicurus' (Epicurus 1926/79, §22, p. 154 (Greek) and p. 155 (English)).
21 "Homo liber de nullâ re minùs, quàm de morte cogitat, & ejus sapientia non mortis, sed vitae meditatio est" (Spinoza (1925), *Ethices*, Pars IV, Prop. LXVII, p. 261).
22 The English translation is adapted from Schiller (1902), p. 306. The German poem, from the group called 'Die Horen' ('The Hours') is, in its entirety:

'Unsterblichkeit'
Vor dem Tod erschrickst du? Du wünschest unsterblich
zu leben?
Leb' im Ganzen! Wenn Du lange dahin bist, es bleibt.
(Schiller 1943—, I, p. 259)

23 Muratov (1924), I, pp. 11—19. [The Russian title of this chapter, 'летейские воды', echoes the Greek, *Λήθης ὕδωρ*, water of oblivion.]

The Bellini painting (c. 1488), which hangs in the Uffizi in Florence, is actually called 'Sacra Allegoria' ('Sacred Allegory'). For the various interpretations of this painting, as well as other information concerning it, see Pallucchini (1962), pp. 84, 86 and 150: that the allegory in question portrays a 'pèlerinage de l'âme' (pilgrimage of the soul) is no longer generally accepted. — Eds.]

24
Era già l'ora che volge il disio
ai navicanti e 'ntenerisce il core
lo dì c'han detto ai dolci amici addio;
e che lo novo peregrin d'amore
punge, se ode squilla* di lontano
che paia il giorno pianger che si more;
(Dante (1970—75) II: *Purgatorio*, VIII, lines 1—6, p. 76).

* [The bell of Compline (the canonical hour that, depending on the season, corresponds roughly to about 6 o'clock), the last divine office of the day. — Eds.]

25
Ne l'ora che comincia i tristi lai
la rondinella presso a la mattina,
forse a memoria de' suo' primi guai,
e che la mente nostra, peregrina
più da la carne e men da' pensier presa,
a le sue visïon quasi è divina
(Dante (1970—75) II, *Purgatorio*, IX, lines 13—18, p. 86).

[26] [Hyrcania was an ancient region on and south of the *Caspian Sea*, whose old name was Mare Hyrcanium. This inland sea is richly stocked with fish. – Eds.]

[27] [When Marx uses the word 'plastic' ("plastic gods of Greek art") in the quotation above, he means shaped or molded, plastic art being any art in which shaping or modelling is done, such as sculpture or ceramics. When Kuznetsov uses this same adjective ("perfect, plastic canons"), he is using it in its figurative sense: plastic, in reference to aesthetic or intellectual conceptions, means formative, procreative, creative (cf. *Oxford English Dictionary, s.v.*). Eds.]

[28] "Dass der Mann im Grossen recht hat mit seiner Ethik ist wohl kaum zu bezweifeln. Andererseits scheint es mir doch nicht den Gegenstand zu erschöpfen, weil die als positiv empfundenen Werte bis zu einem gewissen Grade incomensurable sind und nicht so ohne weiteres addiert und subtrahiert werden können. Angenommen z.B. wir wären überzeugt, dass die Glücksumme bei den Ameisen günstiger liegt als bei den Menschen, wäre es dann vom ethischen Standpunkte aus richtig, wenn die Menschen den Ameisen das Feld räumten?" (Einstein to Solovine, 26 August 1947, in Einstein 1956, p. 86).

[29] denique Democritum postquam matura vetustas
admonuit memores motus languescere mentis,
sponte sua leto caput obvius obtulit ipse.
(Democritus again, when ripe old age warned him that the recording motions of his mind were beginning to fail, of his own free will himself offered his head to death.) (Lucretius 1975, *3*, lines 1039–1041).

[30] his ibi me rebus quaedam divina voluptas
percipit atque horror, quod sic natura tua vi
tam manifesta patens ex omni parte retecta est
(Lucretius 1975, *3*, lines 28–30).

[31] [The English translation is from Pais (1982), p. vi (the page facing the Preface). Ronald Clark, in his biography of Einstein (1971), says that Einstein made this remark in Princeton in 1921 during a discussion of the Michelson-Morley experiment (see pp. 389–390). – Eds.]

[32] ["Du glaubst an den würfelnden Gott und ich an volle Gesetzlichkeit in einer Welt von etwas objectiv Seiendem ..." (Einstein to Max Born, 7 September 1944, in Einstein 1969, p. 204). This was not the first time that Einstein had said this to Born: on 4 December 1926 he wrote: "Quantum mechanics is certainly imposing. But an inner voice tells me that it is not yet the real thing. The theory says a lot, but does not really bring us any closer to the secret of the 'old one'. I, at any rate, am convinced that *He* is not playing at dice" (Einstein 1971, p. 91). ("Die Quantenmechanik ist sehr achtung-gebietend. Aber eine innere Stimme sagt mir, daß das doch nicht der wahre Jakob ist. Die Theorie liefert viel, aber dem Geheimnis des Alten bringt sie uns kaum näher. Jedenfalls bin ich überzeugt, daß *der* nicht würfelt" (Einstein 1969, pp. 129–130).) – Eds.]

[33] "C'etoit une ancienne tradition passée de l'Egypte en Gréce, qu'un Dieu ennemi du repos des hommes, étoit l'inventeur des sciences" (Rousseau 1964, p. 17).

4. THE RATIONALISM OF THE RENAISSANCE

1. REALISM AND THE REVOLUTION OF NONIDENTITY

The evolution of the Heraclitean-Eleatic clash examined in the preceding chapters was closely tied to the evolution of the representations of the abstract and the concrete. A single, fixed substance is associated with the representation of a general form, a general law, a general concept common to all the elements of such a substance at all moments of time. This allows the identification of these elements. On the other hand, the variation of substance in space and time, nonidentity, heterogeneity and motion are associated with concrete definitions which individualize each element.

The general evolution of ancient philosophy and scientific thought consistently demonstrated the indivisibility of the aspects of being just listed. An especially precise demonstration of such indivisibility was the atomism of Epicurus, in which macroscopic laws embracing a set of concrete individual events become the laws of being, if the individual events are to some extent autonomous. Now the representation of being as a synthesis of the identifying, abstract macroscopic aspect and the individualizing, concrete ultramicroscopic aspect has become a physical, very precise, experimentally supported representation. The quantum processes specific to the microworld, which blur the macroscopic world-line, give that line physical meaning; they make it a more abstract concept and, thereby a more concrete, experimentally perceptible, physical concept.

Dialectical philosophy views the process of abstraction as a transition to an increasingly complete comprehension of concrete truth. The concrete is the intersection of multiformity. We define an individual object in an increasingly concrete and complete way by intersecting it with an increasingly large number of abstract multiformities. There is no other way to isolate one given element from other elements identified with it than to refer it to the intersected multiformity. *This* point is isolated from the remaining points on the line because it is a point of intersection with another line. Without the ordering and identifying macroscopic lines, microprocesses do not possess being; they are not individualized.

The evolution of ancient Greek thought moved toward a similar synthesis of the fixed, identical and abstract, on the one hand, with the non-identical, changeable and concrete, on the other. But it also paved the way for the isolation of these polar opposites. Plato contrasts abstract ideas devoid of sensory concreteness with an illusory concrete being. For Aristotle ideas are a real moving force, forming sensorily comprehensible things.

The Frontispiece of this book is a reproduction of the central portion of 'The School of Athens'[1] by Raphael. The figures of Plato and Aristotle may symbolize to some degree the two components of knowledge, the purely logical and the sensory comprehension of the world. This book is dedicated to the synthesis of these two components. The history of rationalism is an increasingly precise demonstration of their indivisibility. There is still another symbol to examine here, namely the painting itself. The culture and, above all, the painting of the Renaissance were the apotheosis of the unity of thought and the sensory perception of the world, an apotheosis that stood in opposition to medieval thought. For Plato and for Aristotle, the demarcation between the world of abstract categories and the world of sensorily comprehensible things had not yet acquired the frozen and absolute form it took on in the Middle Ages.

In the twelfth century, the dominant tendency was Platonic realism, in which the representation of the reality of universals, which are independent of concrete objects, while coming from Plato, was dogmatized in comparison with the Greek original. In the next century, the Aristotelian tradition of the representation of real universals as the essence and moving force that form concrete objects became predominant in medieval realism. Still later, in the fourteenth and fifteenth centuries, a powerful opposing tendency arose: nominalism, which negated the reality of the universals by considering them to be only words — '*vocalia*' or '*nomina*'.

What is the origin of the abrupt cessation of the more profound movement in ancient thought — the development of the conception of knowledge as the *unity* of identifying abstraction and individualizing concrete definitions? It is not difficult to point to the role of social origins. Religious dogma and, subsequently, the entire system of the justification of the church hierarchy were based on realism. The Platonic state was the embodiment of a real universal, independent of the existence of separate, concrete people. Analogously, the church embodied the redemption of mankind from original sin. Just as the real existence of mankind as a universal is what forms the essence of man, so the fall of Adam is transferred to all people, and only the church liberates this real abstract mankind from the burden of sin.

In general, the dictatorship of the church, and the authoritarianism of medieval life as a whole, were connected with the assertion of the autonomous existence of an abstract, levelling Whole, while ignoring all the equivalents of the Epicurean revolt against macroscopic laws.

The rupture of being, the idea of the autonomous existence, first, of the ordering framework of macroscopic laws embodying abstract ideas, and second, of the un-ordered chaos of concrete objects, is a fundamental characteristic of medieval thought. The ancient harmony of abstract forms and the concrete sensory content filling them was exchanged for their opposition and isolation. The perfect heaven of disembodied spirits and the imperfect Earth, the reality of the 'city of God' and the illusoriness of earthly life, the reality of the eternal, of the immutable, and the illusoriness of changes — these isolated poles are characteristic not only of the dominant representations of the world, but also of the attitude toward the world, of man's feelings before the face of the world. During the Middle Ages and the Renaissance, the protest against what exists led to an apologia for chaos, the grotesque, the unexpected, for nonidentity and paradox.

The indivisibility of the two poles — abstract concepts and the macroscopic identification corresponding to them, and individualizing concrete definitions — becomes obvious in motion, in heterogeneity, in development. The clash between the abstract and the concrete is another aspect of the Heraclitean-Eleatic clash between the homogeneous and the fixed on the one hand, and the heterogeneous and the changeable on the other.

Ancient culture was neither homogeneous nor fixed. Stagnant technology was compensated for by the rapid growth of ties with the rest of the world. And the Greek genius learned very early how to find diversity in unity, and unity in diversity. In his *Lectures on the History of Philosophy*, Hegel says that the Greeks, drawing upon materials from Asia, Syria, and Egypt, "have so greatly obliterated the foreign nature of this origin, and it is so much changed, worked upon, turned around, and altogether made so different, that what they, as we, prize, know, and love in it, is essentially their own" (Hegel 1974, **I**, p. 150). To

Brunschvicg's (1949) comment on the genius of the artist who saw in nature the idea animating it (see p. 42 above) it may be added that *only* in nature, in the flow of ever newer impressions did the Greeks see the ideas ordering and uniting this flow: [only in] the flow of nonidentical, really new, really concrete sense impressions did they *see* and grasp ideas, laws, *ratio*. Hegel links this characteristic of the Greek genius with a sense of the beauty of being. Greek aesthetics saw in the beautiful something unified, represented in diversity. Such a definition corresponds to a real aesthetic sense of the world. Hegel writes:

> The stage reached by Greek consciousness is the stage of beauty. For beauty is the ideal; it is the thought which is derived from Mind, but in such a way that the spiritual individuality is not yet explicit as abstract subjectivity that has then in itself to perfect its existence into a world of thought (Hegel 1974, **I**, p. 153).

"The spiritual individuality is not yet explicit as abstract subjectivity" – in other words, the identifying function of reason, the search for general and abstract categories, does not lead to the appearance of 'abstract subjectivity', abstract forms which claim reality while nature is considered in its development and diversity. The development and diversity of nature compel the researcher to see how the identification of concrete elements destroys the wholeness of general concepts, as the universals are explicitly altered under the onslaught of concrete and vital impressions.

Only by acquiring immobility and immutability do universals begin to claim an *a priori* character and an autonomous reality.

Classical antiquity was not frozen. The models of classical art are immortal because they are imbued with the dynamics of the world and its diversity. The classical culture of ancient Greece was not classical in the modern sense, if only because it did not have classical models; it was in this sense the forerunner and those canons that were established only began the tradition without losing their connection with the vital force of nonidentity, non-canonized nature, which continuously reveals its changeability and diversity.

Medieval culture was considerably more stable. This does not mean that it was an entr'acte in history. The Middle Ages were a thousand-year night in a world where work goes on even at night. The Middle Ages were a period of almost continuous growth of economic, technical and intellectual potential which, at the end of the Middle Ages, led to an unprecedented expansion of mankind's horizon.

But the growth in potential — precisely because it was potential — was implicit and seemed (and in many cases was) a conformal transformation: the same forms of production, life and culture were repeated on a broader base; qualitative changes, transformation not of scale but of the forms themselves, were slow, and although each generation could register something new, it could not sense the very dynamic of the transition. Concepts of the canon and repetitions, sacral and fixed criteria hallowed by tradition, were preserved in religion, morality, aesthetics, mores, law, and economics (if only in the concept of 'fair', i.e., traditional profit). This official culture of identity created a breeding ground for the representation of the *a priori*, the sacrality, and independent reality of the results of the identifying, stable, ordering activity of reason.

The official culture stood in contrast to its opposite number, the culture of secularization, a term deriving from *saeculum*: century, meaning 'temporal', 'worldly', 'lay', as opposed to 'eternal', 'spiritual', 'ecclesiastical', and designating the take-over of church property for use by either secular powers or individuals. The essence of the unofficial, anti-canonical, heretical culture of the Middle Ages — a unique spiritual secularization — is the reflection of one such transition from the sphere of the sacred, which is independent of human experience, to the sphere of the temporal, which is dependent upon time and experience.

During the Italian Renaissance spiritual secularization acquired the breadth and depth of a genuine intellectual revolution. It altered viewpoints, representations of the world, and the method of thought itself, for a wide circle of people. No longer restricted to the closed circles of the scholastics, but in the squares of Florence, Rome, Venice and other Italian cities, people argued philosophical problems. But in these arguments the same opposition between abstract identity and the vital concrete diversity of the world still resounded, only now in a new way. In the struggle against a levelling identification, people turned to the real meaning of Greek philosophy. In it, even the most resolute advocates of real general concepts and of the levelling of the concrete, sensory world, did not break with the multicolored, heterogeneous, non-self-identical representation of nature. The path from a dogmatized ancient philosophy to a genuine interest in ancient culture — the movement called humanism — led consciousness from abstract identity to a new representation of the Universe and to a knowledge of it. However tarnished humanism became later on, however much a pedantic, purely literary devotion to old texts developed within its framework, the stream of vital interest in the multicolored concreteness of being was not interrupted until it flowed into the new science in the seventeenth century.

The Renaissance was the renaissance of man, his liberation from the oppressive surveillance of the *civitas dei* and the Scholastic universals that destroyed his individual being. Dante's poetry of genius, the great epoch in painting and sculpture – all this demonstrated the power of human reason: not reason limiting its functions to the identification of being, but reason seeking the truth. "Truth," said Hegel, "is a noble word, and the thing is nobler still. So long as man is sound at heart and in spirit, the search for truth must awake all the enthusiasm of his nature" (Hegel 1975, §19, p. 26). Never before had man been so enthusiastic at the thought of truth; never before had he testified so brilliantly to the health of his soul and spirit.

The general spirit and the general conviction of the Renaissance – in short, its all-pervasive emotional drive – [reflected a belief] in the unlimited possibility of knowledge of objective truth. As in every period in the radical transformation of the world, neither the progressive thinkers nor the participants in the transformation doubted the objectivity of the world, its knowability, its plasticity, or its capacity for change. Thinking about the world was filled with epistemological optimism.

Thought was free. Its freedom seemed an inalienable property, and for the people of the Renaissance the words 'free thought' would have seemed a redundancy.[2] This freedom of thought meant not only freedom from external authority, but also the freedom of genuinely constituted thought from abstract identification.

The Renaissance was the apotheosis of active rationalism, not only seeking reason in nature, but establishing it in society as well. At the end of that epoch, Shakespeare, with the greatest generality and the greatest concreteness (the unity of these criteria became the hall-mark of the age), demonstrated in *Hamlet* the tragedy of reason, a reason no longer satisfied with the rational character of its constructs and longing to restore the *ratio* of the world ("the time is out of joint").

The Renaissance was thus a revolution of non-identity, a revolution against the static reign of universals, an apologia for the individualizing function of reason.

2. THE RATIONALISM OF LEONARDO DA VINCI

More than any other in this book, the content of this chapter, which is devoted to the rationalism of the Renaissance, differs from its title. I opened it with a section on the clash between nominalism and realism, that is, on the basic confrontation of medieval scholasticism. Now, in this the second section, I shall deal with a great fifteenth-century artist, and then in section three, characterize the rationalist tendency in the world-view of Giordano Bruno, a philosopher of the cinquecento [sixteenth century] who died in 1600. Finally, in section four there is an analysis of the ideas of Galileo, a thinker who belongs not to the Renaissance, but to the Baroque. The following considerations serve as a kind of justification for such a departure from the age of the Renaissance, backward to the Middle Ages and forward to the seventeenth century.

If one is speaking of the connection between classical rationalism and science, then from one point of view the Renaissance turns out to be irrelevant. From the twelfth through the sixteenth centuries, there was neither classical rationalism nor science in the modern sense. But [considered] from a historical perspective, we can discover in this period intensive preparations for both, namely, the rationalist ideas that lead to science appear for the moment in artistic creation; the accumulation of empirical materials, which by their nature lead directly to abstract analysis; and the combination of traditional philosophical ideas which leads to a non-traditional outcome in science. The problem of the abstract and the concrete serves as the core of these ideological tendencies; thus their connection with discussions within the framework of the old scholasticism is quite obvious. On the other hand, in the works of Galileo, the genesis of classical science is intertwined with the continuity of humanist traditions. Therefore the influence of rationalism on the origin of science is revealed in the analysis of the ideas of the Renaissance, together with their scholastic prologue and their classical scientific epilogue.

In the culture of the quattrocento [fifteenth century] in its philosophical, scientific and aesthetic tendencies, there is something that recalls ancient Greek science. And so what Hegel said of the Greeks — that "the stage reached by Greek consciousness is the stage of beauty ... it is the thought which is derived from Mind, but in such a way that the spiritual individuality is not yet explicit as abstract subjectivity" (Hegel 1974, **I**, p. 153) – also applies here.

The lines that are drawn from art to science and to rationalist philosophy, which in turn is drawn toward science, are clearly visible in the work of Leonardo da Vinci. We shall examine his natural tendency to move from personal, localized impressions to the general, to space-time diversity.

Paul Valéry, in his essay, 'Leonardo and the Philosophers' (1929), says that the representation of purely localized scientific criteria is opposed to beauty, which is independent of time, place and localized criteria.[3] This aesthetic criterion was transferred to science; the invariance of beauty became the invariance of truth, the invariance of universal laws. For Leonardo, beauty consists in the extension of the individual, in its emergence into the universal in terms of space and time. Here there is a shift of accent: for the Greeks, beauty lies in the local incarnation of the general. And it is even more characteristic of Leonardo that he saw in such a space-time extension of the individual something common to both art and science.

The physics of Leonardo is a qualitative physics; it does not deprive the world of qualitative definitions as do geometry and arithmetic, which "do not encompass anything but the investigation of continuous and discontinuous quantity. They are not concerned with quality, the beauty of nature's creations and the harmony of the world" (Leonardo da Vinci 1956, **I**, § 15, p. 8).[4]

This evaluation of mathematics was intimately connected with the character of the *a priori* rationalist tendency, which led to a representation of mathematics as a suprasensual scheme preceding physical being. The precursor of classical science, the rationalist tendency saw in physical being, in the motion of a particle, the real prototype of mathematical analysis. Leonardo was a representative of this tendency that was oriented toward the future, toward classical science, toward a differential representation of motion. As Giorgio de Santillana writes, mathematics for Leonardo "is not the contemplation of a suprasensible world but the study of the geometrical skeleton of the real one" (de Santillana 1968, pp. 2—3).[5]

What is the relationship between the physics of Leonardo and the physics of Descartes? Valéry says that Leonardo's own words – "Mechanics . . . is the paradise of the mathematical sciences" – express a purely Cartesian thought. In his opinion, the idea of an animal-as-amachine is expressed more precisely by Leonardo than by Descartes: "The dream of creating a mechanical man and hence of achieving knowledge by construction was paramount in his [Leonardo's] thinking" (Valéry 1972, p. 142).[6]

Descartes is really a lineal descendant of Leonardo. But in this connection there is not only a similarity, but a difference as well. The lineage here is historical, and both thinkers retain their historical uniqueness. The early rationalism of Leonardo, with its sensual trappings, leads to a mechanical explanation, but here the mechanics is by no means the geometrized Cartesian mechanics of uniform bodies that are indistinguishable from their spatial locations. This is the mechanics of heterogeneous bodies possessing qualitative differences. Therefore, when Leonardo's painting moves into mechanics, into physics, into philosophy, when it becomes cognition of the world, knowledge of its *ratio*, it does not cease to be painting.

"Leonardo was a painter: *I mean that painting was his philosophy* . . . he regarded painting as a final goal for the efforts of a universal mind" (Valéry 1972, p. 143).[7] His aspiration to see the multicolored generality of the world, with its qualitative predicates, has caused many to liken Leonardo's world-view to that of Goethe. Cassirer, in particular, has said that for Leonardo, just as for Goethe, the "limit of vision is also, of necessity, the limit of conception. Both as artist and as scientist, [Leonardo] is always concerned with the 'world of the eye'; and this world must present itself to him not in bits and pieces but completely and systematically" (Cassirer 1963, p. 157).

But was the 'limit of vision' for Leonardo the limit of comprehension? In any event, it was not *such* a limit as it was for Goethe. Indeed, for Leonardo, mechanical-mathematical methods of knowing nature were within the limits of comprehension.[8] Goethe's 'visibilism' is a protest against the universality of mechanical-mathematical cognition. Leonardo's visibilism is the dawn of such cognition, when it had not yet become clear and well defined, when qualitative distinctions and variations were still retained, and the new science had not yet contrasted it to the almost monochrome picture of Descartes and the already completely monochrome working-plan of Newton.

The pre-Cartesian rationalism of Leonardo failed to meet Descartes's fundamental epistemological *criterion*, namely that of *clarity*. During the seventeenth century, this criterion signified that the objects of cognition could be given as precise a verbal (in particular, symbolic-mathematical) expression as desired. If we not only point to this object or its property but also *name* it, the object or its property loses its uniqueness and becomes the object of classifying, and therefore of conceptual thought. But in addition to losing its uniqueness, in Cartesian philosophy, and in seventeenth- and eighteenth-century rationalist philosophy in general, it loses its colors as well. Leonardo retains the uniqueness and the colors; in fact, for him painting becomes philosophy without ceasing to be painting. But how does one go beyond the limits of the unique in this manner?

For Leonardo, the criterion of clarity is replaced by the criterion of distinctness. Or rather, not 'replaced', but 'anticipated': the criterion of distinctness precedes the criterion of clarity. Here reason uses qualitative definitions, and its cognitive power depends on catching subtle qualitative nuances. Let us emphasize that *reason* uses qualitative definitions. Descartes's quantitative mathematical rationalism must not overshadow the qualitative rationalist tendency of Leonardo for whom the distinctness of nuances became the instrument of reason. As an artist, Leonardo employed very subtle nuances. The demand for subtle nuances was a matter of principle for him. In his words, the mind of the artist, like a mirror, is transformed into as many colors as exist in the objects placed before him.[9] But *what* must the artist, with the help of as many nuances as he likes, express?

Here we approach the major problem of Leonardo's art, one which was for him a major philosophical problem: namely how to go beyond the limits of the unique, how to make the unique element universal. The orientation of this problem toward the future, toward the seventeenth century, toward classical science, results from the fact that the extension of the unique is by no means reducible to its logical position in some *a priori* integral structure. Such an extension occurs in time and space. Thus a logical generalization is transformed into a spatio-temporal one. Unity in diversity becomes identity in the presence of various space-time predicates. Real unity in diversity is the self-identity of a body, which possesses various space-time predicates. Unique events are constituted by the presence of a body in different places at different moments. Self-identity is guaranteed by the continuity of the number of such places and moments — by the *continuous* motion of the body. Thus we see, if not a straight line, then, in any event, a continuous successive line uniting the space-time generalization of the unique with a differential representation of motion, with classical science.

Lenonardo thought that the advantage of painting over poetry lay in the possibility of depicting bodies and events coexisting simultaneously in space. The problem cannot be reduced to the presence, on one canvas, of bodies differing in spatial position. The entire system of coloration, perspective, background, chiaroscuro, the depiction of transparent and semi-transparent media, must demonstrate the connection of the unique *with spatial diversity*. But Leonardo goes further. He demands of painting a transition from the unique, sensual individual to *temporal diversity*. Such a transition reveals the objective *ratio* of the world.

In Leonardo's opinion, "painting is proved to be philosophy because it treats of the motion of bodies and the rapidity of their actions, and philosophy also includes motion" (Leonardo da Vinci 1956, **I**, §7, p. 5).[10] But he goes even further than this. He considers the object of painting and philosophy to be a change in motion, i.e., acceleration. Only a century and a half later were accelerations reckoned as basic processes of the universe, and constant velocities, as unchanging states of being. It is obvious that Leonardo's painting is not static but dynamic. But Leonardo's dynamism is imprinted not only on his artistic practice; it is also found in the *Treatise on Painting*, where Leonardo says, "Painting is therefore philosophy because philosophy treats of the increase and decrease of motion" (Leonardo da Vinci 1956, **I**, §8, p. 5).[11]

V. P. Zubov compares Leonardo's ideas to Lessing's. In *Laocoon*, Lessing says that the painter selects one moment from a series of moments and engraves it.[12] Leonardo considers the task of painting (and philosophy!) to be the comprehension not of a static moment but of a dynamic process. Zubov discusses Leonardo's position:

> In summarizing, let us say again that 'simultaneity' (*in un tempo, un medesimo tempo, in un subito*) is not a moment of existence snatched from the stream of time, but presumes a 'before' and an 'after'; it presumes time to be a form of comprehension of animate, flowing life. Life is possible only where there is a 'before' and an 'after' and a connection between the two. Therefore, time is not only a 'destroyer of things,' but a prerequisite for their true existence (Zubov 1968, p. 263).

At this point we encounter the following difficulty. In many of Leonardo's paintings, the heroes' poses, their facial expressions, the composition and lighting reveal both the genesis and the subsequent development of the depicted situation. But in the portraits of the

Madonna and of women, and especially in the greatest masterpiece of painting of all time, the portrait of *La Gioconda* (Mona Lisa)[13], it is the peaceful pose, and the strange, somewhat unreal, background, and, mainly, the famous smile of the Florentine beauty that rules out a simple connection between 'before' and 'after'. Moreover, one might think that the secret of the smile is of such complexity that Mona Lisa, smiling, states the impossibility of defining her 'before' and 'after'.

But [all] this does not invalidate the rationalist character of the portrait. It is an apotheosis of thought, which not only identifies, establishes, the lawfulness of behavior, and connects 'before' and 'after', but also ascribes to the moment an uncertainty, a *clinamen*, which erodes the simple definedness of life and thus transforms a working-plan into something living, free, and therefore real, concrete, and individualized.

The *Gioconda* is a depiction of thought: it is not only identifying but also individualizing, not only identity but also non-identity. Her smile is not the intervention of the irrational; it is the transformation of rationalism. It is the smile of the Renaissance, revealing that the past universals, the series of things once made identical, do not erase the individual uniqueness of a man and of each moment of his life.

Renaissance art demonstrates this uniqueness. The subsequent development of the connection between 'before' and 'after', already within the range of [Renaissance] science with its differential representation of motion, by no means achieved the precise omniscience of the hypothetical Laplacean spirit, which knows the coordinates and velocities of the particles that comprise the universe. Nature continues to smile at man with the secret smile of the *Gioconda.*

With this reservation, we shall consider Leonardo's 'after' conception. It is the forerunner not only of seventeenth- and eighteenth century science, but of modern science as well. In the nineteenth century, Lagrange's abstract mechanics acquired a more concrete equivalent in the conception of the *field.* It is therefore impossible to compare Leonardo's ideas solely with a mechanical-geometrical scheme, the most general characteristic feature of classical science being not such a scheme, but a somewhat broader conception.

This conception consists of the physically presented connection between an individual body and other bodies, a connection that determines a body's behavior in terms of differential laws. A recent analysis of Leonardo's conception states that his 'philosophy which interprets changes in motion' corresponds to the representation of a field that determines a body's behavior. Such an interpretation is connected with a 'field-like' historical perspective. Valéry, in his *Introduction to the Method of Leonardo da Vinci* (1894), says that "Leonardo's explanation does not *as yet* assume the character of a measurement," but he does assume a concrete physical connection between phenomena. ". . . I believe that nobody else clearly recognized the method during the three centuries after his death, although everyone made use of it — necessarily" (Valéry 1972, p. 56).[14]

Further, Valéry says that action at a distance does not correspond to this assumption of classical science. Although it was given an analytical form, Newton understood the inadequacy of the conception of action at a distance as an explanation of phenomena.[15]

One might add to this that momentary interactions and, correspondingly, concepts of absolute simultaneity contradicted the ideal of classical science — the explanation of phenomena in terms of the interaction occurring not only in space but also in time.

Valéry says that only Faraday returned to the criterion of the physical representability of interaction: "It was Faraday who rediscovered Leonardo's method as applied to the physical sciences."[16] Valéry then quotes the well-known passage from Maxwell's Introduction to his *Treatise on Electricity and Magnetism*: ". . . Faraday, in his mind's eye, saw lines of force traversing all space where the mathematicians saw centres of force attracting at a distance . . ."[17]

To consider Leonardo a 'precursor' of Faraday would mean losing not only the sense of the historical uniqueness of events in the history of science, but also a sense of proportion. Valéry's comparison has an entirely different meaning, which becomes clearer if we return to the evolution of rationalism.

Reason, in comprehending the objective *ratio* of the world, cannot limit itself to the individual. Aristotle wrote:

> If there is nothing apart from individuals, there will be no object of thought, but all things will be objects of sense, and there will not be knowledge of anything, unless we say that sensation is knowledge (Aristotle (1972), Book B. 4, 999^b).

The transition from the individual to the general was realized in an abstract geometrical form, which at times was made absolute, imparting an *a priori* meaning to abstract concepts. At the same time, scientific and philosophical thought attempted to find a mechanism to represent the interaction of individual bodies (that is, a mechanism with models and analogues in sensorily comprehensible processes, such as Faraday's straightforward analogues, or Maxwell's conditional ones). The concept of the physical field for Faraday and Maxwell was, to a certain extent, a synthesis of abstract and model-oriented conceptions.

Among the logical and psychological assumptions of the 'model-oriented', sensualizing accompaniment to (or, to put it more strongly, the sensualizing component of) the rationalist thought of the seventeenth–nineteenth centuries were traditions dating from the Renaissance when, as Valéry expresses it, "explanation had not *as yet* assumed the character of measurement." These traditions go back to the nominalists of the fourteenth century, and even further. But the *quattrocento* [fifteenth century] transformed a narrow stream into a comparatively broad flow. Renaissance art and technology made people's thinking about nature more vivid and 'model-like' by forcing traditional syllogisms to give way. In Leonardo's creative work, art and technology were combined with a direct defense of concrete images as a method for the rational cognition of the world.

3. THE RATIONALISM OF GIORDANO BRUNO

If Leonardo's creative work and world-view approach rationalism from its aesthetic program, then Bruno's approach it from its humanist foundation. But humanist culture was not unified. Giordano Bruno was heir to and representative of the dynamic, sharply critical, militant tendency in humanism which found itself opposed to bookish pedantry. His humanism sought support in national literature and in lively relations with the people while re-examining the legacy of ancient and medieval literature. Moreover it tended to comprehend the world visually. Its past was the rationalist, but multicolored art of the quattrocento, its future, the experimental rationalism of the Baroque. Here, between Leonardo and Galileo, is Giordano Bruno's humanism, which is more than just a humanism: it paves the way for the scientific comprehension of the world and seeks the objective *ratio* of the world, while naming it with traditional humanist pseudonyms.

Before turning to an exposition and analysis of Bruno's ideas, several remarks are needed about the connection between this new ontological rationalism and his own tragic fate.[18]

Galileo renounced his own views but Bruno refused to. Sometimes this difference is explained by the fact that Galileo had already put together his *Discourses and Mathematical Demonstrations Concerning Two New Sciences* (Galileo 1974) in which the ideas of the *Dialogue Concerning the Two Chief World Systems* (Galileo 1967) were developed, thereby excluding [the possibility of] a return to the Aristotelian picture of the world. Moreover, Galileo foresaw the subsequent evolution of classical science. He was convinced that the battle between tradition and science had already ended with science the victor, and that his renunciation was like a rearguard skirmish which could no longer affect the final outcome. And in fact in the *Discourses*, and to an even greater degree in the works of Galileo's students and followers (Viviani, Torricelli, Gassendi) there are almost no polemics; the classical foundations of mechanics are viewed as something experimentally verified, which do not demand further discussion other than a logical, basically mathematical, order and empirical verification.

Einstein wrote:

> My idea of Galileo is quite a different one. There can be no doubt that he was an ardent seeker of truth. But I find it hard to believe that a mature man could find any purpose in overcoming so many obstacles in order to bring a newly discovered truth to the attention of the shallow-minded populace with its paltry interests. Could the task be of such importance as to make him devote the last years of his life to it? . . . For no apparent reason he goes into the lion's den and undertakes a journey to Rome to fight it out with the clerics and the politicians. This does not agree with my idea of old Galileo's inner independence. In any case, I don't think I should ever undertake anything like that to uphold the relativity theory. I should reason that the truth is so much stronger than I am that any idea of mounting a Rosinante and taking up a sword to defend it would seem ridiculously Quixotic (Seelig 1956, p. 210, cited in Kuznetsov 1965, p. 181).

When we discuss Galileo, we shall see the genesis of a new relationship to scientific truth.

Experimental verification came to seem stronger than the propagandistic energy of the thinker. Experiment and mathematical methods carried with them the idea of the absolute indisputability of the truth. But in the sixteenth century there was still a long way to go to reach this point. Bruno wrote his pamphlets and treatises when neither the principle of inertia nor the doctrine of accelerated motion yet existed; nor, on the other hand, was there the telescope or any other instruments which could, strictly speaking, create the 'external confirmation' and 'inner perfection' of classical science, which would make science, as Einstein expressed it, "much stronger than its creators."

During Bruno's time, the force of scientific theory was dependent to a much greater degree upon the personality of its creator, his erudition, talent, and conviction. The defense of a theory was inseparable from the demonstration of these personal qualities, which explains why Bruno was so full of himself, of the self-praise for which later historians, commenting on the sixteenth century, reproached him. By the mid-seventeenth century, this personification of the struggle for scientific truth seemed foreign and strange when considered against the background of an objective exposition, which in Galileo still bore an emotional and personal element absent in Newton.

Bruno could not go beyond an intuitive anticipation of experimental science. Accordingly, he could not attain the objectification of science which later radically altered the relationship between the thinker and his own conception. Galileo could suggest that, with the aid of a telescope, he could convince his opponents of the existence of the moons of Jupiter, the mountains on the Moon, the phases of Venus and the separation of the Milky Way into distinct stars. For Bruno, scientific truth (whose empirical comprehension he could only intuit) was indivisible from his own personality. Thus it is historico-scientific analysis, not the *a priori* subjectivism of 'existential truth', not the moralizing reprimands addressed to Galileo, that can bring us closer to an understanding of Bruno's unique personality, behavior, and fate.

But what was the content and result of Bruno's intuition? It was above all the existence of extended substance in which each extended element depends upon the whole, upon the Universe, and reflects the objective *ratio* of the Universe. For Bruno the problem of substance is inseparable from that of the cosmos. The existence of individual elements in matter is inseparable from the existence of an infinite world. Each finite body is a point in this world, but it does not lose substantial being, it does not become only a point; it retains physical meaning because it is a reflection of the infinite world, a local realization of universal being.

Bruno saw in the inseparability of the two poles — the conservation of self-identical substance, and its changes — the primordial feature of being:

> Time takes all and gives all; everything changes, nothing is destroyed; only that which is one cannot change, is eternal, and can remain eternally one, uniform and identical.[19]

Bruno begins with the self-identity of substance, but he refuses to see in the Universe something immobile, and ascribes real existence to diversity. In connection with this, he sometimes refers to the beauty of the world, which consists in the combination of various elements, in the diversity of the whole.[20] For Bruno, beauty is one of the criteria of reality. It testifies to the coincidence of opposed definitions of the whole. He writes in the dialogue, *Concerning the Cause, Principle, and One*:

> . . . and since all the forms are in it, consequently all definitions are conformable to it, and all contradictory propositions are true (Bruno 1950/78, p. 166).[21]

And in the poem, *De triplici minimo et mensura* (On the Threefold Minimum and Measurement), Bruno speaks of the coincidence of opposites as particularities of the existing:

> Of those things that exist there is nothing that is different such that in another way, indeed in many and even peculiar ways, it does not coincide with that from which it differs and to which it is opposed. . . . It is clear even to the mob of those who philosophize that here all opposites are formed of the same thing because of their common matter. . . . When we look at the universe, there is nothing ugly, evil, incongruous; for diversity and opposition do not prevent all things from being excellent, because they are governed by nature, which, like a choir-master, directs and leads the opposing voices, at the extremes and in the middle, toward one voice for the whole, the best harmony that can be imagined.[22]

It is characteristic of Bruno that he constantly emphasizes the ontological objective quality of the coincidence of opposites. The profoundly dialectical unification of the ontologico-cosmological problem ('What is the universe?') with the epistemological problem ('How is it to be known?') – a unification that does not dissolve the ontological problem into an epistemological one but brings the flexibility and variability of forms and the infinity of knowledge out of the contradiction and infinity of nature – distinguishes him from his predecessors, especially from Nicholas of Cusa.

Of what does the objective *ratio* of the world consist, which is reflected in its every element? Bruno speaks of the 'world soul'. But this is not God; it is something without the ability to think. The rationalist answer to this question, which Bruno hid in the irrational form of the world soul, the substantial form, the 'inner artist', etc., consists in the causal, regular connections between natural processes. The concept of scientific law, as it continually pounds against its shell of Neoplatonic terminology, wants to break out and appear in rational form. The 'world soul' is the Neoplatonic pseudonym for the objective *ratio* of the universe. Without any criterion of reliability, without experiment, guided [only] by polemical enthusiasm, fleeting personal antipathies (more rarely, sympathies), and chance metaphors leading his thought astray, Bruno approaches this *ratio* gropingly by uniting the diverse and often contradictory ideas of antiquity, the Middle Ages and the Renaissance, sometimes supplementing them with fantastic data. In spite of this style of thinking (which was already archaic in the next generation), he does not abandon his main theme — the infinity of substance, the causal harmony of the world, and its rational comprehensibility.

Bruno attacks Providentialism in the name of causal harmony, in the name of the inner connectedness inherent in nature itself and the ordering of its processes. The figure of the Almighty, without whose will not one hair falls from a human head, is the constant subject of his ironies. In *The Expulsion of the Triumphant Beast*, Bruno, slightly masking the Christian God with the name of Jupiter, lists the events with which Providence must be concerned. Here recalls his neighbors from his native village [Nola, near Naples] — real people he remembers from childhood. Mercury, the messenger of Jupiter, relates the events which must be provided for in the village:

That Vasta, the wife of Albenzio, while she tries to curl the hair on her temples, shall burn fifty-seven of them because she overheated her iron, but that she will not burn her head, that she will not swear this time upon smelling the stench, but rather bear it with patience (Bruno 1964b, p. 132).[23]

Bruno is hardly inclined to consider these events as chance. Such an outlook is foreign not only to Bruno's world-view but also to the apologia for man that typifies Renaissance culture as a whole. For the fate of an individual man may turn out to be a vanishing detail in the infinite Cosmos. This view would contradict Bruno's philosophy of nature: the infinite soul of the universe is reflected in each atom, in each detail of being. But this soul of the universe is not a personal god, and its reflection is by no means the will of Providence. The world soul is a pseudonym for the world's causal *ratio*. On this point, Bruno approaches Spinoza's conception. The rationalist tendency in Bruno's world-view becomes especially clear, strange as it may sound, when we try to grasp the meaning of his apologia for irrational knowledge scattered among his works:

I regard one object, which absorbs my mind, and
it is a single visage. I remain fixed upon
one beauty,

which has so pierced my heart, and is a single
dart; by one flame only I burn, and know but
a single paradise. . . .

Love, fate, the object and jealousy are
for me pleasure, torment, content and distress. . . .

Love who shows me so high a truth that
it opens black portals of diamond, enters
its deity through the eyes and by the sight
is born, lives, is nourished and reigns eternally

and makes me perceive how much heaven,
earth and hell conceal. Love brings to light
the true forms of absent things, regains force
and with a sure dart stabs and ever wounds the
heart, uncovers what is within.
(Bruno 1964c, pp. 86—87, 88, 93)[24]

And thus 'love' leads both to Beauty and to Truth. For Bruno, the two are one.

In the dialogue *Concerning the Cause, Principle, and One*, there is the following exchange between Theophilus and Dixon:

Theo. You can judge for yourself. You can raise yourself to the concept — not, I say, of the greatest and highest principle, because that is excluded from our consideration — but to the concept of the soul of the world, inasmuch as it is the act and the potency of all — omnipresent in all. And thence, finally, since there are innumerable beings, everything is one; and the knowledge of this unity is the object and term of all philosophies and all natural contemplations, leaving out in its own limits that contemplation which ascends above nature, which taken to him who does not believe is impossible and nothing (Bruno 1950/78, p. 154).[25]

And thus the goal of knowledge is unified because, in all cases, it consists in the comprehension of the unity of the world. Bruno writes of "that contemplation which ascends above nature". But this phrase means only that nature as a unity rises over nature as an accumulation of individual embodiments of the world soul.

This idea is expressed in an irrational, intuitive form. In modern science, intuition outstrips experiment and mathematical analysis [but] for a very brief period of time, and usually it does not leave any visible traces, as it is soon embodied in experimental or strictly theoretical discoveries which are based upon qualitative calculations. In the sixteenth century, the intuition in Bruno's creative work was a half-century ahead of its embodiment. This was its tragedy. The new intuition, containing in embryonic form the fundamental principles of classical science, appeared without what was the strength of classical science. It was obliged to seek support in erudition, whose inadequacy Bruno already knew, and a foundation in the hypostasization of irrational knowledge into principles: this is the root of Bruno's complicated conflicts and indeterminate differentiations. But be that as it may, the deed was done: the infinity of the universe was proclaimed as an ontological truth.

The Universe's ontological and, moreover, physical character is demonstrated by the fact that Bruno does not speak of infinite space but of infinite worlds. Bradwardine, Palingenius, Patrizi,[26] and Nicholas of Cusa, while ascribing infinite dimensions to space, did not speak of the infinity of worlds. Bradwardine thought that beyond the limits of the finite world lay infinite empty space.[27] Palingenius added infinite space — the domain of God — to the finite universe of Aristotle.[28] For Patrizi, the material of the Cosmos was surrounded by infinite space in which there was light, but no material bodies.[29]

For Nicholas of Cusa, the infinity of the world had epistemological rather than ontological significance, which was reducible to the proof of the limitations upon discursive thought. Therefore he was limited by the idea of the infinity of space, which had pervaded geometry since Euclid.

Bruno says that knowledge reveals in an object the reflection of the infinite 'world soul'. But this thought has an ontological aspect: the real element of matter reflects in its form the Universe as a whole. For Bruno the existence of one element of matter is unthinkable without the existence of its other elements, of matter as a whole, which is the primordial subject of shaping and forming. In this sense, Bruno asks a question of the future which had to be answered not only by the classical principle of relativity but also by subsequent, increasingly remote conceptions of science.

The classical principle of relativity originates in the fact that the behavior of a body — its coordinates and velocity — loses meaning without reference bodies, i.e., other bodies. Without them it is impossible to speak either of the position of a body or of the velocity of change of this position. Newton limited this position by using motion based on inertia, and thought that the acceleration of the body could be determined by relating it not to reference bodies, but rather to local effects — the forces of inertia.

But the ideal of classical science was the reduction of all processes (including the forces of inertia) to interacting or relative positions and movements of bodies. Einstein later called such a reduction Mach's principle. From Bruno's point of view, the absence of other bodies made it impossible to speak not only of the behavior of a body but also of its existence. The being of a body includes something that cannot be reduced to the presence of the body in a given place and its transition from one location to another. In present-day science (in modern cosmology and the theory of elementary particles), there has emerged a representation of a self-coordinated system of interacting particles, whose interaction is the condition for the existence of each of them.

Because of this situation in contemporary science, Bruno's philosophical tendency becomes especially interesting. Of course for Bruno there was and could be no physical deciphering or even a precise expression of this tendency, which was hidden under the irrational formula of being as the reflection of the 'world soul'. This is an anthropomorphic form of a very radical, anti-anthropomorphic idea. The science of the seventeenth century struggled against the anthropomorphic and anthropocentric world view, and advanced the idea of the relativity of space, the equality of the 'human' terrestrial system of coordinates with other systems.

The principle of relativity, beginning with the relativity of the concepts 'up' and 'down', i.e., with the establishment of the spherical form of the Earth, was equivalent to the elimination of anthropomorphism and anthropocentrism from the representation of nature. But in the sixteenth century, such an elimination occurred in an irrational form: individual consciousness was presented with a 'world soul' that it identified with the material harmony of the universe. Nevertheless, individual consciousness stopped considering as absolute the horizon opening before it or the visible canopy of heaven with the sun and stars moving over it.

In the rejection of an anthropomorphic view of the world and in the drive toward an objective knowledge that would continuously overcome the limits of anthropomorphic egocentrism lay the pathos of Bruno's creative work. Hegel says of Bruno:

He sacrificed his personal welfare to the great enthusiasm which filled him, and which left him no peace. It is easy to say that he was "a restless being, who could get on with nobody." But whence did this restlessness come to him? What he could not get on with was the finite, the evil, the ignoble. Thence arose his restlessness. He rose to the one universal substantiality by putting an end to this separation of self-consciousness and nature, whereby both alike are degraded (Hegel 1974, **III**, p. 122).

Let us begin with the last phrase of this apologia for Bruno. The Italian thinker "put an end to [the] separation of self-consciousness and nature". Hegel saw in Bruno's intellectual drive the prototype of his own panlogism, which had not yet reached that level of scientific culture in which nature could be represented as another form of the self-developing spirit. This evaluation, like many other of Hegel's historico-philosophical and philosophical conceptions, can be stood on its head. Bruno really saw in the cognition of nature a transition from individual thought to the 'integral life of thought', but for him this integral intellectual life consisted in the search for an objective harmony of being, its causal *ratio* transforming chaos into cosmos. It was precisely this search that constituted "the grand enthusiasm of a noble soul, which has a sense of indwelling spirit, and knows the unity of its own Being and all Being to be the whole life of thought" (Hegel 1974, **III**, pp. 121—122). Yes, Bruno's drive, his creative work in all its diversity, is a single hymn to cognition, to objective cognition of the external world. It was this very goal that did not correspond to 'scientific culture', since the sixteenth century lacked the powerful methods for cognition of infinite nature that the following century acquired. Hence the stylistic peculiarities of Bruno's creative work, which are so subtly outlined by Hegel:

There is something bacchantic in his way of apprehending this deep consciousness; it overflows in becoming thus an object of thought, and in the expression of its riches. But it is only in knowledge that spirit can bring itself forth as a whole; when it has not yet attained to this point of scientific culture, it reaches out after all forms, without bringing them first into due order. Bruno displays just such an unregulated and multiform profusion . . . (Hegel 1974, **III**, p. 122).

But the order that Bruno needed was not a plan for the self-developing spirit, which Hegel has in mind, but, on the contrary, the conception of nature freeing knowledge from anthropomorphic dreams.

This is so in spite of the conglomeration of anthropomorphic fantasies that Bruno's thought breaches. The central idea, which he opposed to everything finite, is that of the infinite Universe.

Bruno assumed that he could comprehend this infinite universe by a single act of intuition. In this way actual infinity entered the world-picture. For Bruno there was no representation of the infinitely small steps of knowledge about which Galileo spoke, because there were no infinitely small elements in nature itself. Such a representation belongs to the seventeenth century. But what had already been done in the sixteenth century, and done by Bruno himself, was to represent the full epistemological value of a knowledge that embraced infinite nature. However great the elements of irrational comprehension in Bruno's epistemology, as a whole his creative work was a declaration of the power of reason.

The idea of relativity was a direct expression of this power. Man comprehends the world from the motionless origin of coordinates, the motionless Earth. On the other hand, man's activity – active rational activity, the displacement of terrestrial moving objects – includes a set of relative movements. Motion along the surface of the Earth may be viewed as absolute motion, but even the turning of a bar rotating on a lathe demonstrates the equality of the two representations: the instrument moves in relation to the bar, and the bar moves in relation to the instrument. Rational, active activity, work, and the prediction of a result, in a word, production, and productive experience – breaks through the limits of the conception of absolute motion. The concept of relative motion, which grew out of terrestrial mechanics, is transferred to the Cosmos, where the Earth is only one of the bodies, and thus man's reason and labor oblige him to go beyond the limits of immediately empirical anthropocentrism. In this, reason is opposed to both empiricism and *a priori* judgments.

The genesis of the ideas of relative motion was based on more than new ontological and epistemological theses (and for Bruno, these two merge). There was also an essential psychological, emotional aspect. Metaphysical absolutes become sacred concepts. These absolutes do not depend on other objects; they possess an immutable value, significance, and meaning; they are independent of the angle of vision, of the aspect from which they are approached; they turn out to be beyond the range of not only logical criticism but also variable, complex, non-obligatory values in general.

The discreditation of the canonized and sacred absolutes of the Aristotelian world-picture was connected with the 'carnival' tradition of laughter which defeated the sacred principles of the Middle Ages and its culture. The concept of the 'carnival tradition' was introduced by M.

M. Bakhtin in the historical study of the culture, art and artistic literature of the Renaissance, and *mutatis mutandis*, of earlier and later centuries (Bakhtin 1965; 1968; 1984b). This concept may thus be extended to a greater degree to scientific literature because the ideas of classical science were expressed to a greater degree in the stylistic features of the scientific works. That ideas had to be expressed in both logical outline and stylistic features has not only historical-literary but also epistemological significance. This fact, together with literature, characterizes a definite level of knowledge of the world. In returning to the sources of classical science, from Newton to Galileo and from Galileo to Bruno, we encounter an increasing need for definite stylistic features in scientific literature to express new ideas. The text of Galileo's *Dialogue Concerning the Two Chief World Systems* is much longer than Newton's *Principia*, because it must not only expound new concepts but also transform the consciousness of the reader, expunge traditional 'sympathies' from his consciousness, and so on. Correspondingly, Galileo's ideas demand definite stylistic features. For Bruno, concepts of homogeneous space and relative motion, which were ultimately fated to become science, had not yet attained a precise character. Here the psychological aspect of the problem prevails over the logical aspect.

Let us compare Bruno's creative work, that is, his genuinely creative work, not its mere literary form, with the creative work of Rabelais. What strikes one first are the superficial parallels: the use of hyperbole, the grotesque jumble of languages. Then, we begin to observe even deeper analogies: permeating the work of the Italian philosopher and the French writer is an original, grotesque rationalism. Both Bruno and Rabelais faced a common enemy: the strict canons of Aristotelian logic that lay claim to absolute value, the static world-picture, the religious, moral and social norms of the Middle Ages. Their absolute value was destroyed by Rabelaisian laughter, shocking in its earthiness of imagery and its cascades of abuse. All this can also be found in Bruno. But both have another quality – unexpected flights of thought, genre, style; periods that seem alogical, but mask some sort of new logic. All this shatters the system: in Bruno, the system of canonized cosmology, in Rabelais, the entire system of medieval canons.

In the name of what must medieval principles be shattered? What is the positive program of Bruno and Rabelais?

Their program is the independence of reason. Is this a rationalist program? It is difficult to answer this question. In fact, a single, unconditional formula is impossible. Despite all of Bruno's poetic fantasies, his alogical arabesques of exposition, and the very appreciable distance between his ideas and Descartes's classical rationalism, the creative work of the Italian thinker was imbued with an intuitive anticipation of the rationalist picture of the world that was painted in the seventeenth century.

4. THE RATIONALISM OF GALILEO

As mentioned earlier, the inclusion of a section on Galileo in an essay devoted to the rationalism of the Renaissance does not mean the inclusion of Galileo's work within the framework of the pre-history of classical science, or of the scientific representations of the Renaissance. No, Galileo's scientific ideas are not the pre-history of classical science, but its history, its source. Galileo belongs to the science of the Baroque, the science of the seventeenth century. But this is true only if we are speaking of classical *science*. If, on the other hand, we are talking about classical *rationalism*, then Galileo stands only on the threshold and even, if you please, just beyond the threshold of its history.

With regard to the classical rationalism that began with Descartes, Galileo is nearer to the philosophy of the Renaissance than to the metaphysics of the seventeenth century. The search for Galileo's rationalism (and it must be sought out) reveals a rationalism that is indirect, unsystematic, a deduction from astronomical, mechanical and physical conceptions.

The fundamental idea of Galileo's mechanics is the homogeneity of space. A body left to itself does not change its state; it continues to move uniformly. As distinguished from Aristotelian space, where there were isolated points — the natural places of bodies — in Galilean space, all points are equal. The law of inertia expresses the homogeneity of space. According to Noether's theorem, every conservation of physical size corresponds to the homogeneity of some heterogeneity: the conservation of impulse expresses the homogeneity of space, the conservation of energy [expresses] the homogeneity of time. The idea that the conservation of velocity and, consequently, of impulse, expresses the representation of homogeneous space, results from the identifying function of reason. The law of inertia means that the state of motion is subject to some natural law, that what we have is a series of magnitudes whose identity is asserted by the organization of being.

But Galileo does not stop at the homogeneity of space, at the first function of reason, the first definition of rationalism. He speaks of the relativity of motion, which is distinguished from homogeneity, primarily by its logical nature, its relation to the identifying and individualizing function of reason. The homogeneity of space is the equality of its points, in the sense of the identical behavior of particles located at these points. From the homogeneity of space follows the relativity of spatial position and motion. But both have meaning only in relation to *bodies*, not to space itself. Position is the distance to the reference bodies; motion is the change in distance between the moving body and the reference bodies. Relativity is no longer only the identity of the behavior of the particle at points in space. This concept assumes the non-identity of occupied and unoccupied places in space; motion has meaning only in relation to the former, not to the latter. The relativity of position and motion assumes the existence of reference bodies. This is not a geometrical but a physical assertion; the reference bodies are not arbitrarily selected bases of reference in space but real bodies, and without the differentiation between space and bodies, between 'non-being' and 'being', relativity loses physical meaning. Relativity is the expression of the individualizing function of reason.

In formulating the principle of the relativity of motion, Galileo said that during the uniform motion of a ship, bodies in the hold "seem not to move", that in the cabin of a moving ship everything occurs exactly as in the cabin of a non-moving ship, and hence that it is impossible to tell by the events in the cabin whether the ship is moving. This is a rationalist statement, but here rationalism is not reducible to the identification of objects and their unification in a universal. It contains a picture of material bodies that are distinguished from space, sensorily comprehensible, acting on the sense organs, accessible to experimental study. Seen in this light, relativity is a profoundly sensual idea; motion has meaning if it is empirically comprehensible, if it relates to sensorily comprehensible objects.

Thus Galileo's fundamental idea, which gave rise to heliocentrism, expresses in itself the synthesis of the identifying component of rationalism (the homogeneity of space) and its individualizing tendency, which is intimately connected with the sensual accompaniment of rationalism that was typical of the seventeenth century.

The transition from Aristotle's ideas to Galileo's, from peripatetic science to the science of a new age, above all marked the end of absolute identity. Absolute rest, the absolute identity of location, was replaced by motion based on inertia, by the identity of velocity. But this is an 'identity of non-identity': the invariability of velocity means the invariable [i.e., constant] variability of location. This is the point of no return: the Universe has become mobile, and all subsequent development, especially the next step — uniform acceleration, constancy of change of velocity — seems to be a development of Galileo's fundamental idea — the inseparability of being from motion, of immutability from change, of identity from non-identity.

One of the most emotionally charged and at the same time most profound and general remarks in the *Dialogue* is Sagredo's remark concerning the Aristotelian identification of perfection and immobility. Simplicius stated such an argument in favor of the Aristotelian, motionless structure of the Universe, with its fixed center, the Earth. Sagredo replies:

I cannot without great astonishment — I might say without great insult to my intelligence — hear it attributed as a prime perfection and nobility of the natural and integral bodies of the universe that they are invariant, immutable, inalterable, etc., while on the other hand it is called a great imperfection to be alterable, generable, mutable, etc. For my part I consider the earth very noble and admirable precisely because of the diverse alterations, changes, generations, etc. that occur in it incessantly. If, not being subject to any changes, it were a vast desert of sand or a mountain of jasper, or if at the time of the flood the waters which covered it had frozen, and it had remained an enormous globe of ice where nothing was ever born or ever altered or changed, I should deem it a useless lump in the universe, devoid of activity and, in a word, superfluous and essentially nonexistent. This is exactly the difference between a living animal and a dead one; and I say the same of the moon, of Jupiter, and of all other world globes (Galileo 1967, pp. 58—59).[30]

Following this, Galileo makes an interesting point about immobility being equivalent to death, and mobility signifying life. In the Aristotelian tradition, immobility was the fundamental harmony, the basic [form of] existence, eternal immobility, that is to say, eternal life, while motion was the cessation of being. This is in the spirit of the Eleatics who considered substance eternal, immobile, absolutely self-identical.

Sagredo says that the immortality of unalterable matter is equivalent to the absence of life:

Those who so greatly exalt incorruptibility, inalterability, etc. are reduced to talking this way, I believe, by their great desire to go on living, and by the terror they have of death. They do not reflect that if men were immortal, they themselves would never have come

> into the world. Such men really deserve to encounter a Medusa's head which would transmute them into statues of jasper or of diamond, and thus make them more perfect than they are (Galileo 1967, p. 59).[31]

Immobility is the murdering head of Medusa; it makes immortality a synonym for nonbeing. This idea anticipates constructions as remote as that of the abstract, homogeneous, immobile 'pure being' that is identical to 'pure nothing'. Like every powerful and general conception of the new age, it finds an antecedent in ancient philosophy. It was Heraclitus, who unlike the Eleatics, connected being not with immobility but with motion, with the destruction of absolute identity, with non-trivial identity. At the very onset of the concepts of space, time, identity and non-identity, people began to understand that absolutely identical time is compressed into a moment, while absolutely identical space is compressed into a point, into non-being.

Now the mobility of being, its inseparability from motion, is often called dynamism, which is not entirely precise: *δύναμις* is not motion but force. For the modern representation of motion, such a usage is legitimate: motion is generally accelerated motion; it occurs in a force field. Motion based on inertia is changed or accelerated by the action of forces, and the body left to itself is an abstraction; the body's uniform motion is only a component of real motion. But in the history of science this view came after a plan of the universe in which there were no forces, and all motion, including the orbital motions of the planets, was considered motion based on inertia. This was precisely the celestial mechanics of Galileo's *Dialogue*. It was rapidly transformed into a dynamic scheme. Newton made the concept of force the fundamental concept of science. Therefore in Galileo, and generally in all cases where one image of motion is being contrasted with a later image of motion under the influence of forces, it is better to use the term 'kinematics'.

The difference between Newtonian dynamics and Galileo kinematics is very radical. Dynamics led to the concept of the field. As we shall see later, it exerted a powerful influence on philosophy and was one of the sources of the conceptions of Leibniz, Kant and Hegel as well as one of the sources of French materialism. It rebuilt the system of categories in political economy. However, with regard to the principal thread of the development of rationalism, the inclusion of the concepts of non-identity, change, and motion within the basic ideas of rationalism, kinematics and dynamics were distinguished only by the degree of their influence. Motion in philosophy (as opposed to natural science) — at least in the philosophy that broke away from Cartesianism — was the general basis of all change, not only the change of displacement, but the change in all real derivatives of place, plane, and state (velocity, acceleration, etc.) as well. All forms and all classes (orders of derivatives) of motion were generalized by the inclusion of nonidentity among the fundamental categories of rationalism.

In the system of categories that Galileo introduced into mechanics there was no concept of force. But Galileo did take a great step in the direction of dynamics. If the central concept of the *Dialogue* is inertia, then the central concept of the *Discourses and Mathematical Demonstrations Concerning Two New Sciences* is continuous uniform acceleration. Acceleration for Galileo is not the conservation of a state (i.e., uniform motion) but a change in state (i.e., an event). It occurs at every point and at every moment, and obliges us to trace motion from point to point and from moment to moment. This gives rise to a differential representation of motion, and a new concept of infinity (although it is close to Archimedes's views) as an infinite number of infinitely small segments of a finite distance, and infinitely small intervals of time in a finite interval.

The phrase "*Eppur si muove*!" — attributed to Galileo — is apocryphal. But all of Galileo's activity after the trial in 1633 is symbolized by this phrase. In a symbolic sense, this phrase might refer not only to the Earth, but to the Universe as a whole, to being as a whole, — and not only as a whole but in all its details and elements as well.

The differential representation of motion transformed motion into infinitely small phases. Within limits, predicates of motion become predicates of points and moments. Local definitions of the 'here and now' appear, which describe the behavior of a body 'there and then', the potential behavior of a body. These are already familiar to us; their prototype was the aesthetic program of Leonardo. Such a fundamentally new relationship between a point and a segment, between a moment and a temporal interval, attains a precise form in the conception of virtual displacements. Galileo established the conditions for the equilibrium of a system at a given moment by examining the virtual displacements of the bodies comprising it. For example, an unbalanced lever of the first kind can be balanced if forces inversely proportional to the displacement of its arms are applied to them. The principle of virtual displacements means defining statics in terms of dynamics; but at the same time dynamics, motion, is defined in terms of local situations in the 'here and now'. In one of the following chapters we shall see what a powerful influence the local criteria of motion and the filling of the world-picture with local events exerted on both the style of scientific thought and culture as a whole.

A local, individual event is determined by motion as well as by forces which ultimately connect the entire Universe into a whole. Bruno's idea that the individual reflects the whole is embodied in Galileo's mechanics in a way that is by no means mystical, that has already lost its Neoplatonic birth-marks, and is not entirely rational. Two hundred years later, after a very interesting evolution, this idea would be expressed in Hegel's doctrine of 'true infinity', which does not counter each of its elements with their illusory completion, but is present in each element, determines each element. Only, unlike Bruno's, Galileo's infinite whole is no longer the infinitely extended Universe, but an infinity that results from an infinite division into increasingly smaller parts.

Galileo's epistemological platform, his views of the knowability of the world and the certainty of knowledge, are connected with the differential representation of motion. Let us cite an excerpt from the *Dialogue*, which figured in Galileo's trial by the Inquisition in 1633. In discussing the question of the certainty of knowledge, Salviati says:

> . . . human understanding can be taken in two modes, the *intensive* or the *extensive*. *Extensively*, that is, with regard to the multitude of intelligibles, which are infinite, the human understanding is as nothing even if it understands a thousand propositions; for a thousand in relation to infinity is zero. But taking man's understanding *intensively*, in so far as this term denotes understanding some proposition perfectly, I say that the human intellect does understand some of them perfectly, and thus in these it has as much absolute certainty as Nature itself has. Of such are the mathematical sciences alone; that is, geometry and arithmetic, in which the Divine intellect indeed knows infinitely more propositions, since it knows all. But with regard to those few which the human intellect does understand, I believe that knowledge equals the Divine in objective certainty, for here it succeeds in understanding necessity, beyond which there can be no greater sureness (Galileo 1967, p. 103).[32]

Let us examine the content of Salviati's remarks. At first glance, there is an almost trivial idea here, in any event, one that is not too different from the traditional view: concrete truths are certain, close to immediately certain first-hand impressions. But other truths, beyond the limits of first-hand impressions, are not known to us at all or are not known with certainty. They are innumerable. Therefore our reason comprehends an infinitely small portion of integral truth, so small as to amount to nothing. And thus knowledge is a transition from one certain truth to another, a transition which can never approach the full comprehension of the truth.

Such an epistemological conception seems to leave no room for the idea propounded by Bruno and many other thinkers of the Middle Ages and the Renaissance: that knowledge may go beyond the limits of the finite, and intuitively grasp the infinite truth. This idea relates primarily to extensive knowledge, to the comprehension of an infinitely large Universe. Galileo's conception apparently ignores intuitive knowledge of infinity; for him knowledge is the infinite sum of certain partial statements.

Now let us turn our attention to the last phrase of the passage. In the few truths that are comprehended by reason, reason approximates a certainty equal to that of divine knowledge, because here human knowledge, in comprehending partial truths, "succeeds in understanding necessity", which is the embodiment of the highest certainty.

Necessity means the connection between one given truth and another, this connection being continuous and unbroken; otherwise necessity is no longer guaranteed, and it becomes chance, complexity and uncertainty. Necessity is established as a transition from partial comprehension to general, and then finally to universal comprehension. Such a conception, so logically precise in Spinoza, attained precise physical form for Newton: knowledge of the truth is knowledge of the distribution and interaction of masses. We can find the distribution of masses with certainty by referring to forces and the distribution of forces by referring to the distribution of masses. But both Newton and Galileo understood that complete knowledge of the distribution of masses was unrealizable. Much later Laplace ascribed such complete understanding of necessity to a conditional higher reason — knowledge of the coordinates and velocities of all the particles in the Universe.

In any event the phrase that "[human] knowledge [of partial truths] equals the Divine in objective certainty" as far as the discovery of their necessity is concerned, shows that Galileo's epistemology is far from the linear linking of sure statements. Each step taken by knowledge compares the discovered truth with a more general one; each empirical statement is impossible without the speculation, prediction, and expansion of experimentally obtained truth and assumptions about its extension, about the series of identical statements.

We have before us a rationalist conception. It is characterized by a certain expansionist tendency. Galileo's thought includes Bruno's notion of the intuitive understanding of infinity and an anticipation of Newton's knowledge of dynamics, which seeks in forces the connecting link between each kinematic situation. Only in Galileo, as opposed to Bruno, the microworld is infinitely subdivided and therefore complex in any minimal region. Its complexity no longer lies so much in the reflection of the infinitely great whole as in the possibility of subsequent unlimited division.

Nevertheless each separate empirical statement is compared with the whole, which is infinite, with the general differential law, which has infinitely many applications, and, in any finite segment, governs the infinite number of local events.

For Galileo, rationalism became not only kinematic and infinitesimal; it became paradoxical as well. Let us recall the subject of our second chapter on 'Paradoxes of Being', and introduce certain supplementary characteristics of scientific paradox. The mechanism of paradox in science — what Bohr called the madness of science — stems from the constant conflict and the complex nonlinear interaction of the concepts of speculation with the empirical. Empirical data are, to some degree, identified, and they make possible the introduction of some abstract concept to represent a set of identified statements. But to some extent (sporadically in classical science, continuously in nonclassical science) the process of identification changes the meaning of the statements and leads to discrepancies between separate statements and general concepts, between individual concepts and fundamental concepts. The 'regularity' of an event means that it occurs in the same manner as every other event included under their common law. If one takes metrical statements, then 'regularity' means that each value of the variable corresponding to the given event converts the equation expressing the general rule, the general law, into an identity. The paradoxical event destroys such identity. It remains paradoxical until a new rule, a new law, a new set of identified events, is found.

The history of science may be viewed as a series of transitions to new paradoxical schemes. The first of these general transitions was the representation of the Earth as a sphere and the isotropism of space.

This contradicted the 'regular' identical behavior of heavy bodies; heavy bodies uniformly fall vertically in directly observed space where, in an absolute sense, the sky is up and the surface of the Earth is down. Next came the geocentric scheme: all heavy bodies fall in directions that cross the immobile center of the Universe, the center of the Earth. This seemed a paradoxical static scheme. Heliocentrism is a paradoxical kinematic scheme, while relativistic cosmology which ascribes non-Euclidean properties to the Universe, is a geometrical paradoxical scheme that has become the paradox of being.

In this evolution, Galileo's *Dialogue* was a profoundly paradoxical declaration of a picture of the world that contradicted the 'obvious', 'common sense', and all the other pseudonyms for tradition.

It was in fact tradition, and traditional concepts, that were contradicted, not direct observation. Geocentrism and the absolute character of the coordinate system of the Earth were not the directly observed picture of the world. Yes, and does this term 'directly observed' make sense? Even the primitive representation of the flat Earth was not a direct observation, but the result of hypothetical, i.e., identifying generalization: "the surface of the Earth and further, beyond the horizon, remains flat". Heliocentrism, in the form given it by Galileo was not a conflict of reason with experience; it was a conflict within reason, the inner conflict of rationalism.

In Galileo's time, the generalizing potential of science and philosophy was sufficiently high. The paradox of the rationalist comprehension of the world led to a new representation of the character of reason, to a new style of rationalist thought. It was not yet based on experiment, but was itself already experimental by nature. Not only traditional syllogisms but also thought experiments figured in it, which had also been encountered earlier; they illustrated the traditional syllogisms. But now they were no longer illustrations, but starting points for new logical constructions. Language remained one of the roots of rationalist thought, but the relation of thought to language was altered: there was a weakening of humanist excursions into the past; there was less influence from Latin with its elaborate, fixed figures of speech that absorb scholastic constructions. To an ever greater extent the Italian language became the source of rationalist associations. Science, drawing new associations from it, especially those connected with production (let us recall that the opening words of the *Discourses and Mathematical*

Proofs on the Two New Sciences single out the Arsenal at Venice)[33] also influenced language, and Galileo became one of the masters of Italian prose.

In speaking of language, we must mention a characteristic peculiar to Galileo — his *silence*. Of course, this is not a feature of language, but it is directly related to language and its philosophical and logical functions. For the Renaissance thinkers in the strict sense of the word, including Bruno, silence was the murder of truth. Only in language, literature, oral discussions and correspondence, could an idea obtain verification of its clarity, lack of contradiction, content. For the Baroque thinkers, as previously mentioned, truth had different, primarily experimental criteria. Not only real experiments but also logically irreproachable thought experiments gave truth such certainty ("[human] knowledge equals the Divine in objective certainty") that the problem no longer consisted in the advocacy of truth but in its development and application.

"Silence is the lot of the despairing and the convinced", wrote Galileo. Silence had long been the lot of the despairing. It now became the lot of the convinced. The phrase occurs in a polemical pamphlet that was written at a moment when Galileo had decided to abandon the polemics directed against the Aristotelians. The pamphlet ends with the sentence: "Truths that have found confirmation . . . can by no means be contradicted with arguments".[34]

Galileo's stand is connected with his rationalism — no longer literary and humanist but experimental, claiming an adequate comprehension of objective *ratio*, objective truth. There is a connection between this new type of rationalist thought and Galileo's recantation — an event which provoked a flood of trivial, moralizing reprimands addressed to the thinker.

NOTES

1 [Executed between mid-1510 and late 1511 in the *Stanza della Segnatura* (papal judicial tribunal) in the Vatican, from a cartoon now in the Biblioteca Ambrosiana, Milan (see Raphael Sanzio 1972). The 'School of Athens' ('*Scuola d'Atene*') is one of a cycle of four wall frescoes designed to illustrate the Platonic view that "*universalia sunt ante rem*, that the things of this earth are only embodiments, however incomplete, of general ideas or principles" (Gombrich 1972, p. 88), these being the Good, the True, the Beautiful, represented in the *Stanza* by the four ceiling figures of Philosophy and Theology (the True), Poetry (the Beautiful) and Justice (the Good). Standing beneath the ceiling figure of Philosophy, who has on either side of her the words *causarum*

cognitio (knowledge of causes) and who holds two volumes, one marked "*moralis*", the other, "naturalis", the 'School of Athens' has at its center (shown in the Frontispiece) *Plato* (natural philosophy) on the *left*, holding in one hand the *Timaeus* (dialogue concerned with creation and the nature of the universe) and pointing upwards with the other, probably to the knowledge of causes or essence of things (Gombrich 1972, p. 91); and *Aristotle* (moral philosophy) on the *right*, holding the *Ethics* in his left hand, and stretching out his right between heaven and earth "to suggest that the universal does not exist outside the concrete from which it is abstracted by the reasoning intellect" (Redig de Campos 1952, p. 17; 1965, p. 15). See also Gombrich (1972); Pope-Hennessy (1970), pp. 98—106, 137—139. Redig de Campos (1965) identifies the philosophers (and their models) who are grouped in the foreground and to either side of Plato and Aristotle. – Eds.]

[2] Much later Hegel, in his *Logic*, spoke of this characteristic of 'freedom of thought': ". . . this exertion of thought is *my* act. If this be so, the real nature is a *product* of *my* mind, in its character of thinking subject – generated by me in my simple universality, self-collected and removed from extraneous influences – in one word, in my Freedom" (Hegel 1975, §23, p. 35). See also below, pp. 351—52.

[3] "But just as we are still rather firmly attached to the idea of a pure science rigorously developed on the basis of *local* evidence, having properties that may be extended indefinitely from identity to identity, in the same fashion we are still half convinced of the existence of a single *Morality* and a single *Beauty*, both independent of times, places, peoples or persons" (Valéry 1972, pp. 116—117). ("Cependant, de même que nous sommes encore assez attachés à l'idée d'une science pure, développée en toute rigueur à partir d'évidences *locales* dont les propriétés pourraient s'étendre indéfiniment d'identité en identité, ainsi sommes-nous encore à demi convaincus de l'existence d'une *Morale* et de celle d'une *Beauté* indépendantes des temps, des lieux, des races et des personnes" (Valéry 1938, pp. 127—128)).

[4] ". . . queste due scientie nó s'estendono se nó alla notitia della quantita continua et discontinua ma della qualita nó si travaglia la qual é bellezza de l'opere de natura et ornamento del mondo" (Leonardo da Vinci 1956, **II**, §15, f. 7v).

[5] "ce n'est pas la contemplation d'un monde suprasensible, c'est la recherche de l'ossature géométrique du réel" (de Santillana 1952, p. 44).

[6] "La mécanique . . . est le paradis des sciences mathématiques. . . . L'ambition de l'automate, de la conaissance par la construction, était souveraine en lui" (Valéry 1938, pp. 151—152).

[7] "Léonard est peintre": *je dis qu'il a la peinture pour philosophie*. . . . il le regarde comme une fin dernière de l'effort d'un esprit universel" (Valéry 1938, pp. 152—153).

[8] See Luporini (1953), pp. 153—154; Zubov (1968), pp. 161—163.

[9] "The painter's mind should be like a mirror, which transforms itself into the color of the thing that it has as its object, and is filled with as many likenesses as there are things placed before it" (Leonardo da Vinci 1956, **I**, §71, p. 48). ("L'ingegno del pittore vol esser à similitudine dello specchio il quale sempre si trasmuta nel colore di quella cosa ch'egli ha per obbietto e' di tante similitudini s'empie quante sono le cose che li sono contraposte" (Leonardo da Vinci 1956, **II**, §71, f. 32r)).

[10] "Si prova la pittura essere filosofia perche essa tratta del moto de corpi nella prontitudine delle loro attioni e la filosofia anchora lei s'estende nel moto" (Leonardo da Vinci 1956, **II**, f. 4r).

[11] "donque la pittura è filosofia perche la filosofia tratta de moto aumentativo e diminutivo . . ." (Leonardo da Vinci 1956, **II**, f. 4r).
[12] ["from the eternally changing reality, the artist can take only one moment, and the painter can take that one moment only from one point of view" (Lessing in Zubov 1968, p. 261). ("Kann der Künstler von der immer veränderlichen Natur nie mehr als einen einzigen Augenblick, und der Maler insbesondere diesen einzigen Augenblick auch nur aus einem einzigen Gesichtspunkte, brauchen . . .") (Lessing 1893, p. 19).]
[13] [The portrait of Madonna Lisa, the third wife of Francesco del Giocondo (*c.* 1503–06; Paris, Museum of the Louvre) – Eds.]
[14] "Ici, l'explication ne revêt pas *encore* le caractère d'une mesure. . . . et il me paraît que, pendant les trois siècles après sa mort, cette méthode n'a été reconnue par personne, tout le monde s'en servant, – nécessairement" (Valéry 1938, p. 112).

[It may be helpful here to read Leonardo's own words as they appear in Valéry:

> The air is full of infinite lines, straight and radiating, intercrossing and interweaving without ever coinciding one with another; and they *represent* for every object the true FORM of their reason (or their explanation). (L'aria e piena d'infinite linie rette e radiose insieme intersegate e intessute sanza ochupatione luna dellaltra rapresantano aqualunche obieto lauera forma della lor chagione.) In Valéry (1972), p. 55.

By "'*field-like*' historical perspective" (italics ours), Kuznetsov is echoing a point made by Valéry about Leonardo:

> Leonardo's explanation does not *as yet* assume the character of a measurment. It consists merely in the emission of an image, a *concrete mental relation between phenomena* [italics ours] – or, let us say to be exact, between the images of phenomena (Valéry 1972, p. 56). (Ici, l'explication ne revêt pas *encore* le caractère d'une mesure. Elle ne consiste que dans l'émission d'une image, d'une relation mentale concrète entre des phénomènes, disons, pour être rigoureux, entre les images des phénomènes (Valéry 1938, p. 112) – Eds.]

[15] [Cf. Valéry 1972, p. 57; 1938, p. 112 – Eds.]
[16] *Ibid.* ("Mais il était réservé à Faraday de retrouver dans la science physique la méthode de Léonard.")
[17] Maxwell 1954, **I**, p. ix. Valéry 1972, p. 58.
[18] The next ten pages present a brief exposition of several evaluations contained in the concluding chapters of my book, *Dzhordano Bruno i genezis klassicheskoi nauki* (Giordano Bruno and the Origin of Classical Science, 1970a), Chapter 5, 'Beskonechnost' Vselennoi' (The Infinity of the Universe), Chapter 6, 'Atomistika' (Atomism), Chapter 7, 'Otnositel'nost' ' (Relativity), Chapter 8, 'Stil' (Style).
[19] "Il tempo tutto toglie e tutto dà; ogni cosa si muta, nulla s'annichila; è un solo che non può mutarsi, un solo è eterno, e può perseverare eternamente uno, simile e medesmo" Bruno 1969, p. 22 [The passage is from the Dedication of the *Candelaio* (literally, candle-maker, but in the play the word seems to allude to homosexuality), a satiric comedy first published in Paris in 1582, whence Bruno had fled in order to escape punishment for heresy in Italy. The Introduction, by G. B. Squarotti, to the above cited Italian edition of the play, discusses the *Candelaio* in terms of its intel-

lectual, cultural and social criticism, which Bruno effects, on the one hand, by means of very rich, destructive, linguistic inventions, and, on the other, by the use of sixteenth century, Neapolitan 'underworld' characters living outside of the decadent and no longer adequate structure that he is attacking. An English translation of the play by J. R. Hale, called *The Candle Bearer*, can be found in *The Genius of the Italian Theater*, edited by Eric Bentley (New York: The New American Library, 1964), pp. 194—314. The passage cited above occurs on p. 199. Our translation is in accord with the French of Michel's *La cosmologie de Giordano Bruno* (1962), p. 80 — Eds.]

[20] See note 22 below for the passage in *De triplici minimo et mensura*. [For an explanation of this title see Michel (1973), pp. 80—81 — Eds.]

[21] ". . . e perché tutte le forme sono in esso, conseguentemente tutte le diffinizioni gli convengono; e per tanto le contradittorie enunciazioni son vere" (Bruno 1958, p. 329).

[22] "Nihil eorum quae sunt ita differt, ut et aliquo pacto imo etiam in pluribus atque praecipuis non conveniat cum eo a quo differt et cui contrariatur. . . . Hinc contraria omnia propter communem utriusque materiam in eodem genere consistere, etiam vulgo philosophantum est manifestum. . . . Hinc ad universum respicienti nihil occurret turpe, malum, incongruum; neque etenim varietas atque contrarietas efficit quominus omnia sint optima, prout videlicet a natura gubernantur, quae veluti phonascus contrarias voces extremas atqua medias ad unam omni (quam possimus imaginari) optimam symphoniam dirigit et perducit" (Bruno 1879—91, Vol. 1, part 3, p. 272).

[23] "Che Vasta, moglie di Albenzio, mentre si vuole increspar gli capelli de le tempie, vegna, per aver troppo scaldato il ferro, a bruggiarne cinquanta sette; ma che non si scotte la testa, e per questa volta non biastemi quando sentirà il puzzo; ma con pazienza la passe" (Bruno 1958, p. 634).

[24]
Un oggetto riguardo;
Chi la mente m'ingombra, è un sol viso.
Ad una beltà sola io resto affiso,

Chi sí m'ha punto il cor, è un sol dardo,
Per un sol fuoco m'ardo,
E non conosco piú ch'un paradiso. . . .

Amor, sorte, l'oggetto e gelosia
M'appaga, affanna, contenta e sconsola. . . .

Amor, per cui tant'alto il ver discerno,
Ch'apre le porte di diamante nere,
Per gli occhi entra il mio nume, e per vedere
Nasce, vive, si nutre, ha regno eterno;

Fa scorger quanto ha 'l ciel, terra ed inferno.
Fa presenti d'absenti effiggie vere,
Repiglia forze, e col trar dritto, fere,
E impiaga sempr' il cor, scuopre l'interno.
(Bruno 1958, pp. 963, 964, 969—970).

[25] "*Teofilo*. Giudicate voi. Possete quindi montar al concetto, non dico del summo ed ottimo principio, escluso della nostra considerazione; ma de l'anima del mondo, come è atto di tutto e potenza di tutto, ed è tutta in tutto; onde al fine (dato che sieno

innumerabili individui) ogni cosa è uno; e il conoscere questa unità è il scopo e termine di tutte le filosofie e contemplazioni naturali: lasciando ne' sua termini la piú alta contemplazione, che ascende sopra la natura, la quale a chi non crede è impossibile e nulla" (Bruno 1958, p. 308).

26 [Palingenius, or Marcellus Palingenius Stellatus is the pseudonym of Pietro Angelo Manzoli (or Manzolli), fl. 1528. He is the author of the poem *Zodiacus vitae* (Venice, 1531). For more information about Francesco Patrizi, see Paolo Rossi, 'Francesco Patrizi: Heavenly Spheres and Flocks of Cranes', in *Italian Studies in the Philosophy of Science*, edited by Maria Luisa dalla Chiara, *Boston Studies in the Philosophy of Science*, vol. 47 (Dordrecht, Boston: D. Reidel, 1981), pp. 363—388 — Eds.]

27 See Michel 1973, p. 154; Zubov 1960; Murdoch 1960; Clagett 1959, pp. 232—234.

28 Michel 1973, pp. 155—156; Koyré 1957, pp. 24—27.

29 See Patrizi 1587, col. I, pt. II. See also Zubov 1965, p. 199.

30 "Sagr. Io non posso senza grande ammirazione, e dirò gran repugnanza al mio intelletto, sentir attribuir per gran nobiltà e perfezione a i corpi naturali ed integranti dell'universo questo esser impassibile, immutabile, inalterabile etc., ed all'incontro stimar grande imperfezione l'esser alterabile, generabile, mutabile etc.: io per me reputo la Terra nobilissima ed ammirabile per le tante e sì diverse alterazioni, mutazioni, generazioni etc., che in lei incessabilmente si fanno; e quando, senza esser suggetta ad alcuna mutazione, ella fusse tutta una vasta solitudine d'arena o una massa di diaspro, o che al tempo del diluvio diacciandosi l'acque che la coprivano fusse restata un globo immenso di cristallo, dove mai non nascesse nè si alterasse o si mutasse cosa veruna, io la stimerei un corpaccio inutile al mondo, pieno di ozio e, per dirla in breve, superfluo e come se non fusse in natura, e quella stessa differenza ci farei che è tra l'animal vivo e il morto; ed il medesimo dico della Luna, di Giove e di tutti gli altri globi mondani" (Galileo 1897, p. 83).

31 . . . Questi che esaltano tanto l'incorruttibilità, l'inalterabilità etc., credo che si riduchino a dir queste cose per il desiderio grande di campare assai e per il terrore che hanno della morte; e non considerano che quando gli uomini fussero immortali, a loro non toccava a venire al mondo. Questi meriterebbero d'incontrarsi in un capo di Medusa, che gli trasmutasse in istatue di diaspro o di diamante, per diventar più perfetti che non sono. (Galileo 1897, p. 84).

32 ". . . l'intendere si può pigliare in due modi, cioè *intensive* o vero *extensive*: e che *extensive*, cioè quanto alla moltitudine degli intelligibili, che sono infiniti, l'intender umano è come nullo, quando bene egli intendesse mille proposizioni, perchè mille rispetto all'infinità è come un zero; ma pigliando l'intendere *intensive*, in quanto cotal termine importa intensivamente, cioè perfettamente, alcuna proposizione, dico che l'intelletto umano ne intende alcune così perfettamente, e ne ha così assoluta certezza, quanto se n'abbia l'istessa natura; e tali sono le scienze matematiche pure, cioè la geometria e l'aritmetica, delle quali l'intelletto divino ne sa bene infinite proposizioni di più, perchè le sa tutte, ma di quelle poche intese dall'intelletto umano credo che la cognizione agguagli la divina nella certezza obiettiva, poichè arriva a comprenderne la necessità, sopra la quale non par che possa esser sicurezza maggiore" (Galileo 1897, pp. 128—129).

33 "Frequent experience of your famous arsenal, my Venetian friends, seems to me to open a large field to speculative minds for philosophizing . . ." (Galileo 1974, p. 11).

("Largo campo di filosofare a gl'intelletti specolativi parmi che porga la frequente pratica del famoso arsenale di voi, Signori Veneziani . . ." (Galileo 1898, p. 49).

[34] ". . . il tacere . . . è da disperati e convinti; . . . il contradire con ragioni alle verità dimostrate . . . è del tutto impossibile . . ." (Galileo 1894, p. 691). [The Italian says literally that "silence is with the despairing and convinced". — Eds.]

5. DESCARTES

1. THE ILLUMINATION AT ULM

In his concluding characterization of Cartesianism and its place in the history of philosophy, Kuno Fischer begins with the following definition of a philosophical school (Fischer 1887, pp. 466—469). A school states conclusions drawn from several first principles. The elaboration and refinement of the link between first principles and conclusions, including partial revision of these conclusions, represents the evolution of a school. The transformation — partial or complete — of the first principles takes an idea beyond the school's boundaries; new principles answer the fundamental questions in a different way. If the questions themselves also change, then there is a transition not only to a new school, but also to a new epoch in philosophy. Of course, such a hierarchy of transformations must not so much define the position of Descartes, as emphasize the epochal character of Kant's ideas. Fischer relates the evolution of Cartesianism within its own boundaries, which are those of a self-identical school, to the activity of Descartes's and Malebranche's students. The partial transformation of first principles he ascribes to Spinoza, the complete transformation, to Leibniz, and the rejection of the questions that the old and new principles answered, to the work of Kant.

Such an historico-philosophical scheme is unsatisfactory for many reasons, only one of which it is necessary for us to touch upon here: the question of the modern, nonclassical historical perspective. Nonclassical science destroys in an obvious manner each rigid differentiation between first principles and conclusions. Einstein, as we know, spoke of a physical theory's external confirmation (correspondence to observation) and its inner perfection (natural deduction from the most general principles). These criteria merge in the dynamics of scientific progress, in continuous dynamics, that is, in nonclassical science. The paradoxical results of experiment lose their paradoxical quality during the transition to principles that are paradoxical from the traditional point of view. Such a transition, such a revision of first, fundamental principles occurs often enough in nonclassical science to make their theoretical fluidity apparent. Accordingly, modern philosophy cannot develop within the framework of schools, which claim inflexibility of fundamental principles that produce only partial conclusions from these propositions in the ever-changing flow of time. But this is not enough.

In modern science, it is no longer possible to be sure that experiment will not oblige us to reject the question asked of nature, will not reveal the senselessness of such a question, will not force us to replace it with another. The accelerators that impart energy in the hundreds of millions of electron volts to elementary particles will perhaps force us to abandon the search for a space-time picture of events in ultramicroscopic regions. Therefore, changes not only in the fundamental answers, but also in the questions themselves (what Fischer considered the boundaries of epochs) now become a continuous component of philosophical thought. Let us emphasize that we are dealing with philosophical thought; answers and questions relating to being as a whole change.

Nevertheless nonclassical science demonstrates the continuity of scientific and philosophical thought. Strange though it may seem, our epoch, the most dynamic known to the history of the world, feels an affinity with classical science and classical philosophy, with the philosophy and science of the Renaissance and the Middle Ages and, perhaps to a greater degree, with ancient thought. But not with a school of philosophy. Modern science, with its continuous gradient of change, with the expansion and deepening of the questions it asks of nature, with its specific goal to include in these questions both the metagalaxy and elementary particles, is close to the critical, pivotal moments in the history of human thought when new problems arise, new principles are expressed, new directions appear. From a non-classical historical perspective, from the perspective of modern science, its history is presented as a history of revolutions. But these revolutions do not resemble the cataclysms of Cuvier. They did not erase from the slate what had previously been written on it. They deepened and modified several recurring ideas. In the history of thought, just as in physics, there can be no motion without that which moves, no change without the self-identical subject of change.

Among the most radical revolutionary upheavals in the annals of human thought is the philosophy of Descartes — the transition from the prehistory of classical rationalism to its history. As seen from the perspective of modern science, Descartes's fundamental ideas may be defined and evaluated once again — in particular: his fundamental position, *cogito ergo sum*; the criterion of clarity — the possibilities for the precise mental and verbal representation of real objects, and the style he uses in connection with this criterion; his physical ideas — the identification of matter and space, the concept of straight-line inertia, the theory of the motion of bodies, as well as his mathematical conceptions.

In re-evaluating Descartes's philosophical and physical ideas from a modern historical perspective, the light shed by several biographical problems changes — such problems as his initial refusal to publish the new conceptions and his subsequent literary activity, his initial thirst for personal contact and his subsequent search for solitude. Let us touch upon these problems briefly.

The most important component in the life of every great thinker — that is, his inner intellectual and moral evolution — reflects his thoughts about the world and about knowledge of the world, especially his representations of the empirical and speculative components of knowledge. From this point of view, a great epistemological model shows through the *peripeteia* of Descartes's life.

The point at which the course of a life culminates — the moment when a twist in the course coincides to the greatest degree with the evolution of human thought, when biography merges with history — is fixed very precisely for Descartes. It was November 10, 1619, and the following night. This date is just as important for epistemology as the night of January 7, 1610 is for astronomy (and for epistemology as well!), when Galileo first directed a telescope at the starry sky.

According to some of the most recent evidence, in a lost diary of Descartes there was an entry:

> November 10, 1619, filled with enthusiasm, I discovered the foundations of the marvelous science.[1]

This enthusiasm, while reminding us of Bruno's 'heroic enthusiasm', had a somewhat different character.

At this time, the winter of 1619—20, Descartes was in the army of Duke Maximilian of Bavaria, then quartered on the banks of the Danube at Ulm and one year later victorious [against Frederick V, the Elector Palatine and king of Bohemia] in the battle of the White Mountain near Prague during the Thirty Years' War.

What was the "marvelous science" whose foundations were discovered in 1619? It is difficult to say: perhaps these are mathematical discoveries, perhaps a 'universal mathematics', i.e., a conception going beyond the limits of mathematics (Asmus 1956, pp. 48–52). In any event, it is difficult to think that the illumination at Ulm did not contain Descartes's fundamental epistemological ideas which occurred to the thinker, together with separate mathematical conceptions. These conceptions are directly connected with Descartes's epistemology, and we shall consider November 10, 1619 as the birthday of a new epistemology, of new criteria for the certainty of knowledge, of everything which was later expressed by the formula *cogito ergo sum*.

This formula was the starting point of a complex and detailed system of epistemological concepts and criteria, of physical principles and models, thought experiments, hypotheses, mathematical constructions, etc. But the representation of the idea that thought itself, independent of content and objects, demonstrates the reality of the thinking subject, that in all doubtful conclusions in thought, the doubt itself is certain – this representation could come to Descartes as a sudden illumination. Emotional enthusiasm could hardly appear without the solution of the problem of the certainty of knowledge, which for Descartes had persisted from his earliest years.

The illumination at Ulm may have been sudden, but the groundwork for it had already been laid. It was the result of prolonged, purely intellectual efforts, of prolonged and agonizing meditation and, at the same time, of a great and intense empirical comprehension of the world. Following two years of solitude in Paris in the faubourg St.-Germain, Descartes experienced long years of wandering and constant (at times anxious, at times peaceful) contact with people. In the *Discourse on Method*, Descartes described these years and the inner impulse which forced him to spend "the rest of my youth in travel, in frequenting courts and armies, in mixing with people of various dispositions and ranks . . ." (Descartes 1954, p. 13). Such personal contact continued during Descartes's military service, interrupted by considerably prolonged periods of solitude, and then was maintained in Paris until he emigrated to Holland.

His motives for this emigration shed light on the correspondence between the inner logic of the development of science and philosophy and, on the other hand, the external, in particular historical, conditions. In the spring of 1631, in his famous letter to Jean-Louis de Balzac[2] to explain why he preferred to stay in Holland rather than in his native land or in Italy, Descartes spoke mostly of the possibility for solitude. It is necessary to note, however, that his solitary reflections came after lengthy and intense empirical impressions. Descartes *reflected* on what he *had seen* earlier; the evolution of his relationship to the world corresponded not only to the rationalist ideal of pure knowledge, but also to the sensualist accompaniment of rationalism without which rationalism loses its connection with science and, in essence, its meaning.

But all this refers to the immanent, inner impulses of Descartes's creative life, which to such a great degree – and herein lies his greatness – corresponds to the phylogenesis of science, to the historical evolution of knowledge! The inner impulses are interwoven with social conditions and their influence on this history of thought. In listing the advantages of Holland in the letter to Jean de Balzac, Descartes asks, "In what other country can one enjoy a liberty so complete . . .?"[3] Thus it turns out that solitude, as a condition for intensive philosophical reflection, depends on social, historically rooted conditions, in other words, external isolation from the environment turns out to be dependence on the environment. Having fallen away from Catholicism, Holland was for many seventeenth century thinkers a haven where they found freedom (although a very ephemeral freedom, as Descartes found out soon enough).

The alternation and intertwining of the urge toward personal contact and the urge toward solitude in Descartes are undoubtedly connected with the complex structure of rationalist thought, with the factual indivisibility of the speculative and the empirical comprehension of the world. The essence of the Cartesian conception of knowledge is connected with Descartes's characteristic attitude toward the presentation of his ideas and the publication of his works. A little later we shall dwell in more detail on the paramount importance of the criterion of clarity for Descartes's epistemological, physical and mathematical credo. This criterion connects Descartes's original thesis – *cogito ergo sum* – with the bases of Cartesian physics, the reduction of the predicates of matter to spatial predicates, and, in its turn, the endowment of space with physical properties. For Descartes, the world of spatial predicates represents something comprehensible on the basis of the clarity inherent in the very process of thought. The clarity of the spatio-mechanical picture guarantees its certainty. The fact that such a picture can be expressed in words, in the embodiment of *cogito*, has for

Descartes not a didactic, but an epistemological significance. Hence the inner need for the presentation of philosophical and physical doctrines, which was initially a drive to give them a foundation; only later, when the system was shown to be complete, did the need acquire a didactic coloration. At the root of the matter lay the epistemological tendency – the criterion of *clarté*, the representation of clarity as the guarantee of truthfulness. This tendency not only influenced Descartes's attitude toward the preparation and publication of his works, but also determined their style and, to a great extent the style of the science and culture of the subsequent period.

Of course external influences were essential. Galileo's trial was the immediate reason for Descartes's refusal to publish the *Traité du monde*. But his didactic temperament did not oppose this refusal – at least not in 1630 or for several years thereafter. Descartes understood that it was impossible to publish the *Traité* without conflict with the Church, but he was easily reconciled to this. What was for the seventeenth century a new attitude toward a scientific system, one which had increased the representation of the system's ineluctable inner cogency, what distinguished Galileo from Bruno, Galileo's *Discourses* from his *Dialogue*, Galileo's students from their teacher, had already gone as far as it could go. Thinkers of the first half of the seventeenth century depended upon this inner cogency, while a new surge of polemics (this time clearly anticlerical) was still far off. To this idea of inner cogency, Descartes added the criterion of clarity which was intimately connected with the presentation of the new doctrines. Then, when Descartes thought that the bases of the new teachings seemed complete, he began to seek the publication of his works with greater energy.

Nevertheless, even as late as the period following the publication of his *Principles of Philosophy* (1644), Descartes continued to speak of his antipathy toward publication. He wrote to Pierre Chanut on November 1, 1646:

> . . . M. Clerselier has written me that you are waiting to read my *Meditations françaises*, so that you can present them to the Queen of the country in which you are living [Queen Christina of Sweden]. I have never had sufficient ambition to want people of that rank to know my name, and if I were as wise as savages are convinced that apes are, I would never have become known to anyone as a writer of books: for savages imagine that apes could talk if they wanted to, but that they refrain from doing so in order not to be forced to work, and since I have never had the same prudence to keep me from writing, I do not have as much leisure and rest as I might have had I the mental disposition to keep quiet.[4]

But the presentation of ideas on paper always seemed and really was for Descartes the *experimentum crucis*, the test of true clarity and thus of the truth of these ideas.

2. THE CRITERIA OF CERTAINTY

Let us leave these biographical questions, and turn to the fundamental problem of the relation of Cartesian rationalism to science, namely that of the problem of the transition from the certainty of doubt, to the certainty of the objective being of the world.

The concept of the criterion of clarity was the first link in such a transition. Doubts of long-standing about the certainty of the sensory world and the certainty of science led Descartes to a great and completely unexpected advance in thought, to its radical transformation, to a change in the very concept of certainty. Descartes did not find that sensory feelings guaranteed their own truth. They cannot be distinguished from dreams. Man, as the bearer of sensory impressions, cannot be convinced of their reality and certainty, cannot know whether what he sees exists outside of his consciousness. And here, as a result of his sudden illumination, Descartes found himself on another path, one which no longer led to agnosticism and, even further down the path, to solipsism. Man doubts and must doubt everything — everything except the fact of doubt itself. This fact appears to him absolutely clearly. Accordingly, thought possesses complete certainty — a certainty not of first impressions and not of conclusions, but of the very fact of thought itself. Descartes liberates thought from concrete images and arrives at an abstract statement of thought. Thought itself also possesses absolute certainty. Accordingly, the being of the thinking and doubting subject is also certain: *cogito ergo sum*.

But this is not yet the first step in the direction of science. It was not the certainty of doubt and thought that led Descartes beyond the limits of traditional thought. He was led there by the second statement: *only* thought possesses absolute and direct certainty. Thus the negative problem is resolved: Descartes was beyond traditional thought about the world; this traditional thought was left behind. Now the positive problem begins. Descartes ascribed a not-to-be-doubted character to doubt itself, to the thought and being of the subject of thought. Here, turned in upon itself, doubt disappears. It is opposed by the immediate clarity of the very fact of thought, of *cogito* itself, which is abstract, independent of its object and of the object of doubt. Now Descartes began to seek in the object of thought and doubt, in the concrete images filling *cogito*, something analogous to thought in clarity. In the end he arrived at mathematics as the universal method for knowing the world, and at the world itself as an accumulation of the parts of space — moving, possessing physical being, but having no properties other than spatial ones.

Now what was only the starting point of a new path — *cogito ergo sum* — becomes the real starting point of a really new path. Verhaeren can turn to Descartes with this explanation of *cogito*:

> "Penser, même doubter que l'on pense, c'est être.
> Première! au jour intérieur, cette fenêtre.
> L'idée éclot innée, elle se scrute, insiste;
> L'infini se conçoit: donc il existe,
> Et Dieu ne trompe pas l'homme sur l'univers.
> Mais l'âme humaine encore gothique
> Maintient le corps que rongeront les vers
> Ainsi qu'un instrument sous son doigté mystique."
>
> "To think, even to doubt that one thinks, is being.
> First! This window that faces the inner light.
> The idea hatches, innate, scrutinizes itself, persists;
> The infinite is conceived: thus it exists,
> And God does not deceive man about the universe.
> But the still gothic human soul
> Sustains the body that will be eaten away by worms
> Like an instrument beneath its mystical fingering."[5]

The certainty of doubt must be transformed into the certainty of a positive picture of the world. The idea "scrutinizes itself"; it finds in the contemplation of its own self the criterion of certainty and, further, "[it] persists", it turns its attention towards the body and here the "still gothic" soul finds objective certainty.

Let us attempt to examine more closely the content and meaning of this criterion of clarity which leads Descartes away from initial doubt and the statement of its certainty in a positive cosmological picture. Does this criterion retain its meaning and content during the transition from a concrete series of representations to an abstract statement of thought?

In such a transition, as in every act of abstraction, we ignore the differences between the concrete images which replace one another in the flow of thought, keeping only a homogeneous series of images which possess abstract being. The verb *to be* acquires another meaning; it is already impossible to say that the subject (in this case, the image, concept, deduction) 'is one thing or another', it simply 'is'. 'Is' is no longer a *copula* uniting the subject with the predicate; it is now the predicate itself. Thought (later we shall see that it is thought within the limits of common sense, but not within the limits of reason) identifies its objects, relating them to various classes and ascribing general predicates to them. Now this activity of thought is applied to itself. Thought seems to split in two; it becomes its own object and obtains a general identifying predicate of being which eliminates differences. Does this general predicate of being, this 'is' as a predicate, add anything new to what was contained in the concrete 'is' as a *copula*, in concrete statements like 'the subject is something', 'Bucephalus is a horse'? Yes, it does. And thus it limits the sum of 'is' as *copulas*, carves out of the set of such copulas those which possess real being. Copulas of the type 'the subject is something', i.e., 'the subject possesses a certain predicate', can participate in thought, but these statements can refer to imaginary, negative, wittingly conditional statements. 'Is' as a predicate of being eliminates these conditional, unreal statements and unites those remaining. It seems to be the confirmation of a separate statement, a supplementary affirmation of its reality.

Not long before Descartes, this supplementary affirmation had played an essential role in philosophical and theological confrontations (the problem of the ontological proof of the existence of God), and now it proved to be related to thought and to its subject. In order to understand where it leads (and this is precisely what interests us here), we shall dwell not on the historical sources, but on the modern illustrations of such supplementary confirmation. Let us take as one such illustration the principle of least action in the form given it by Lagrange. A particle can move in various ways, along various trajectories that differ infinitesimally from one another. All these trajectories represent geometrical images, and only the trajectory characterized by the lowest value of the function of action is a physical image which has physical meaning.

Being as a predicate, 'is' not as a copula but as a logically conclusive statement, being when applied to thought itself, eliminates the uncertainty of sensory knowledge that troubled Descartes. If sensory cognition in philosophy and in physics was the verification of truth and the criterion of real being, then for Descartes this cognition became the most suspect source of knowledge. The transition from 'is' as a copula to 'is' as a predicate is no longer the result of empirical experience, but the result of the exclusion from thought of the concrete images of empirical origin. This abstract identification of the concrete links of thought, this transition from thought as a set of various concrete statements to thought as such, guarantees the certainty of being. It is possible to doubt the certainty of *what* a man thinks, i.e., the certainty of the content of his thought, but man cannot doubt that he *thinks*, cannot doubt the certainty of the fact of doubt itself.

In this way, Descartes separates the problem of truth from the problem of sensory cognition. But in this process truth and certain being create a monopoly of nonextended substance, of spirit. How are the concepts of truth and being to be moved beyond these limits? Descartes performed an operation on extended substance somewhat analogous to the one he performed on thought: He removed the properties of extended substance that are the object of thought as a set of concrete statements and retained in extensiveness only its extensiveness, just as he retained in thought only thought itself, only its undoubted existence. Why does the image of a body lacking properties other than extensiveness, an image identical to a portion of space, seem *clear* to Descartes? Why does he ascribe certain being to it? When one examines a body as a portion of space, it means that one isolates only the object of cognition and rejects other statements. Here the answer to the question 'What is a given object?' is reduced to the statement 'the given object *is*'. And 'is', in this case, means that the representation of the object is clear, that is, it lacks empirically comprehensible properties. Clear thought, which states its own being with certainty, confronts an object, which does not demand from thought anything that goes beyond the limits of clarity. Clarity thus ceases to be a predicate of thought characterizing the being of the thinking *subject* and becomes the predicate of the object of thought, the clear representation of what is found outside of thought but which is comprehended by thought. Here all that is known of the object is that it has a certain volume, that it possesses extensiveness and thus can serve as the object of thought *as such*, without losing its self-certainty. During this process, thought not only loses its self-certainty, but also ascribes it to the object of thought.

Cartesian physics and certainly its basis — the identity of extended substance with extensiveness itself — was an attempt to return to thought its 'objectness', as it still retained its clarity, its self-certainty, everything that substantiates *cogito ergo sum*.

But in such objectification, the thought that figures in *cogito ergo sum*, while not losing its certainty, does nevertheless change its character. It becomes thought about extended objects and thus thought about quantities. In order that the objects of thought not deprive thought of clarity, they must possess only purely spatial predicates. This is the function of clarity as a criterion leading from *cogito ergo sum* to mathematics and to mathematical science. What is the meaning of clarity for the transition to experiment and to experimental science?

Descartes's *cogito ergo sum* contradicts *esse est percipi*. Berkeley identified being with thought, while Descartes sought in thought the criteria for the investigation and discovery of objects existing with certainty outside of thought, independent of it, extended. In their extensiveness (i.e., in an attribute foreign to thought), which is not dimmed by illusory representations of heterogeneous matter distinguished from space, Descartes finds the differentiation of the objective from the subjective. But if the extended world remains an object of certain thought, this means that the objective *ratio* is its ruler — the extended equivalent of nonextended *clarté*. The world is made up of categories of clear thought, that is why it is not chaos, but cosmos. The chaos of events becomes cosmos, because causality reigns within it. This causality is mechanical: true, not illusory events consist of the motion of bodies, and they are connected through the impetus which some bodies receive from others. It is precisely such causality that is revealed by thought, retaining its self-certainty as well as producing a certain picture of the extended world. Thus thought moves from *cogito ergo sum* to *cogito ergo sunt*, from the undoubted existence of the subject of cognition, to the undoubted existence of the objective extended world and to the acknowledgment of the certainty of science.

However, the rationalist path from *clarté* in thought to *clarté* in nature was not limited to the mathematical discovery of order and objective reason in the cosmos. It also led to experiment. Order and reason in nature consist in conformity to law, the repetition of phenomena, the identical behavior of bodies belonging to the same class, i.e., bodies possessing identical predicates in unaltered conditions. But such identity, like the real homogeneity of matter, is hidden from cognition by a complex interweaving of events which complicate the identification of conditions and of the behavior of bodies. In order to disclose the identity of the behavior of bodies (and ultimately, the homogeneity of matter and its identity with space), it is necessary to bring the real conditions as close as possible to some defined scheme and then observe how events occur: will they occur uniformly or will they change because of a previously known change in other conditions? Such identification of conditions, such elimination of nonidentity, lies at the basis of experiment. Rationalism does not eliminate experiment; the illusion of a purely speculative comprehension of the world is not characteristic of rationalism in general nor of the evolution of rationalism toward science in particular. The search for the real rationalization of processes, the creation of identical conditions, the discovery during observation of identical results, homogeneity, conservation, and symmetry, are a necessary component of the rationalist approach to nature.

In Galileo rationalism was already to some degree experimental. For Descartes experiment (of course, not so much experiment itself as a penchant for experimentation, for the representation of the problem and the meaning of the experiment) is connected with the basic rationalist tendency, with the search for *clarté* in nature. But the influence of experiment on the evolution of rationalism is unclear, and the concept of experimental rationalism somewhat exaggerates a real, but not very distinct tendency. The opposite tendency — *rationalist experiment* as opposed to experimental rationalism — is more distinct, and we may speak of it as an obvious and characteristic feature of seventeenth-century science, which began at the turn of the century with Gilbert.

In the seventeenth century, experiment was no longer purely empirical (if, generally speaking, the concept of pure empiricism has any real meaning). The overwhelming majority of experimenters started from a kinetic plan; they sought in experiment a scheme, freed from complicating details, for the displacement of matter. This matter (at least in the kinetic experiments) was not considered homogeneous; non-Cartesian, dynamic properties were often ascribed to it, but somehow or other noncausal schemes for experiment, which in principle excluded rational comprehensibility, became rare and usually did not influence the development of science. Reports of experiments with a magnet dipped into a red substance (the satisfaction of the 'king of stones' with royal purple must have been expressed in the intensification of the attraction of the magnet) was out of date much earlier than Kircher stated.[6]

But why did the rationalist comprehension of the world demand experiment? This is the fundamental feature of classical rationalism. The need to experiment stems from the very essence of rationalism and sheds light on its position in history, its connection with sensualism, the evolution of classical rationalism, and its points of contact with nonclassical science.

Later we shall dwell more than once on this problem, which is really central and acquires a very precise character in the analysis of classical rationalism from the vantage point of nonclassical science. Here we can limit ourselves to a reference to Jean Hyppolite's marvelous comment: "If there were only ideas, there would not be any ideas."[7] Thought that operates only with ideas may discover identity in nature but cannot discover nonidentity. Without nonidentity, without unexpected and previously unrepresented interventions of experience, observation, and experiment in the linear chains of identical events, we would see in nature only these series. Or rather, we would not see even them. Identity without its destruction, rational series without empirical interference, would transform the world into a single moment and a single point, and the existence of logical constructions would become impossible.

Therefore, for as long as science has existed, empirical cognition of the world has served as a condition for its rational cognition, and experiment serves as a new and higher form of the synthesis of both. It would be more correct to speak not of the synthesis, but of the complementarity — as Bohr used the term — of the two poles of cognition which exclude each other, but which lose meaning without each other.

3. UNIVERSAL MATHEMATICS

The complementarity of the speculative-logical and empirical aspects of cognition of the world explains the relation of rationalism not only to experiment but also to mathematics. As with experiment, we must limit ourselves to some preliminary and fragmentary remarks before going on to a more detailed analysis. The problem of identity and nonidentity, of logic and empiricism in mathematics, is best examined in the context of the genesis of the differential calculus. But first, a few observations to help us understand the Cartesian conception of space and motion.

Descartes's mathematical ideas stem from the same criterion of clarity as the Cartesian conception of experiment. For Descartes, mathematics is the embodiment of clear, by no means tautological thought directed toward the discovery of new truths. The physical equivalent of *such* thought is the motion of a body, a self-identical but changing position in space. This is nontrivial self-identity, which is retained during a change in spatial predicates. The representation of motion corresponds to mathematical thought. For Descartes this was the first and most important element of universal mathematics — the science that, like mathematics in a narrow sense, uses purely quantitative concepts to investigate nature as a whole. Only such investigation can give certain knowledge. It begins with certain intuition, with the immediate conviction of the spirit of its own existence, which does not demand a logical outcome. After this comes deduction, which retains the initial certainty because it uses clear concepts, as clear as the original *cogito ergo sum*. These are the quantitative concepts of mathematics. But if science as a whole is transformed into mathematics, then mathematics becoming universal in its turn, is transformed into science as a whole, the science of an extended world. First and foremost, the most abstract theory of pure quantities — algebra — acquires a spatial meaning and merges with geometry. Hence geometrical problems may be solved by means of algebraic equations, but because of the analytical geometry created in this manner, algebra also acquires, in principle, the continuous possibility of spatio-geometrical interpretation. It becomes a doctrine of space.

Nor does the matter end here. Space has been substantialized, it has become matter, it is filled in. The doctrine of space is the doctrine of matter, of Cartesian matter devoid of attributes other than extensiveness. Universal mathematics becomes physics. Speculative thought comprehends the material world and acquires a substantive character. Thought as such, thought as it figures in the initial intuition of *cogito ergo sum*, becomes rich in content and is now characterized by concrete objects. Descartes wants to retain initial certainty. By no means does he reject experiment; without it, thought would have been reduced to the traditional tautological dialectic. Descartes demands of experiment that it find under the deceptive surface of phenomena the certain movements of the parts of space. Then experiment does not destroy the certainty of cognition which is operating with purely quantitative concepts.

The physics of Descartes is a great attempt to retain the certainty of thought when it is directed not toward itself, but toward the extended world. This was his initial intention, and in the possibility of such a retention lay the great illumination at Ulm in 1619. For Descartes the thought of the certainty of the direct intuition of *cogito ergo sum* would not have illuminated his entire future course without the thought of a universal science. Such illumination was not even the thought of a universal science lacking in certainty: The *coup de foudre* at Ulm was the thought of the possibility of a certain, universal science.

Nor was thought that was dependent on subjective appraisal in its concern with a certain *object*-knowledge, with an adequate picture of the extended world, the sole basis for Cartesianism. Historically, it was a new and seminal idea for philosophy and science. Augustine spoke of the certainty of the existence of the thinker, even of a spirit gone astray. Descartes later discovered a foreshadowing of *cogito ergo sum* in Arnauld. The idea of an universal science was expressed more than once, but Descartes rejected the certainty of the empirical comprehension of the world on which these ideas were based.

With Cartesianism, philosophy became science. The criterion of certainty led philosophy to science, and science, within the framework of Cartesianism, to philosophy. In this relationship, Cartesianism differs from the previous stage when philosophy and science went their separate ways, and from the subsequent stage when they parted again, this time to a considerable extent. For Descartes, rationalism became science. The point is not the mathematical and physical discoveries (including the very fundamental ones of coordinate representation and linear inertia) for which Descartes deserves credit. The point is that the principal epistemological problem of certain cognition was resolved in Cartesian philosophy constructively by means of the elaboration of a universal science that claimed the certainty of its conclusions.

Such a claim was based on the fact that the primary, certain intuition – *cogito ergo sum* – went beyond the limits of an anthropological problem, beyond the limits of thought as such: it is a transition to object-thought, to thought about the world. When Descartes speaks of method and insists on the definition of the possibilities and conditions for certain cognition, we are dealing with the certain cognition of the extended world.

It is a movement of thought into the world, a search in the world for what is similar to thought, for what makes the world comprehensible. It is a search for an answer to the question of the knowability of the world, which Einstein considered the most difficult: "The eternal mystery of the world is its comprehensibility" (Einstein 1936a, p. 351).[8] The constructive answer reveals the *ratio* of the world in its certain picture. This is the path from the primary intuition of *cogito ergo sum* to the deduction *ergo sunt*.

In order to grasp this *sunt*, thought, logic, the word, must go beyond its boundaries while remaining itself. We have already cited Hyppolite's statement about ideas that cannot survive without the complementary pole of cognition. Hyppolite says further: "We cannot go beyond Logos, but Logos goes beyond itself while remaining itself."[9]

Hyppolite says this about Hegel but it can also be applied to Descartes. For him, Logos goes beyond its primary anthropological limits, while retaining its certainty and finding beyond these limits the rationally comprehensible world of matter, which is identical to space.

The transition from the anthropological *sum* to the naturalist *sunt* somewhat alters the criteria of being. In the first, Descartes substantiates the certainty of being by direct intuition and by theological reference to the certainty of what man knows from God. In the second, we have not intuition but deduction. The criterion for the truth of the results of deduction is its clarity. Unlike intuition, deduction — the motion of ideas, the transition from one idea to another — is subordinate to the very general condition formulated by Hyppolite: "If there were only ideas, there would not be any ideas." Deduction demands an empirical accompaniment. In rejecting empiricism as such in [his notion of] certainty, Descartes, with his representation of experiment, admits the empirical accompaniment into the sphere of the certainty of cognition.

In reality such an accompaniment is necessary to guarantee not only *sunt* but also *sum*. Here we leave the ground for the exposition and interpretation of Descartes's ideas. It is a question of *cogito*, which guarantees *sum*. The very old problem of the theory of being, a problem still not eliminated and to all appearances pervasive, is the problem of change and immutability. It became fundamental for Heraclitus who considered change and motion the stuff of being, and for the Eleatics who maintained that being is immutable. To a certain extent, Democritus went beyond this clash by ascribing to being — atoms — the conservation of immutable predicates as well as continuous change of position in space: motion. We shall soon see what role such a combination of immutability and motion played in Descartes's physics, i.e., in the problem of *sunt*. But in the problem of *sum* as well, apparently, it is impossible to do without the complementary poles of constancy and change.

Let us examine more closely the process of thought: *cogito*. When we deal with the purely abstract statement of thought (*cogito*), we do not notice the above-mentioned polarity. But does such an abstract statement provide the basis for the conclusion, *ergo sum*? No, this conclusion may be drawn if the thinking process indicates the existence of the subject of thought. Such a subject must possess a certain immutable substratum of psychic activity, immutable conserving images, representations, memories, logical norms, concepts. *I*, as the subject of thought, must possess this immutable substratum; then the concept of *I* has meaning. But such a trivial identity of *I* is insufficient in and of itself. Not only *cogito ergo sum* but *cogito* itself loses meaning if the subject possesses immutable, identical states of consciousness. Let us imagine the immutable contents of consciousness. In this case thought disappears, as in the case when consciousness is completely varied, and the self-identity of the subject of consciousness is compressed into a moment. Nontrivial self-identity is the condition of being, for both the object of thought and its subject. Thought identifies separate images but their identification and the creation of classes, into which the identified images and concepts enter, are impossible without identity. In this sense the subjects of thought somewhat resemble photons; they disappear when stopped.

Change, motion, nonidentity as a component of nontrivial identity are brought into consciousness from without. They impart to thought its substantial character. Therefore *sum* follows from *cogito* if the pure self-contemplation of the spirit, convincing it of the certainty of being, is accompanied by a change in concrete states of consciousness. These two poles — thought, which is homogeneous, lacking in content and empirical sources; and thought, which finds differences, nonidentity, which changes its contents, moves — are mutually exclusive. Moreover, each of these poles loses being without the other, becomes an illusory concept. At present, such a relation of the poles does not seem paradoxical; nineteenth-century classical philosophy made it fundamental and twentieth-century nonclassical science embodies it in precise physical schemes. However, in a preliminary and unclear form, it was already hidden in the rationalist philosophy of the seventeenth century.

4. BEING AND SPACE

The purpose of this book is to investigate the course of rationalism which leads to science, classical science, then to its contradictions and dynamics, and thus to nonclassical science which has partly resolved, partly deepened the classical contradictions. The cardinal contradiction, the cardinal *aporia* in Cartesian physics, stems from the identification of matter with space, rendering geometry physical and physics geometrical. Descartes ascribed clarity to physical conceptions and likewise to certainty because only spatial definitions occur in these conceptions. But lacking all properties other than extensiveness, the bodies are indistinguishable from the surrounding bodies, and as a result, another criterion of certainty disappears.

The development of physics after Descartes, along other, no longer Cartesian paths, was directed toward the resolution of or escape from this *aporia*. Dynamic properties were ascribed to matter, and thus bodies were divided from space, which once again became 'nonbeing' in the Democritean sense of the word. Soon, however, it became clear that empty space, 'nonbeing' possesses dynamic properties. The concept of the field of force came into being. Having filled space with hypothetical ether, the field of force was made into 'being'. But the ether could not perform the same task of precisely representing a body. This task now consisted in determining the body's location by means of reference bodies. The ether could not serve as such a reference body. Therefore it ultimately disappeared from the world-picture. This has already happened within the framework of nonclassical science. The world was represented by science as a framework of world-lines. What are these world-lines — are they geometrical or physical objects? What distinguishes them from geometrical objects, what gives them physical being? The answer — again not an exhaustive one — was given as a result of a new twist in nonclassical thought. It turned out that in the microworld, Cartesian *clarity* in the representation of an object (the conviction that it is present at a given point) excludes *precision* in representation (the isolation of this point from surrounding space).

We have skipped ahead and rendered modern concepts somewhat 'archaic', or rather 'Cartesianized' them. Now it will be easier to 'modernize' Cartesian concepts. These words are given in quotes because in reality there is no such thing as rendering archaic or modernizing: it is not the answers but the questions that are juxtaposed;

each answer reveals the question posed and most often re-posed to the future. It is precisely this kind of analysis that allows the history of science to "find not the ashes, but the fire of the past."

Let us return to the epistemological sources of the Cartesian representation of space and matter. In his conception of experiment, Descartes contrasted it with the direct flow of sensory impressions, which bring to thought uncertain, perhaps illusory, elements. Experiment reveals the certain 'solid remains' of the empirical; it deprives experience of its color and *desensualizes* it. As was already mentioned, for the seventeenth and eighteenth centuries, the typical experiment discovered under the multicolored surface of phenomena the monochrome, or better still, the colorless dislocation and redislocation of particles of homogeneous matter. At worst, experiment should have decreased the number of colors in the world-picture, but the ideal of cognition as a whole remained the Cartesian ideal – the complete reduction of qualitative differences to the purely spatial. It is amazing how much the change of the *empirical* basis of science, a change in the problems and character of experiment, depended on the *rationalist* ideal.

Thought performs a task analogous to that of experiment. It desensualizes cognition in an ideal manner, eliminating *concepts* that are connected with experience but foreign to thought, and are therefore foreign to a certain picture of the world. The ontological content of science must possess the same degree of clarity and precision as thought itself. This is the criterion of the selection of predicates for matter. By means of this criterion, Descartes distills a certain world-picture from the set of qualitative and dynamic characteristics, a set suspect in its possible illusoriness, and arrives at the abstract being of the extended world. This world exists outside of the cognizant spirit because it is an extended world as distinct from the nonextended subject of knowledge. But in this extensiveness is everything; extended substance does not have other predicates.

In Descartes's conception, thought itself attains clarity and precision as abstract thought. It becomes substantive and objectified; it passes from *ergo sum* to *ergo sunt* because it finds in the purely spatial world an object just as abstract. But a very difficult *aporia* lies in wait for this well-proportioned and non-contradictory construction. The thought that figures in *cogito ergo sum* is abstract thought as such. The philosopher, in this case, is not interested in *what* man thinks, it is sufficient *that* he thinks. Concrete images and deductions are uncertain, and only the simple statement of thought, its content aside, is certain. It possesses undoubted clarity, but does it possess the precise isolation of each link, of each psychic act (if we eliminate the content of this act), of a separate image, representation, or conclusion? Here thought comes to resemble the Cartesian extended substance; it consists of bodies, but each body is only continuous extensiveness and cannot be separated in a precise manner from the surrounding extended substance. Descartes's *cogito* is only thought continuing in time, and just as his concept of matter runs the risk of being transformed into space and only space, so Descartes's thought, lacking in content, may turn out to be only temporal duration, time and only time.

Descartes fills thought with substantive as well as certain [i.e., reliable] representations and conclusions. Thought acquires precision without losing clarity. But these certain images, representations and conclusions themselves represent images of matter identical to space, representations of bodies as parts of space, and conclusions relating to such parts. At this point the *aporia* enters physics. Suspended above it is the problem of the differentiation of the body from the surrounding substance, the problem of the individualization of the body.

This problem is the worst one for Cartesianism. The individualization of the body in any philosophical or physical system that admits substantial qualitative differences is not a complex matter. But what differentiates a body that is part of a *homogeneous* substance from its other parts? When we think of the body as a part of a homogeneous substance, we include this body in an imaginary, infinitely fine film, but this is an obvious transition from Cartesian clarity and certainty to an undeniable illusory representation. We shall now see how Descartes gets out of, or tries to get out of this clash between clarity, which demands a purely spatial homogeneous substance, and precision, which demands the individualization of the bodies. Before this, however, we must discuss the problem of filling space.

The filling of space in Descartes was by no means a return to Aristotle's space. The association of Descartes with the peripatetic tradition provoked Voltaire to ironic laughter, but in reality there was no such association. Aristotle's space was filled with heterogeneous matter which possessed properties that differentiated one body from another, and, at the same time, matter as a whole from simple extension. For Descartes, space was filled with matter which, in its turn, was not differentiated from space. But what does the physical being of space consist of, what does its *filling* signify?

This question belongs to a number of pervasive problems which every epoch receives from the preceding one, gives it a particular but not exhaustive answer, and then readdresses the same questions to the future. The subsequent, new answer permits a more precise definition of the meaning and significance of the previous one. In this case, it is natural to compare the Cartesian filling of space with the conception of the field. As already mentioned, the premise of the conception of a field was the rejection of Descartes's purely kinetic view and the inclusion of dynamic representations in the world-picture with the inclusion of the concept of force in physics. But science goes in spirals, and in the nineteenth century people began to understand by the field a material medium which through its influence not only determines the behavior of bodies, but exists in their absence as well.

This last circumstance is important in explaining the relation of field theory to the general evolution of the doctrine of space. Democritus ascribed reality to empty space, to 'nonbeing', because it represented the accumulation of the possible positions of the atom. But he ascribed 'being' only to space that was actually filled at any one time, that is, to the atom occupying this space at any given moment. Descartes ruled out this possible being: for him space is already filled with matter and in this sense it possesses actual and only actual physical being. The theory of the field of force (or what replaced it) in the physics of action at a distance ascribed only possible being to the field: if a particle appears at a given point, then force acts on it; the particle receives a real impulse, and this testifies to the reality of the field. In Faraday's theory this is all different. For him the field already exists in actuality, even without the presence of the test particle at a given point. The field seems to await the particle, but possesses actual being even without it. In the wave theory of light this actual being loses its connection with the possible. The field does not simply exist and await the particle in order to demonstrate its existence; rather, here it reveals its existence in the most direct manner: light is the most universal and important agent acting on the sense organs — after all, people see it.

The photon theory meant that once again space was not continually filled with something actually existing. Light — the flow of photons, the electromagnetic field — exists only where photons are found. This representation lasted only a short time. Then it turned out that space is filled everywhere after all — with probabilities of a meeting with a photon. Possible being was transformed into *probable* being, probability being precisely defined. But this was only the beginning. Space possesses not only possible being; the wave function indicates the probability of its transformation into the actual. It possesses *virtual* being; even in the absence of 'real' particles, several ultramicroscopic events take place. Of course, a completely new concept does not fit into traditional reminiscences and differentiations. But the question posed in the past still confronts physics today, although of course in a very modified form. What is meant by the filling of space? What distinguishes the substance filling space from space itself? What distinguishes physical objects from geometrical objects? This question cannot be passed over now, just as it could not be passed over by Descartes in the seventeenth century. We shall see how Descartes answered it.

Descartes's answer is simple. A body is individualized insofar as and when it moves. Motion performs this function of individualization if it occurs in reference to the bodies contiguous to it. We shall see later whether this condition is sufficient. But it is necessary. Motion, according to Descartes, is change of location. The location of a body is the surface of the bodies in contact with it. Here the significance, for the Cartesian theory of motion, of the problem of the individualization of a body is readily apparent. Motion is the property of a body that isolates it from the homogeneous medium. Therefore, only motion in reference to the adjacent surface from which the body must be separated by motion is essential for philosophy, for the resolution of the fundamental philosophical question, for the guarantee of substance, object, precision of thought, in other words, for Descartes's fundamental task. Thus we can see how important Cartesian relativism is for Descartes's *entire* conception.

Descartes ascribed two meanings to the concept of motion: one is the real meaning, the other, the generally accepted meaning. *Motion in the real sense* is motion in reference to adjacent bodies. Without such motion, the boundary of a body disappears; the body loses individuality and ceases to exist. In Descartes's opinion, the kernel of a nut really exists as a separate body if it is displaced relative to the shell. *Motion in the generally accepted sense* refers to the reference body, which can be far away. We consider it motionless. A ship, carried along in the flow of a river, is motionless in the real sense and moves in the generally accepted sense relative to the banks.

What does the entire Cartesian theory of motion mean as a whole?

Primarily it means that Descartes had found an objective equivalent for clear thought, an objective *ratio* of the world. He found it in the totality of moving bodies. In contrast to the motionless system of natural places in peripatetic cosmology, the objective *ratio* of the world now consisted of the motion of bodies – an ordered motion maintaining immutable velocity and immutable direction. All these properties of world harmony, world *ratio*, are inseparable from repetition, immobility, identity. Without this there is no being. Therefore, the concepts of repetition, immutability, and identity are inevitably basic to every picture of the objective world, to every conception of being. In this sense, what Descartes did cannot be rejected. He introduced the criterion of clarity from *cogito* into nature, identifying it with homogeneous, extended, and only extended, substance. But rationalist science could not develop without concepts expressing nonidentity, the heterogeneity of being, and the irreducibility of empirical data to the identifying deductions of reason, hence the empirical accompaniment to rationalist thought. The idea of such an accompaniment was expressed more than once – sometimes in the form of a conception of the limits of rationalism which separate it from the irrational. In the seventeenth century, immediately after Descartes, rationalism overstepped these boundaries, but beyond them it met nothing that could be called irrational in principle.

5. THE LIMITS OF CARTESIAN RATIONALISM

In the next chapter, which is devoted to several aspects of Spinoza's philosophy, we shall analyze the passage of rationalism across the initial Cartesian boundaries. Here, however, we shall touch upon that indefinite, almost intuitive sense of the limitations of Cartesian rationalism which existed during Descartes's lifetime and beyond, into the middle of the seventeenth century. This is the sense of limits which grew from Cartesianism itself, and not from any anti-Cartesian tendencies. We have seen that this sense accompanied rationalism even in its prehistory, when rationalism itself was expressed not so much in precise representations of paths, aims and possibilities for the rationalist knowledge of the world and of the *ratio* of the world itself, as in the emotions, inclinations and sympathies of man. Now when rationalism had attained in Cartesian physics a more precise form, one intimately connected with positive scientific knowledge, the intuitive sense of the limits of rationalism ceased to be sense and became thought — rationalist thought. Reason no longer pushed against the boundaries which emotion, aesthetics and morality placed in front of it. Reason itself induced its emotional accompaniment. It criticized its own inadequacy, its limitations at a given moment as compared to its unlimited potential. The contradiction between the limitation and the limitlessness of reason became an internal, and at times a tragic, contradiction for rationalism. For Galileo, the infinity of nature and the limitation of already acquired representations of nature (extensive knowledge) were not a tragic dissonance. Galileo was confident of the absolute certainty of the method which leads to the successive expansion and refinement of acquired knowledge. The source of his optimism was his confidence in the universal effect of this mechanical method; it withstood his most difficult personal trials.

The contradiction between the limitation of cognition and the limitlessness of its object, between the limitation of human existence and the infinity of the world, between the power of reason and the defenselessness of man, became a tragic *inner* contradiction of rationalism in Pascal's world-view. Pascal could not become reconciled to Montaigne's individualistic scepticism. He wanted objective and generally significant knowledge. But Descartes's objective and generally significant physics seemed dry and soulless to Pascal. Nature, in which there are no predicates other than extension, did not answer moral and emotional questions; it did not explain why mortal man thinks of eternity, why man's limited knowledge embodies his striving and ability to grasp and resolve problems of the infinite universe; it did not create the foundation for moral principles and the world of human feelings.

Let us emphasize once again that Pascal's drama was the inner drama of rationalism. Thought itself had to go beyond the framework of the colorless world of Cartesian physics and find the moral and emotional equivalents of science. Pascal sought a way out in Jansenism with its mystical perception of God, a vision which limited the scientific comprehension of the world. But he could not go beyond the limits of rationalism, and for Pascal the very defense of Jansenism and the battle against the Jesuits was profoundly rationalist.

Pascal spoke of man as a great and insignificant creation. His greatness lies in thinking. Man thinks of his sad fate, of death, weakness, loneliness; and however insignificant this weak, defenseless, momentarily living creature, the thought of this weakness is a great thought. Thinking is the peak of human activity and will remain man's great distinction.

The inner rationalist character of the clash between frail and limited human existence and the infinite being of nature rends the soul of the thinking man, hence the tragic character of this clash. The limitation of rationalism by external, purely *a priori* or purely empirical concepts, not subject to rational comprehension, was not the source of tragedy. But for Pascal, rationalism strives to grasp everything, and the impossibility of reconciling and resolving the fundamental contradiction of human existence becomes extremely difficult for the thinker.

Two aspects of this fundamental contradiction do not disappear from Pascal's consciousness: the limitation of human existence in time — man's death — and his limitation in space. Before him is eternity in time and infinity in space. Pascal writes:

> 'I do not know who put me into the world, nor what the world is, nor what I am myself. I am terribly ignorant about everything. I do not know what my body is, or my senses, or my soul, or even that part of me which thinks what I am saying, which reflects about everything and about itself, and does not know itself any better than it knows anything else.
>
> 'I see the terrifying spaces of the universe hemming me in, and I find myself attached to one corner of this vast expanse without knowing why I have been put in this place rather than that, or why the brief span of life allotted to me should be assigned to one moment rather than another of all the eternity which went before me and all that which will come after me. I see only infinity on every side, hemming me in like an atom or like the shadow of a fleeting instant. All I know is that I must soon die, but what I know least about is this very death which I cannot evade' (Pascal 1966, §427 [Brunschvicg §194], pp. 157–158).[10]

It would seem that these problems of the individual, the atom — of the shadow which is deprived of real being in the face of spatial and temporal infinity — were resolved by the philosophy and culture of the Renaissance. Renaissance art, in the person of Leonardo (and not only Leonardo), saw in the individual the limited and fleeting reflection of infinity and bore witness to this reflection. Renaissance philosophy, by representing the atom as a reflection of the infinite soul of the world, kept it from disappearing in the infinite world, from being reduced to nothing, to zero. Why then, in the middle of the seventeenth century, was the fate of the individual — and above all man's position in the infinite world and in its infinite evolution — once again posed as a fundamental, unresolved problem in an unprecedentedly sharp and tragic form?

After Descartes it was no longer possible either to pose or resolve this problem as other than a problem of reason within the framework of rationalism. For Leonardo, the path from the individual and limited to the general and infinite was aesthetic. The *beautiful* was the synthesis of the concrete, individual and general, a path from the individual to the general. For Bruno, individual consciousness comprehended its connection with the infinite, and thus overcame the rift between the mortal and the immortal, the finite and the infinite, through the intuitive romantic drive of 'heroic enthusiasm'. Now it was necessary to connect the "one corner of this vast expanse" with the most immeasurable space, and the "brief span of life allotted to [one]" with eternity, to connect the precise and accurate causal connection with the framework of classical rationalism. But to do this rationalist thought had to comprehend not only homogeneous and abstract space and time, but also space and time that was filled, heterogeneous, that consisted of an infinite number of modes — in other words, space and time that were extended but not reducible to the extension of substance. In the picture of the infinite Universe, where the infinitely small spatial and temporal cells cannot be ignored, the "one corner" and the "brief span" do not disappear before infinity; they retain their individual being and enter the network of causal connections; they reveal their necessity for the whole and are represented as modes of infinite nature.

The task of reconciling the individual with the infinite Universe, and the transitory and spatially limited being of man with the infinite being of nature, is examined in a general way by Spinoza, in a more concrete way by classical science, and in a significantly more precise and explicit way by nonclassical science. From this point of view, Pascal's ideas were a search, a demand, a problem, which was resolved inconclusively by seventeenth-century philosophy as it persisted with Bruno's intuitive 'heroic enthusiasm' within the framework of rationalism.

Awe before infinity and the foreboding that before it the individual being of man would be reduced to nought also showed up in seventeenth-century astronomers. After the publication of the *Starry Messenger* [*Siderius nuncius*], Kepler, in a letter to Galileo, told how the first news of the book's contents inspired in him the fear that Galileo had demonstrated the infinity of the world.

> In the first place, I rejoice that I am to some extent restored to life by your work. If you had discovered any planets revolving around one of the fixed stars, there would now be waiting for me chains and a prison amid Bruno's innumerabilities, I should rather say, exile to his infinite space. Therefore, by reporting that these four planets revolve, not around one of the fixed stars, but around the planet Jupiter, you have for the present freed me from the great fear which gripped me as soon as I had heard about your book from my opponent's triumphal shout (Kepler 1965, pp. 36—37).[11]

But this 'great fear' did not go too deep. When it became clear that the 'Medicean stars' discovered by Galileo revolve around Jupiter, that everything occurs within the solar system, and that there was no basis for the conclusion of the infinity of the world, Kepler calmed down at once. But even before he learned this and although he found the prospect of an infinite Universe bleak, infinity did not arouse in Kepler a tragic sense of the illusoriness of human existence before the immeasurable Universe.

Kepler was convinced that an *a priori* harmony rules in nature. The motion of the planets in their orbits is maintained by immutable laws, which connect their velocity at each point with their distance from the center of revolution. But what do the orbits themselves depend on, what are the primary conditions of motion, what is the structure of the solar system? Kepler drew a plan of the solar system by deducing the mean radii of the orbits from *a priori* geometrical ratios, and he spoke of the predetermined, mystical 'music of the spheres'. For Descartes, an *a priori* harmony guaranteed the being of the world and the immutability of the laws of the universe. Pascal had no such confidence. He sought refuge in mystical contact with God, in the apologia of Christianity, but his tragic sense of the disharmony of being persisted. The disharmony of the finite and the infinite of which Pascal spoke in the passage cited above demanded for its reconciliation with being a rationalist solution, without reference to predetermined, *a priori*, rationally incomprehensible forces.

This was what history demanded. Pascal's agonizing twists of thought expressed an objective search for a conception of the world in which each finite element of being would become an element of the infinite Universe, not its mystical reflection, but rather a participant in a universal connection that was rationally comprehensible and left no room for a predetermined *a priori* scheme, for forces standing above or outside of nature. And left no room for God. It was precisely because of this that Pascal's tragedy was a stage in the history of thought. Not having resolved the doubts and demands of thought, his religious, mystical drives expressed essentially the expectation of a rationalist, *universal* scheme of being. In order to resolve the clash between finite human existence and infinite nature, it was necessary to free nature from God. This was what Spinoza did. At the same time, man became free. Free in a new sense, unknown to the past. He is free because his actions depend not on external impulses, but flow logically from his essence, like geometrical theorems from axioms. Man's freedom becomes conscious; man no longer views infinite nature as something foreign, he is joined to it in the process of adequate cognition — as Spinoza said, in the comprehension of nature creating its own self, and free from external forces. Such cognition is connected with *amor intellectualis*, with the emotional drive toward truth, toward its rational comprehension.

The transition from Descartes to Spinoza is a transition across the boundary of rationalism, but not across a general boundary separating rationalism from irrationalism. It is a transition from one historically limited form of rationalism to another, more general form. In the Introduction to this book, it is called ultrarationalism. Such a term makes sense when we compare Spinoza's rationalism with that of Descartes. Ultrarationalism is the rational comprehension not only of the behavior of physical objects, but of their being as well. Previously this was viewed as *a priori* and not subject to rational analysis. Spinoza overstepped this boundary. Thus the infinite Universe, foreign to man as something that was *a priori* in opposition to him, is humanized and becomes the object of rational human cognition, not only in its behavior but also in its being. Awe before infinity, the sense of the limitation of human existence, gives way to the sense of merging with nature that was discussed in our chapter on the philosophy of Epicurus.

NOTES

[1] A very interesting historico-philosophical and psychological analysis of the illumination at Ulm is contained in the short biography of Descartes written by Paul Valéry for a collection of excerpts of Descartes's writings ('Une vue de Descartes' in Valéry (1961), pp. 7—51; an English translation, 'A View of Descartes' can be found in Valéry (1968), pp. 36—68.

[The source that Valéry cites for the biographical information contained in his sketch is Adrien Baillet's *Vie de Descartes* (1691). It describes the illumination at Ulm in part as follows:

> Il [Descartes] nous apprend que le dixiéme de Novembre mil six cent dix-neuf, s'étant couché *tout rempli de son enthousiasme*, & tout occupé de la pensée *d'avoir trouvé ce jour là les fondemens de la science admirable* . . . (Baillet 1691, p. 81) — Eds.]

[2] [Jean-Louis Guez de Balzac (1597—1654) was a French man of letters who had a great influence on the development of classical French prose. On May 5, 1631 Descartes wrote to him from Amsterdam:

> Mesme vous deuez pardonner à mon zele, si ie vous conuie de choisir Amsterdam pour vostre retraitte & et de le preferer, ie ne vous diray pas seulement à tous les Conuens des Capucins & des Chartreux, où force honnestes gens se retirent, mais aussi à toutes les plus belles demeures de France & d'Italie, mesme à ce celebre Hermitage dans lequel vous estiez l'année passée. Quelque accomplie que puisse estre vne maison des chams, il y manque tousiours vne infinité de commoditez, qui ne se trouuent que dans les villes; & la solitude mesme qu'on y espere, ne s'y rencontre iamais toute parfaite. (. . . you must forgive my zeal if I invite you to choose Amsterdam for your retreat and to prefer it not only to all the cloisters of the Capuchins and Carthusians to which many honorable people retire, but also to all the most beautiful places in France and Italy, even to that renowned hermitage where you stayed last year. A country house, no matter how well furnished, always lacks an infinite number of conveniences and comforts that one finds only in cities; and even the solitude that one hopes for in the country is never complete (Descartes 1969, pp. 202—203).) — Eds.]

[3] "Quel autre pays où l'on puisse iouyr d'vne liberté entiere . . .?" (Descartes 1969, p. 204).

[4] . . . Monsieur Cl(erselier) m'a écrit que vous attendez de luy mes Meditations Françoises, pour les presenter à la Reyne du païs où vous estes. Ie n'ay iamais eu assez d'ambition pour desirer que les personnes de ce rang sceussent mon nom, & mesme, si i'auois esté seulement aussi sage qu'on dit que les sauuages se persuadent que sont les singes, ie n'aurois iamais esté connu de qui que ce soit, en qualité de faiseur de liures: car on dit qu'ils s'imaginent que les singes pourroient parler, s'ils vouloient, mais qu'ils s'en abstiennent, afin qu'on ne les contraigne point de trauailler; & pource que ie n'ay pas eu la mesme prudence à m'abstenir d'écrire, ie n'ay plus tant de loisir ny tant de repos que i'aurois, si i'eusse eu l'esprit de me taire (Descartes 1972, p. 535).

[5] [Emile Verhaeren (1855—1916) was the foremost Belgian poet to write in French. The lines quoted here are from 'Les livres', one of the poems in the cycle entitled 'Les flambeaux noirs' (the Black Torches). This cycle, published in 1890, is the last group of poems in a trilogy, the first two groups of which are 'Les soirs' (1887) and 'Les débâcles' (1888). Baudouin (1924) writes that these poems, which are central to the poet's oeuvre, "were a product of extreme mental anguish" (p. 106) and that in 'Les livres' "Verhaeren gives expression to the distress that arises in a modern mind owing to the secular evolution of philosophic doctrines, and their crazy dance in such a mind" (p. 149). Significant studies of Verhaeren include Corell (1927), Zweig 1914), Baudoin (1924) and Starkie (1927) — Eds.]

[6] [Athanasius Kircher (1602 or 1601—1680), a German scholar and mathematician, wrote five books on magnetism, the most important of which is *Ars magnesia* . . . (Würzburg, 1631). For more information, see the article on Kircher by Hans Kangro in the *Dictionary of Scientific Biography*, edited by Charles Gillispie (New York: Scribners, 1970—1980), vol. VII, pp. 374—378 — Eds.]

[7] "S'il n'y avait que des idées, il n'y aurait pas d'idées" (Hyppolite 1953, p. 131).

[8] ". . . Das ewig Unbegreifliche an der Welt ist ihre Begreiflichkeit" (Einstein 1936b, p. 315).

[9] "Nous ne pouvons pas sortir du Logos, mais le Logos sort de soi en restant soi . . ." (Hyppolite 1953, p. 131).

[10] "Je ne sais qui m'a mis au monde, ni ce que c'est que le monde, ni que moi-même; je suis dans une ignorance terrible de toutes choses; je ne sais ce que c'est que mon corps, que mes sens, que mon âme et cette partie même de moi qui pense ce que je dis, qui fait réflexion sur tout et sur elle-même, et ne se connaît non plus que le reste.

"Je vois ces effroyables espaces de l'univers qui m'enferment, et je me trouve attaché à un coin de cette vaste étendue, sans que je sache pourquoi je suis plutôt placé en ce lieu qu'en un autre, ni pourquoi ce peu de temps qui m'est donné à vivre m'est assigné à ce point plutôt qu'à un autre de toute l'éternité qui m'a précédé et de toute celle qui me suit. Je ne vois que des infinités de toutes parts, qui m'enferment comme un atome et comme une ombre qui ne dure qu'un instant sans retour. Tout ce que je connais est que je dois bientôt mourir; mais ce que j'ignore le plus est cette mort mème que je ne saurais éviter" (Pascal 1958, vol. II, §427, p. 22). "Brunschvicg §194" refers to the numbering used in the edition of the *Pensées* edited by Léon Brunschvicg (in three volumes, Paris: Hachette, 1904). This passage occurs in volume II, p. 106 of the Brunschvicg edition.

[11] "Primum exulto, me tuis laboribus nonnihil recreari. Si circa unam fixarum discursitantes invenisses Planetas, iam erant mihi, apud Bruni innumerabilitates, parata vincula et carcer, immo potius exilium in illo infinito. Itaque magno in praesens me liberasti metu, quem ad primam libri tui famam ex opponentis mei triumpho conceperam; quod quatuor istos Planetas, non circa unam fixarum, sed circa sidus Iovis, ais discurrere" (Kepler in Galileo 1892, p. 119).

6. SPINOZA AND EINSTEIN

1. SUBSTANCE

In this book rationalism and its evolution are discussed from the point of view of both their influence and dependence on science. From this, and perhaps not only from this perspective, the basic difficulty in the rationalist conception of cognition lies in the isolation of reason from the object of cognition: the basic course of the evolution of rationalism is the search for a bridge between reason and the extended world. The strictly physical *aporia* of Cartesian physics stems from the epistemological difficulty just indicated. Descartes seeks in the world for that which, by its clarity, exists on the same level as reason. He finds this objective equivalent of reason in extension and then faces a new difficulty: a homogeneous extended substance cannot be divided into individualized objects. Its division is purely mental, and thus Descartes comes back to nonextended substance — to thought.

Bridges between the extended and nonextended world have not been built. Dualism cannot explain the existence and evolution of objective knowledge. Malebranche posits the isolation of two substances. Sensation cannot arise from motion. Ideas do not penetrate from the outside into thought. Malebranche entrusts the establishment of links between reason and the extended world to God.

But here is the beginning of a more radical twist in rationalist thought. Spinoza unites thought and extension. They are different *attributes* of one and the same — extended and thinking — substance. These attributes express the constant, inalienable properties of substance. So far we are dealing only with attributes — there is still no living, changing fabric of finite determinations that replace, contradict, and negate each other. Such a fabric consists of modal determinations. These are modes, properties which can or cannot be ascribed to substance, and this 'or' creates the entire diversity of the world.

Now the pervasive *aporia* of rationalism moves to another plane. It is no longer a question of the unavoidable rupture of substances. This is no longer sufficient. It is necessary to extract from the constant, unchanging, inalienable attributes of substance that element of negation which makes the world nonidentical to itself, or rather nontrivially identical, changing in space and in time. Spinoza knows about negation as a condition that determines and limits modes, but the difficulty lies in extracting negation and nonidentity from inalienable predicates.

How is it possible to link the unity of substance and the unchangeability of its attributes with the contradictory kaleidoscope of modes?

This is not a new problem. The heterogeneity of the extension of the world was always an obstacle to purely rational understanding. Such an understanding identifies the world, and already for Descartes the objective, extended equivalent of thought could not be divided into bodies because of its homogeneity. Therefore the heterogeneity of modes expresses in itself the heterogeneity of the contemplated world — the constant difficulty of rationalist epistemology. Delimination: substance, attributes, modes will be the natural framework for analysing this difficulty. Therefore, before turning to these three fundamental concepts of Spinoza, something must be said about the subsequent course of rationalism.

It is difficult to build bridges between thought and extension, even when there are no longer different substances but different attributes on the opposite shores. In this case the problem itself can be changed. Leibniz individualized the concept of substance; he ascribed substantive character to individual elements of being. For him extension had already become not an attribute but a mode — the result of the dynamic predicates of a monad. Kant transferred extension as a whole to the sphere of reason and made it an *a priori* category of contemplation. Hegel looked at thinking itself from a different point of view; he brought purely identifying thought to the level of the understanding and used the term 'reason' for the more complex thought that included negation and contradiction. He began to consider the very heterogeneity of the objective, extended world as another type of being of the heterogeneous, contradictory spirit.

However, no matter how the problem of understanding the extended world by nonextended reason was modified, to a certain extent its solution will always revert to Spinoza. The idea of a unified — extended and thinking — substance will always be, if not the solution to the problem, then the most general program for a solution. Looking ahead again, let us mention that, considered from such a 'programmatic' point of view, Spinoza's teaching concerning a substance is not, as the

Kantians often affirm, the peak of dogmatic philosophy (if this very definition still holds up after an analysis that 'questions' the components of the development of philosophy and science). Spinoza's teaching is full of unresolved clashes directed to the future.

Spinoza's substance changes the epistemological meaning of the very concept of substance. For Descartes, substance is something possessing *a priori* predicates, something lasting, constant, constant in space (and therefore homogeneous) and in time. Only the modes of substance change. This is thinking substance, thought: *cogito*, which proves the existence of a thinking subject and is homogeneous in its essence; only the content of thinking changes. Matter is abstract extension that possesses the epistemological criterion of clarity. It becomes heterogeneous when it succeeds (and if it succeeds) in introducing the boundaries of bodies into space. But in both cases this intervention of heterogeneity pushes rational thought to its sensual sources: for thinking and for space the problem of being filled is the problem of the empirical roots of cognition.

For Spinoza, substance is not the last and final link of the analysis, but the sum total; it is unified and *for this reason heterogeneous.* It includes all determination: outside of substance there is nothing. Substance possesses not only constant and inalienable properties — attributes — but also modal, changing properties, which are also the properties of substance, its modes. Therefore the cognition of substance does not lead to the last, final links in the analysis.

Rationalist thought prior to Descartes and for Descartes himself is motion along a straight line on a limited plane. It consists of a linear sequence of identifying deductions of the type: 'if A is B and B is C, then A is C'. This leads to a particular final concept, one which does not need and does not admit any further continuity. For Spinoza rationalist cognition is motion on a spherical plane; it never reaches limits and can continue infinitely. The ascent of thought to substance is not an ascent to ultimate abstraction, but an ascent to the highest concreteness. Spinoza's substance is not only the substance of nature, it is nature itself. A limited and transient mode cannot be understood without the representation of a heterogeneous but unified Whole; in the mode is embodied a single substance with its infinite attributes.

Spinoza's thought breaks out of the traditional clash between nominalism and realism, the traditional juxtaposition of the abstract and the concrete. Universals are real, no longer the abstract ideas of Plato, deprived of a sensual form but concrete multitudes; of them, the most concrete and richest in attributes is nature in its heterogeneous integrity.

Such substance — no longer abstract and homogeneous, but rich in concrete attributes — serves as the object of causal analysis. We shall not dwell on 'substance'; it does not interrupt the series of questions, 'why?'. This question also relates to nature as a whole: 'why is nature thus and not otherwise?' The answer must be causal. What makes a causal answer possible if there is nothing outside of nature?

2. *CAUSA SUI*

The possibility of a causal answer exists because Spinoza introduces a new, more general and broader understanding of cause. The universe itself is the cause of its own existence — *causa sui*. The concept of *causa sui* finds a very clear physical equivalent in *contemporary* science. Today the idea of the interaction of the Universe with itself as a cause for its existence does not seem paradoxical to anyone. The very concept of interaction with the self, of 'self-action', has become customary in quantum electrodynamics and in elementary particle theory. The electron radiates and absorbs virtual photons, and the mass of the electron to some extent depends on this process. Moreover, we can imagine that the particle undergoes virtual transmutation, is transformed into a particle of another type, and then regenerates; a new particle is transformed into a particle of the initial type. Such a scheme (already mentioned in our introduction) may now seem arbitrary, but it no longer seems paradoxical. The same can be said concerning hypotheses which examine the effect of the Universe as a whole on each particle as the cause of its being *as* a particle of a determinate type, with a determinate rest mass and a determinate charge.

Such a self-coordinated system of particles, where the existence and substantive predicates of each particle are explained by the existence of others, at first seems to go from Pontius to Pilate and reminds us of a person who lifts himself up by the scruff of his neck. But this first impression is the result of the traditional limitation of causal analysis using the world of modes. Traditional thought seeks the cause of the behavior of each physical object in the behavior of *other* objects and, as a rule, does not develop a causal analysis of the sum total of objects, it does not ask why this sum total exists. The object of traditional causal analysis is modal attributes — coordinates, speeds, accelerations, and all quantities that are combinations of these. But elementary particle physics persists in searching for a causal explanation not of modes but of substantive predicates, a causal explanation of not only behavior, but also the existence of particles. This is the natural transition to the existence of the Universe as a causal problem.

The transition from a causal explanation of the *behavior* of bodies composing the Universe to the causal explanation of their *being* is the basis of Spinoza's philosophy and distinguishes it from Cartesianism. Corresponding to this transition is the generalization of the concept of causation, the concept of *causa sui*, the inclusion of freedom in the definition of cause — *causa libera*, free cause as opposed to *libera voluntas*, and the image of nature not only as created (*natura naturata*) but also as creating (*natura naturans*). Before returning to these fundamental concepts, we must say a few words about their physical equivalents.

Modern elementary particle and cosmological theories surround each of the abovementioned concepts with a mass of very concrete physical associations, which are sometimes precise and clear, sometimes intuitive and vague. Therefore Spinoza's categories begin to seem like souls which searched for physical embodiment for three centuries but did not find it. This is only partly true. First of all, each conception seeks its — increasingly concrete and broad — physical embodiment in the future. Secondly, Spinoza's concepts had their physical equivalents in classical science, but these equivalents were not explicit. On more than one occasion we have taken various approaches to the distinction between classical and nonclassical science. One characteristic of nonclassical science is its constant or nearly constant return to the most general principles, and to a review of these principles in solving particular problems. Using Spinoza's concepts, one can say that nonclassical science, in solving problems directly related to modes, raises the notion of attributes. The theory of relativity, which investigates a purely modal problem, namely the constancy of the speed of light in moving systems, could not fail to raise the problem of space, that is, the problem of an attribute. When classical science studied modes (for example, the position, velocity and acceleration of bodies), it did not have to deal with attributes and the problem of substance as a whole. Scientific thought posed these problems in the course of *generalizing* modal problems, but this generalization was not organically linked to positive knowledge.

The transition from causal constructions relating to modes (where an external cause can be found) to a causal explanation of the being of nature as a whole occurred not in everyday science (as it did in nonclassical science), but in the moments of its generalization. But in the final analysis particular causal conceptions pushed thought to a causal explanation of the being of the Universe as a whole. To be precise, these impulses, which went from science to philosophy, had to lead to the conception, *causa sui*. Other explanations — above all creationist ones — did not come from science. The idea of *causa sui* is a direct generalization of particular causal statements. If the causes of modal events are located inside nature, inside extended substances, then when dealing with this substance, we must refer to it, but we do not have to limit ourselves to the concept of external effect. It is necessary to introduce internal effect into the concept of cause; nature develops one way and not another because of its internal structure. Hence, determinism loses its link with the idea of external coercion; cause can exist in the internal nature of the object: No one or nothing outside of the triangle, says Spinoza, "can bring it about that it should not follow from the nature of a triangle that its three angles are equal to two right angles . . ." (Spinoza 1982, Pt. 1, Prop. 17, Scholium, p. 44).[1] In this absence of external impulses, in this immanent effect of nature on itself consists the free cause — *causa libera* — of being. Nature, which acts in this way, is productive nature, *natura naturans*. When we examine only the result of this constant effect of nature on itself, we see created nature, *natura naturata*.

3. *NATURA NATURANS*

The relation between *natura naturans* and *natura naturata* is one of the most important historico-philosophical problems connected with the interaction of classical rationalism and classical science. The cognition of *natura naturans* is the cognition of nature in its wholeness and unity, the cognition of the harmony of the universe, of objective *ratio* which defines all that occurs in nature as *causa libera*. Obviously, the cognition of *causa libera* must have a mathematical character; the analogy of this concept to the geometrical necessity of the equality of the sum of the triangle's angles to two right angles penetrates into the very essence of *causa libera*. The cognition of *natura naturata* is the cognition of created nature. Spinoza considers it inadequate. He operates with general concepts, with universals. These concepts, which Spinoza calls 'transcendental terms' ('being', 'nothing'), and generic concepts ('man', 'horse') result from obscuring and concealing the differences between individual notions. However, empirical cognition is also inadequate, for it encompasses only that aspect of external objects directed to the sense organs. For Spinoza, inadequate cognition is only a negative definition, a lack of true adequate knowledge. Such a lack is revealed in the movement toward true knowledge. "Indeed, just as light makes manifest both itself and darkness, so truth is the standard both of itself and falsity" (Spinoza 1982, Pt. 2, Prop. 43, Scholium, p. 92).[2] True, adequate cognition is cognition of *natura naturans*; it is of the highest concreteness, not fragmented by generic concepts, but united by a general causal link, by necessity. This necessity is analogous to mathematical necessity; it joins individual and transient elements of being to eternity, for the necessity that the sum of a triangle's angles equal two right angles is an eternal necessity. "It is in the nature of reason to perceive things in the light of eternity" (Spinoza 1982, Pt. 2, Prop. 44, Cor. 2, p. 93).[3]

There is no need to mention here the early equivalents of the most recent epistemological ideas. In order to clarify the link between the rationalism of Spinoza and science, we must emphasize something else. What is the significance of the 'form [or light] of eternity' for the cognition of dynamics, nonidentity, and the paradoxical quality of the world? Does it include adequate cognition, cognition of *natura naturans*, the ascertaining and explanation of the living diversity of modes, or does Spinoza's scheme reduce this diversity to substantive identity?

It is particularly this aspect of the question that concerns Hegel in his evaluation of Spinoza's philosophy, an evaluation which is relevant for our time. We shall examine several statements of Hegel in his *Science of Logic*, where the substance of Spinoza converges with the immovable being of Parmenides.

In Book One of *Science of Logic*, in which he introduced the concept of becoming, Hegel included a long note about the limitation and isolation of the abstract concepts 'being' and 'nothing'. Here he stated that being by itself, as a true, unmodified origin, cannot develop in any other direction. "It is just as impossible for anything to break forth from it as to break into it; with Parmenides as with Spinoza, there is no progress from being or absolute substance to the negative, to the finite" (Hegel 1969, Bk. 1, Remark 3, pp. 94—95).

We shall encounter this concept of 'being' as a pure, totally non-mediated beginning again. But here it is already clear that the divergence between Spinoza and Hegel includes their understandings of both *beginning* and *mediation*. For Hegel, 'being' is the beginning, because it is not mediated by anything. Therefore nothing can follow from nothing and thus the beginning "is at the same time the end" (Hegel 1969, Bk. 1, Remark 3, p. 94). For Spinoza, being can be the beginning since it is mediated; because it is *causa sui*, it is rich in mediations, determinations (including finite determinations), the highest concreteness. The mediation of being is not external, but internal, in itself. As we approach being, substance, we find ourselves at the beginning, we meet a *natura naturans* that is ready for the ultimate mediation, for creation, for differentiation, that is full of modes. In this sense, for Spinoza, *the end is at the same time the beginning.*

Later Hegel turns to the problem of the rise of diversity, difference, nonidentity and determination from the utmost abstractions, which are absolutely unique, absolutely indeterminate. Here he expresses a very important thought: these indeterminate abstractions make sense only in the presence of determinate diversity. This remark is made in regard to the abstraction of absolute, empty space, which is particularly interesting both for the history of science and the clarification of the natural scientific sources of the problem of being. Absolute, empty space like absolute, empty time, does not exist.

> . . . there is empirically present no such space and time which is not spatially and temporally limited, or whose continuity is not filled by manifoldly limited determinate being and change, so that these limits and changes belong, unseparated and inseparable, to the nature of spatiality and temporality (Hegel 1969, Bk. 1, Remark 3, p. 98).

Analogously, pure consciousness does not exist; it is filled with determinate sensations, representations, desires. Here Hegel seems to point directly to the *aporia* of homogeneous extended substance and to the homogeneous thinking substance of Descartes.

However, the very abstract concepts of indeterminate, pure being, of absolute space and time, not filled by thinking, already assume determinate being — a space filled with finite areas, a time filled with events.

> The demonstration of the empirical nullity of empty space, and so forth, is not, however, what we are concerned with. Consciousness by making abstraction can, of course, fill itself with such indeterminates also and the abstractions thus held fast are the *thoughts* of pure space, pure time, pure consciousness, or pure being. It is the

thought of pure space, etc. — that is, pure space, etc., *in its own self* — that is to be demonstrated as null: that it is as such already its own opposite, that its opposite has already penetrated into it, that it is already by itself the accomplished coming-forth-from-itself, a determinateness (Hegel 1969, Bk. 1, Remark 3, p. 98).

If, thus, pure being already assumes determinate being, then why cannot diversity, which is rich in determinate concreteness, serve as a beginning? In the third section of the second book of *Science of Logic* ('Actuality'), Hegel partially reveals the roots of his conception of beginning as pure, indeterminate being. He speaks of *causa sui* and other fundamental concepts of Spinoza, and continues:

These notions, profound and correct as they are, are *definitions*, which are *immediately* assumed at the outset of the science. Mathematics and other subordinate sciences must begin with something *presupposed* which constitutes its element and positive foundation. But the absolute cannot be a first, an immediate; on the contrary, the absolute is essentially *its result* (Hegel 1969, Bk. 2, § 3, Chapter 1, Remark, p. 537).

Of course, in thinking, there is the richness of attributes: filled space and time, nature and history, the result of the successive transition from abstractions to higher concreteness. Mathematics (in this context mathematics figures as a type of apparatus for the comprehension of the truth, of coming close to the absolute) and "other subordinate sciences" start from abstract premises. But the absolute — the highest concreteness — is not a premise, but a result of the transition to the concrete.

This occurs in this manner in cognition. In reality, the concrete is the premise for initial abstractions ('Introduction to the Critique of Political Economy' in Marx 1913, § 3, pp. 292—305: 'The Method of Political Economy'). As an initial concept, Spinoza introduces nature as a heterogeneous, multiform, concrete multitide, that is, as if he were not using initial pure abstractions. In this sense he diverges from mathematics and 'other subordinate sciences'. But Spinoza's substance is not unformed chaos, it is divided into modes, divided by abstract concepts.

Perhaps an analogy from the natural sciences will help to present these very subtle controversies more clearly. Let us return to space as natural science's most important initial concept. The space of Newton is homogeneous and empty. Classical science as a whole builds on this abstract foundation a complex, concrete building. It rejects this primary abstraction, fills it with bodies, with their interactions, with events, and arrives at highly concrete physical diversity. The space of Einstein is different. It presents itself at once as determinate, possessing a different curvature – depending on the sum total of masses, energies, and impulses distributed in space. In Einstein's theory the very first abstractions are no longer abstractions in the traditional sense, they *manifestly* depend on real concrete conditions. Einstein's initial definition is non-Newtonian: "absolute space, in its own nature, without relation to anything external, remains always similar and immovable" (Newton 1934, Def. VIII, Scholium, p. 6).[4] Einstein's initial definition is space, fused with time; the curvature of space-time depends on physical conditions. Here abstractions serve as the source of analysis, forming a concrete picture of space filled with events, and as a result of such filling, it is determinate, heterogeneous, consisting of finite modes.

A nonclassical historical perspective allows us to see the dynamism of Spinoza's substance, which Hegel did not see. Let us concentrate on this problem.

For classical thought, dynamism, motion, time, and change in time supplemented rest, three-dimensional, simultaneous, 'instantaneous' space. The universe of peripatetic science is, above all, the motionless framework of the center and the boundaries of the world. To make the framework movable and dynamic, we must add the words 'plus time' to this scheme. Classical science did just this. Newton's system, with its simultaneity, with the illusion of the real being of the world in one and the same given moment for the entire world, did not fully part with the motionless Universe of the Aristotelians. It limited the immovable Universe to a given moment and added to this momentary, simultaneous, but real Universe: 'plus time'.

When Einstein excluded instantaneous action at a distance and absolute simultaneity from the scientific picture of the world, the classical scheme 'instantaneous static harmony *plus time*' yielded its place to another scheme – a four-dimensional scheme which no longer demands 'plus time', which does not simply contain time as an independent reality in relation to space but deprives both time and space of reality, preserving reality for the space-time continuum.

The harmony of Einstein's world is the framework of four-dimensional world-lines. This framework is by no means an instantaneous scheme of the universe; it is not statics, which demands a supplementary 'plus time'. Time, change in time, and dynamics already exist in this scheme. The immutable structure of world-lines describes change, motion. This is a dynamic scheme.

What is the substance of Spinoza? As *causa sui* it gives rise to itself, it is *natura naturans*, i.e., it creates, renews itself, creates ever newer links of *natura naturata*, and again appears in these links as *natura naturans*. Hegel sees in this substance something motionless, analogous to a three-dimensional world supplemented by dynamism 'and time!' He includes initial, pure being in the chain of the universal process, identifying it with '*nothing*' and proceeding from it to becoming. But it is not necessary for being, which is heterogeneous *causa sui*, includes modes and is dynamic, to be in conflict with '*nothing*' in order to become dynamic, just as it is not necessary, given a four-dimensional scheme of the universe, to have the addition of time, changeability, evolution, and the addition of 'plus time'.

This is not only an analogy. Hegelian pure being is the philosophical generalization of abstract categories, freed from concrete content, which Hegel identifies with '*nothing*'; this was a radical break with doctrines about being. One of the sources of such an abstraction was empty space, deprived of bodies, which possessed a certain reality in classical science and lost this reality in the framework of nonclassical science. Spinoza's conception of being is different: it is being which constantly gives rise to and renews itself; it is a philosophic reflection of space filled with temporal events. In classical science there already existed such a four-dimensional space; everybody knew that real events reveal themselves in time and space. Einstein's theory asserts only (and herein lies the break with the classical bases of science) that the four-dimensional continuum cannot degenerate into a three-dimensional one; that space, taken outside time, cannot be the arena of real processes; that the velocity of the real physical motions of self-identical objects is neither instantaneous nor, generally speaking, greater than the speed of light.

Viewed from the historical perspective of relativity, the classical conception of motion represents a four-dimensional scheme, but one which is capable of regeneration, of transformation into a three-dimensional one. Such a scheme describes real physical processes filling space. Spinoza's conception is that of a filled world; it corresponds to the classical four-dimensional scheme, i.e., to that which clearly appears in classical science when viewed in the light of a nonclassical historical perspective.

As a whole, the evolution of science demonstrated the inseparability of matter and motion. Classical science made use of the concept of material bodies at rest and gave this concept an absolute sense. Relativistic physics deprived rest of an absolute sense. Now quantum physics and, even more clearly, quantum-relativistic conceptions (elementary particle theory) considers even mass at rest the result of motion and of interaction.

This evolution of science has a certain corresponding tendency in the development of classical rationalism — a distinctive philosophical introduction. When seventeenth-century rationalist philosophy repudiated nonspatial, nonmaterial forces which bring inert matter into motion, *when matter became all*, it began to figure in philosophical constructions as moving, heterogeneous in space and time, multiform and evolving. This was the substance of Spinoza. The world, comprehended by reason, turned out to be the world of multiform and evolving matter. Rationalism became the doctrine of matter.

4. ATTRIBUTES

Natura naturans, nature creating, which appears to Spinoza sometimes under the pseudonym of god, sometimes under its own rational name, is nature in all its unity and diversity. Here there do not yet exist differentiations, generic concepts, determinations which are always negation (*determinatio est negatio*). Cognition must introduce these differentiations, then once more rise up to nature as to the highest concreteness, and finally it appears as an infinite multitude of determinations. The first step towards differentiation of a unified nature, which has so far been deprived of determinations, is the differentiation of attributes. "By attribute I mean that which the intellect perceives of substance as constituting its essence" (Spinoza 1982, Part 1, Definition 4, p. 31).[5]

Thus reason has already encroached upon being. The process of regulating initial concreteness has begun. But it does not follow at all from this that attributes are forms of comprehension, peculiar not to nature but to reason. No, reason does not introduce attributes into nature but finds them in her; they are real, they are found outside of reason (*extra intellectum*) and express that which is eternal and always present in substance.

Spinoza examines two attributes of substance: thinking and extension. But he repeats over and over again that the number of attributes is infinite. This assertion gave birth to a great many puzzles which were already discussed by Spinoza's pupils during his lifetime and have cropped up constantly right down to the present. We shall not touch on the controversial aspects of this question and shall limit ourselves to an attempt to explain the infinite plurality of attributes.

This attempt is connected with the problem of the filling of space – the Achilles' heel of Cartesianism, the *aporia* that has pervaded physics from antiquity to the present. Spinoza says that reason knows substance in two of its attributes: extension and thinking. But here, there immediately arises the problem of the content of thinking and the filling of space. Descartes wants to ascribe content to thought, to force it to reflect another substance – extension; thinking comprehends the motion of discrete bodies. But the very discreteness of bodies, their individualization, remains a simple reflection of thought; they are bodies deprived of physical being if they are parts of a homogeneous substance. Leibniz subsequently ascribed dynamic properties to discrete bodies, but in so doing he essentially rejected the problem of the filling of space. For him this problem is neither substance in Descartes's sense nor attribute in Spinoza's sense, but a secondary concept – the result of nonextended forces. In the atomism of the eighteenth and later centuries, dynamic properties, especially weight, were ascribed to particles. Weight turned out to be proportional to mass and, practically speaking, became the quantitative measure of matter as distinguished from space, i.e., the measure of Democritus's 'being'.

The concept of filled extension was not and could not be a generalization of the dynamic tendencies we have discussed. It was only a question, a conviction that attributes of substance are not limited to thinking and extension, it was a question without any dogmatic metaphysical overtones. But for all its extremely undefined form, this question, this conviction was the question of the filling of space.

Let us imagine that substance does not possess any attributes other than thinking and extension. Then both of these attributes are wasted: thinking loses the object, and extension the subject. Thinking occurs, but what is this thinking about? We return to the Cartesian *cogito* – abstract thinking. An analogous question concerning extension remains unanswered: what possesses extension? This would be a return to the space of Descartes, filled with and indistinguishable from matter and therefore filled with bodies that cannot be physically distinguished from one another. But Spinoza does not return to Cartesian *aporia*. His substance possesses other attributes, and it is they which create the subject in the sentence '*something* possesses extension'. This is matter distinguished from space. 'Something', possessing extension, possesses innumerable attributes. Man cannot attain this infinite object other than in spatial representation, but this does not mean that the attributes of substance can be reduced to extension. The infinite number of attributes signify the infinite complexity of the object: no matter how much we reduce it to quantity, form, to grouping and regrouping, something will always remain which possesses size and form, the subject of grouping and regrouping. This is the very general conception of being as the unity of Democritus's 'nonbeing' (the spatial-geometrical 'predicates', the behavior of bodies, their coordinates, and various combinations and derivative spatial definitions) and Democritus's 'being' (the bodies themselves as they are distinguished from space, the 'subject' in physical statements).

The principal content of thought and the principal filling of space allow a more accurate evaluation of the role of the mathematical method which Spinoza used in both the exposition and, as can be demonstrated, the construction of his system. The chain of definitions, theorems, corollaries and scholia are not a mathematical apparatus, nor are they the external form of exposition which was fashionable, widespread and, at times, ridiculed in the next century. Having eliminated purpose, chance, and nondetermined will, Spinoza wants to base a whole system of cognition on the concept of necessity, namely that necessity which figures in mathematics, for example, the necessity for a circle to have equal radii. This necessity is not like a sequence in time — the radii do not become equal after the circle appears. In this instance necessity is not realized in time; it has an eternal character, and is no more dependent on time than the necessity of a theorem which flows from an axiom's contents. But we see already that such a necessity, reigning throughout nature as if it were purely logical, does not at all signify the unchangeability of nature in time and in space. Space and time do not disappear, they do not yield their place to extratemporal and extraspatial logico-mathematical necessity; it is as if they fused with it.

Spinoza's mathematical method is a cognitive method for the objective world, which, moreover, cannot be separated from the object of cognition. Brunschvicg says that analytic geometry serves as the foundation for Spinoza's theory of cognition.[6] But Spinoza's mathematical method not only unites thinking and space [but also] demonstrates their unity, thus bringing the whole theory of cognition closer to analytic geometry. Geometry itself becomes the ideal scheme of the cognition of being, an adequate cognition, which, as we know, rises not to abstract universals but to a rich, an infinitely rich concreteness, to *natura naturans*. The geometrical scheme becomes the scheme of the universe.

If we look for the prototypes of the relativistic picture of the world, where the *ratio* of the world is the framework of world-lines, a four-dimensional geometric scheme, our thought must turn not only to Descartes, and not as much to him as to Spinoza. For Descartes, geometry did become physics, but a physics indistinguishable from geometry. For Spinoza, geometry is transformed into a broader and more general theory, which is no longer only axioms, theorems and definitions relating to space. Space is only one attribute of substance, and Spinoza's geometry is extended to the world as a whole, to substance as a whole. The contemporary conception of abstract spaces correlates these spaces not only with extension, but also with practically any diversity. Spinoza's geometrical method reveals necessity — the highest truth of being — no longer for *any* diversity, but for all being as a whole; it replaces or embodies the world soul of Giordano Bruno. And if the geometrical apparatus of the theory of relativity is sometimes called physical geometry, then Spinoza's mathematical method is epistemological geometry. It is not only a method or one of the methods of reason, it is *the* universal method of reason which leads to the adequate cognition of being as a whole.

5. MODES

The third of Spinoza's fundamental concepts of philosophy — the concept of modes — seems to be closest of all to positive science, observation, and experiment. The mode is the state of substance, dependent not on internal essence, but on external effects. It is the finite forms of the infinite attributes of substance. The initial determination of mode is its finite character. The mode is something determinate and insofar as determination is negation, then the concept of mode includes the concept of boundary, limit, beyond which the given mode disappears. Thinking — an attribute of substance — is infinite, but it consists of individual modes replacing each other — ideas. Extension — the other attribute of substance — is unlimited, space is infinite. But space includes determinate and, consequently, limited finite bodies.

Substance consists of individualized, limited, determinate, finite objects; these are its states, its modes. They are deprived of freedom and determined from the outside; Spinoza calls them *causati*.[7]

But this is only Spinoza's first thesis. Further on there is a new turn of thought, one that is extremely important for the problem of being, namely the concept of infinite modes. An individual finite mode is limited by others. One body is limited by others; one idea in a train of thought is replaced by another. This process of limitation is in itself unlimited. Therefore modes, when added up, form unlimited, infinite states of substance. Spinoza calls them infinite modes. Thus, side by side with a single, undifferentiated attribute of substance, there also arise infinite but differentiated, modes with individualized elements. Nonidentity, difference, and modification enter into the doctrines about substance. Without this there could be no modes; modes are differentiating, determinate states of substance. But now substance appears to us as something heterogeneous. If modification of an attribute is inevitable, then it exists as endless modification. One motion is replaced by another and determined by another, but this is another motion, in its turn a determined one. "This mode is determined by another mode, another reason, the last – again by another, and this series is infinite." This series of modes form in their totality *natura naturata*, created nature.

The terms *natura naturans* and *natura naturata* have a long history. They radically change their meaning from the Neoplatonists and early commentators of Aristotle to the modern era. For Spinoza, the opposition and inseparability of the concepts of creating and created nature acquired such a deep and general sense that it is already easy to see the logical connection between these concepts and the clashes in science.

Natura naturans is the totality of a substance's attributes. *Natura naturata* is the totality of its modes. A continuous attribute, deprived of determinateness (for our problem we must keep extension in mind as an attribute), is transformed into an infinite diversity of determinate, finite modes. According to Spinoza, without such a shift, the concepts of substance and attribute become meaningless, for the fundamental idea of Spinoza's philosophy is the identity of creating and created nature. Spinoza expresses this idea with the formula: "god's power is his very essence" (Spinoza 1982, Pt. 1, Prop. 34, p. 56).[8] "God's power" is nature created, "God's essence" is nature creating. This identification means that apart from nature there is nothing real in the world, that created nature is at the same time creating; that nature is *causa sui*, the creation of nature is found in its very essence.

Such an interpretation of Spinoza's philosophy was a way out of rationalism into science. Long before Feuerbach science went from *deus sive natura* to *aut deus aut natura*. But this shift stemmed from the very essence of Spinoza's philosophy. In particular the concepts that demonstrate the inseparability of *natura naturans* from *natura naturata* demonstrate most clearly the uniqueness of the choice made by science between god and nature and the logical impossibility of any other choice. If modes and their totality — created nature — are opposed to creating nature, and if creating nature is opposed to created nature — to modes, to the finite which enters into empirical experience — then the problem of the relation of God to the world remains meaningful. But if these poles are fused together, then nothing is opposed to nature, and rationalism makes a complete break with all forms of creationism and turns to science. All connection between God and the world remains behind, in particular all modifications of Neoplatonic emanation. Lying ahead is the basic problem of the shift from indeterminate and undifferentiated attributes of substance to its determinate finite modes and the infinite plurality of finite modes.

The shift from seventeenth-century metaphysics and classical rationalism to classical science, from Descartes and Spinoza to Newton, was a shift to these infinite pluralities. Seventeenth-century metaphysics focussed most of its attention on creating nature, which appeared under the pseudonym of 'God'. It culminated in the philosophy of Spinoza, which made this pseudonym simply a pseudonym. After Spinoza and beginning with Leibniz, who was not only the last of the great seventeenth-century metaphysicians, but also one of the first naturalists of the eighteenth century (which, in this respect, began in the latter half of the seventeenth century), the center of gravity of theoretical processes moved to *natura naturata*, to created nature. But the problem of the origin of *natura naturata* no longer appeared in an explicit form in science, and in this sense classical science was isolated from classical philosophy. The shift had distinct historical roots, but here there was actually a logical prerequisite — the liberation of extended nature from the phantom of external nonextended forces. This liberation was the essence of Spinoza's philosophy, namely the sense that all categories unite *natura naturata* with *natura naturans*.

Classical science concentrated its attention on created nature, but in an inexplicit way its basic tendency continued to link created nature with creating nature. As we shall see later, the problems of infinite modes, and all that links the attributes of substance with its modes, the infinite with the infinite, lay on these main paths. In order to prepare for such a 'neo-Spinozist' analysis of classical (and then of nonclassical) science, we shall return to one of the problems about modes in Spinoza's doctrines, namely, the problem of their individualization.

In the individualization of finite modes, there is a shift from the attribute to infinite diversity. Space could serve as a geometric prototype of the infinite mode (that is, the infinite number of finite modes), which has a certain system of coordinates. Space, in which points do not have parameters and are indistinguishable, is an attribute, like extension. When a point is assigned some value by coordinates we get a finite mode; the infinite diversity of these modes becomes an infinite mode. But extension is an attribute of substance, and therefore Spinoza would not have been satisfied with a purely geometrical example. Actually Spinoza, in reply to a question of Tschirnhaus about the meaning of the concept of the finite mode, wrote that motion and rest serve as an example of a finite mode as it relates to extension,[9] in other words, what we call the behavior of a body at a given point in space. The contemporary equivalent of the infinite mode that corresponds to extension is parametrized space, which includes the physical field, whose points are distinguished physically, that is, by various experimentally observable processes.

This contemporary interpretation of Spinoza's mode seems, at first glance, to contradict the definition of a mode as the finite state of substance. At issue is the point and infinitely small values. It would be easy to clear up this problem by referring to the fact that finite values of coordinates, speeds and other magnitudes can be either a finite object or a finite state. But the problem is really more complicated than that. The differential calculus and differential ideas of motion (the subject of one of the following chapters) allows us to consider finite volumes of space as infinite pluralities of various infinitely small volumes, each of which is characterized by various finite values of geometrical and physical magnitudes. This means that each element of extended substance can potentially be represented as an infinite diversity of physically different modes.

The individualization of modes is the condition for the transition from the indeterminate *natura naturans* to the determinate *natura naturata*. But for Spinoza, this aspect of the matter — the nonidentity of modes, their individuality and determinateness — is less emphasized than the other side — the unity and indistinguishability of substance. In this sense Spinoza continues Bruno's line of thought: what is important for him is the dependence of modes, individual states, and individual acts on nature in its undifferentiated unity. But Spinoza takes a radical step forward by decisively eliminating universal reason as something standing over nature, something personal. The objective *ratio* of the world, which is found in nature itself, is all that exists. The *ratio* of the world is the processes which take place in it. Cognition of this *ratio* is cognition of substance, of extended substance. Adequate cognition finds this substance by examining modes as links in the necessary, causal harmony of the world which does not permit any deviations.

Spinoza was further from positive science than Descartes or Leibniz. While there were some Cartesians and Leibnizians to be found among eighteenth century scholars, there were, with few exceptions, no Spinozists. But, by contrast, Spinoza's philosophical position in the seventeenth century ran closest to the broadest and deepest currents of science. For Spinoza, in particular, rationalism became the doctrine about extended substance and only that. Reason no longer dictated its laws to nature, did not seek something in nature to correspond to its ideals and generally ceased to oppose itself to nature. Rationalism, in short, became science.

If science, beginning with the seventeenth century, found principles that could be generalized and modified (but not discarded) and still remain valid in their field as lawful approximations, then seventeenth-century rationalism arrived at concepts that were a generalization and anticipation of historically invariant principles of science. The basic concept of rationalism, as it existed in Spinoza's philosophy, was the concept of a creating nature inseparable from created nature, inseparable from modes, which were finite, determinate, changing, susceptible of quantitative analysis and experimental assertions. From here on, the study of modes became the adequate cognition of substance. Science sought in modes, in non-identity, in change, something unified. This means that cognition, acting experimentally on nature, provokes changes of modes; in juxtaposing their changes with mathematical images, it finds in nature *causa sui*, and no longer preserves anything beyond these methods — neither *a priori* constructions of reason, nor purely empirical assertions.

Thus Spinoza's doctrine about the modes of substance reflected that aspect of rationalism concerned with mathematics and experiment and their synthesis, that is with the flow of increasingly certain information about the world made possible by the development of science. Spinoza's conception is tied to the conviction of the certainty of this information. This certainty is guaranteed by necessity, determination, the singleness of all that results from creating nature. Here the word 'result' means something analogous to a mathematical deduction of the type 'the theorem results from the axiom', with the difference that the train of mathematical deduction is broken up into axioms, while *natura naturans* is not the last link of a causal chain. This last circumstance must be emphasized as it is very important. Creating nature [*natura naturans*] cannot be the last link of the causal chain because it is also created nature [*natura naturata*], in that it gives birth to itself in the process of infinite renewal as *causa sui*. It is particularly this sense of *natura naturans* — nontraditional, paradoxical and deeply antidogmatic in the highest degree — that must be borne in mind when we speak of the conception of modes. In the origin of modes as finite objects, what role is played by this unity of creating nature, its lack of determinations, limits, and finite objects? How is it that this infinite differentiated element turns out to be a plurality of determinate and finite modes, created by nature? In other words, how is it that undifferentiated substance turns out to be differentiated, consisting of finite objects?

Contemporary science grew out of a continuous effort to understand the unity, identity and indissolubility of (1) continuous substance without finite elements and determinate limits and of (2) *that very* substance as a plurality of discrete objects. Although contemporary science has moved very far toward its resolution, the identity of the continuous substance of a wave field with the discrete substance in its corpuscular representation remains its deepest *aporia*. Certain [i.e., authentic] representations of being now embody the idea that corpuscular determinations, classical modes and the significance of dynamic changing particles are already contained in the continuous substance of a wave field as probabilities, that this field is a field of probabilities and that the concept of probability is a substantive category. But unlike most of the laws of the state, the laws of science can be applied retroactively: they force one to re-evaluate the past in order to see in it what was previously not visible.

6. DOES SPINOZA'S GOD PLAY DICE?

From an historical perspective, Spinoza's work ceases, to a significant extent, to justify the traditional evaluation; it no longer seems dogmatic. It contains unanswered questions that are often more important for the development of science than definitive answers. One such question concerns the mechanism of the links between undifferentiated *natura naturans* and *natura naturata*, which consists of determinate modes.

Therefore it is entirely natural to call on Spinoza as an arbiter in the debate between Einstein and Bohr. In 1937, Bohr visited Princeton. As usual, his argument with Einstein began with the foundations of quantum mechanics. The debate took a somewhat unexpected form; the debaters discussed what Spinoza's position would have been regarding the question 'does God play dice?'[10]

Spinoza's intervention in the debate between Einstein and Bohr could have fundamentally altered the two opposing physical conceptions. We can imagine that the appeal to Spinoza forced both Einstein and Bohr to shift the debate to the level of the most general controversies, which, by their very generality, are opposed to the concepts introduced by Spinoza. It would doubtless be interesting to write a hypothetical dialogue between Einstein, Bohr, and Spinoza, but it is not given to us mortals to guess the words of immortals and we should not put our own conjectures into their mouths. However, several guesses — not hypothetical texts, but guesses concerning the logical links between the conceptions of Spinoza, Einstein and Bohr — flow with a certain probability from a comparison of these concepts.

First a few words about the relationship of the theory of relativity to Spinoza's identification of *natura naturans* and *natura naturata*. The theory of relativity is derived from the metrical properties of space as a given; it explains the dependence of these properties on the movement of a reference system and this whole range of problems within the boundaries of *natura naturata*. But this explanation relates to the theory of relativity in its finished form. We reach a different conclusion if we remember those paths of generalization and the course of the development of the theory of relativity that Einstein noted and which are the fundamental directions in the development of science today.

The theory of relativity was a great victory for the principle of causation in its 'modal' form. The starting point of the theory of relativity was to complete the construction of the edifice of classical science, by placing itself within the framework of Spinoza's *natura naturata*, that is, the sum of modes, the totality of impulses and motions linked to one another in a simple way. This is what was done. The special theory of relativity ruled out immediate [instantaneous] action at a distance.

The general theory of relativity put an end to another concept which in its essence contradicted both the universal causal dependence of modes in Spinoza's *natura naturata*, and the ideal of classical science. Empty space as a whole, that is, something not counted as a mode, no longer serves as the cause of inertial forces, which must be explained in terms of a theory constructed in the spirit of the concept of *natura naturata*.

And where are we with regard to the conception *natura naturans*, to substance, to *causa sui*? We shall now see that it was precisely this problem that pushed Einstein forward to the unification of macroscopic relativistic laws with ultramicroscopic, ultrarelativistic laws.

As we already know, Einstein thought that one defect in the special theory of relativity was that its laws were not deduced from the ultramicroscopic structure of reference bodies (in Schilpp 1949, p. 59). The point is that the metrical properties of reference bodies must be states of a substance and must not result from geometric postulates, postulates of pseudo-Euclidean, four-dimensional geometry (in the case of the special theory of relativity) and of non-Euclidean geometry (in the case of the general theory of relativity). Rather, these postulates themselves, which describe the behavior of reference bodies, must result from the attributes of substance, from that which distinguishes a physical object from space or from space-time.

Did Einstein see his way to a physical representation of *natura naturans*, of self-acting substance, whose states served as the metrical properties of scales and clocks?

Such a path goes through quantum mechanics. It includes in the picture of the world a representation of indeterminate processes and their transformation into determinate, modal meanings of dynamic variables entering into *natura naturata*. The first hint at the existence of metrically unobservable processes consists in the interaction of the body being measured with the measuring apparatus. The measuring apparatus — the body by means of which the measurement is carried out — reacts on the body being measured. In a letter to Maurice Solovine, Einstein speaks of such reaction and then offers a conclusion that is unexpected in its generality: "If we do not sin at all against reason, we will never get anywhere."[11]

Does quantum mechanics sin against reason? It is not blind to the effect of the measuring apparatus; on the contrary, quantum mechanical constructions begin with the establishment of such an effect. The 'sin' consists in allowing effects that are uncontrolled in principle, in allowing metrically unobservable processes. But Einstein speaks of something else, of a deeper 'sin' against reason — the sin of *ignoring* the effect of the measuring apparatus on the body being measured. Quantum mechanics is guilty of a somewhat analogous liberalism. It takes into account uncontrolled changes in the behavior of particles, but ignores such changes when dealing with the apparatus. The postulate of the interaction of classical bodies with guaranteed localized elements (for example, openings in a motionless diaphragm) or with a guaranteed reaction to an impulse (a diaphragm with a very small door) allows quantum objects to acquire space-time localization and metrically registrable impulse-energetic interactions, and thus to return to the world of *natura naturata*. But still a sin is a sin, and after the introduction of quantum mechanics, the classical picture of world-lines that depend on impulses which one body (one modal extension) gives to another body (another modal extension) lost its self-evident absolute meaning.

Quantum mechanics examines the effect of macroscopic bodies on the *behavior* of a particle, on its position and impulse. What if the effect derives not from separate bodies, but from a universal field, perhaps from the Cosmos as a whole and determines not the behavior but the existence of particles? Then the 'sin against reason' acquires a completely new meaning. This is the betrayal of rationalism. But which rationalism? The rationalism of Descartes? Yes! The rationalism of the simple causal link of modes? Yes! But if we speak of the rationalism of Spinoza, which is the basis of the doctrine concerning substance, of Spinoza's *ultrarationalism*, then what we have before us is not a betrayal, but an apotheosis. 'The sin against reason' is the sin against reason which perceives being as a system of impulses and lines of behavior which are determined by these impulses in a simple way. Accordingly, it is a sin against reason embodied in *natura naturata*, but a sin in the name of *natura naturans*, creating nature, *causa sui*, substance.

Now we can understand the meaning of that reply which, as we said before, we would not venture to put into Spinoza's mouth.

Apparently, Spinoza would have agreed with Einstein: the lawful regularities of nature are governed not by probabilities, but by the events themselves, which determine the behavior of each particle from point to point and from instant to instant; they determine world-lines, and a causal world disintegrates into world-lines.

But Spinoza would also have agreed with Bohr: the scheme of determinate world-lines does not exhaust the deepest attributes of nature, the revolt of the particle against a determinate world-line. This revolt against determinate impulse-energetic interaction perhaps testifies to the fact that the substance of the particle is not reduced to its determinate space-time and impulse-energetic behavior. The God of Spinoza is not *natura naturata*, but *natura naturans*, substance. Although the modes of substance have a rigorously determined character and consist not of determined probabilities but of determined events, it by no means follows that substance itself is like this. Perhaps Spinoza's God, unlike Einstein's, still plays dice.

It would be possible to compare Spinoza to the famous wise man who, deciding a lawsuit, said to the plaintiff, "you are right, my son", and to the defendant the same thing. When the wise man's wife protested against the settlement with its mutually exclusive versions, he answered "and you too are right". In some sense Spinoza would have approved of the opposing conceptions in the Einstein-Bohr rivalry, just as to some extent he would have agreed with their incompatibility. But the most important matter in the arbiter's decision must consist of the development and the generalization of the conceptions presented to him for judgment.

Can the conceptions of Einstein and Bohr be developed, generalized, modified? To do this it would seem necessary to turn to a new field of physical experiment, to a new series of physical phenomena. But from the contemporary point of view, such a shift was in fact a necessary part of the Princetonian appeal to Spinoza's ghost. For here it was not a question of a spirtual seance, but a physical debate, and Spinoza's name could be only a symbol for the search for new physical facts. In 1937 there still could exist a notion that the debate could be resolved with logical arguments and historical reminiscences. But even then, the question raised by Einstein and Bohr began with the words "if Spinoza had known quantum mechanics . . ." Today this question would begin "if Spinoza had known about virtual particles, the vacuum, relativistic quantum effects. . . ."

In the 1970's, if we are dealing with a physical, and not an historical-scientific problem, the appeal to Spinoza's ghost can mean only one thing. The physical equivalent of the idea of *natura naturata* is a system of world-lines and impulse-energetic interactions, which determine events — space-time localization and the exchange of impulses and energies. The uncertainty of both is the physical equivalent of the irreducibility of *natura naturans* to classical modes. Where does the correct solution to the problem lie? Where is the physical equivalent to Spinoza's god, Spinoza's substance — *causa sui*?

The first, classical approach to nature provides a system of determined events, where there is no game of dice. This is *natura naturata*. These are classical modes. Farther on is the quantum world. And farther still, or more accurately, deeper still, lies the world of ultrarelativistic effects about which we know very little. Perhaps by means of accelerators on the order of 400—1000 billion electron-volts that can penetrate into the region on the order of 10^{-13}—10^{-15} cm and 10^{-24}—10^{-25} sec, we shall find out more about it. But one particular aspect of ultrarelativistic energies and the ultrarelativistic world can now be guessed at with a certain probability, one already illustrated earlier in this book by a purely hypothetical scheme, namely the transmutation of particles on the order of 10^{-13} cm and 10^{-24} sec. Space in these ranges is not metrical, and here distance is neither a function of coordinates nor the magnitude that characterizes the motion of a self-identical particle. Rather, space is the condition for the existence of such a particle, the condition for its motion. Here space is not a mode, but an attribute of substance.

We can link the ultrarelativistic regenerations [transmutations] of a particle (the fundamental ones not derived from a metrical formula) to the self-action of substance, to *causa sui*, by means of a hypothetical, quasi-physical scheme that illustrates only the principal possibility for bringing contemporary concepts close to Spinoza's. Let us assume that regenerations in all directions at a distance of 10^{-13} cm in the course of 10^{-24} sec possess an equal probability and that they depend on the effect of the Metagalaxy as a whole on each of its elements, on each particle. This is essentially the self-action of the Metagalaxy, of *causa sui*. Such self-action has nothing to do with Mach's principle in that we are not dealing with the impulses on a particle but with initial spatial attributes.

There are a few other equally hypothetical assumptions regarding the mechanism of shifting from substantive extension, i.e., from ultrarelativistic regenerations, to modal extension, i.e., to motions across differing distances at different speeds, which depend on impulses.

Besides the self-action of the Metagalaxy, we find local fields in the world: gravitational fields of celestial bodies and systems, and other fields of force. These local fields are responsible for modal motions. They destroy the symmetry of the probabilities of regeneration and create instead a dissymmetry. Therefore, a particle in a local field, to a certain extent (depending on the reference system), goes beyond the initial symmetrical dispersion of the directions of elementary displacements. It acquires a macroscopic, non-zero order trajectory (over a given period of time, after a large number of random movements, the particle turns out to be more or less distant from the starting point) as well as a macroscopic non-zero order velocity. (However, it does not exceed the velocities of the elementary displacement $10^{-13}/10^{-24} = c$, i.e., the 'velocity' of light. The word 'velocity' here is in quotation marks because in the range indicated, there is no motion in a modal sense.)

Does Spinoza's God play dice? This question did in fact interest Einstein and Bohr. We could now give this question an experimental character. Since, with the shift to higher energies, we penetrate an even smaller space-time range, we can imagine that energies on the order of 400—1000 billion electron volts reveal a range where we collide directly with the attributes of substance, with *natura naturans*. But an experimental conversation with the God of Spinoza requires considerable resources, it may not occur in the near future and, most importantly, it will not occur without new concepts: high-energy physics will probably answer a series of questions by denying their physical sense. Therefore it may not be superfluous to check whether the question 'does God play dice' still makes sense when referring to the God of Spinoza, to *natura naturans*.

This question takes on a physical meaning if the roll of the dice — the result of experiment — can be observed, even in principle. Then it is possible to answer the question about whether God plays dice in either the affirmative or the negative. If the regularities [in the sense of conformity to law] of being determine the roll of the dice in a certain [i.e., reliable] way, then there is no game of chance, God does not play dice. But if the regularities of being determine only the probability of the result of the experiment, if God knows in advance only the chances of each fall of the dice, then God does play dice.

This whole dilemma seems inapplicable to nonmodal extension, to *natura naturans*, as illustrated by the scheme mentioned earlier. Is the probability known for spatially undetermined regenerations? Probability is at issue only if we introduce the coordinate axes, direction, symmetry and dissymmetry of directions into space. As long as all this is not introduced, we can speak only of regenerations of the particle without characterizing directions. But regeneration by itself is not only possible, not only certain [i.e., reliable], it also represents in itself *natura naturans*, substantive being, the existence of substance. In Spinoza's language, it is impossible to raise the question about whether God is gambling his own being in the dice game. The point, however, is whether *natura naturans* has a physical meaning without *natura naturata*.

As regards the modal motion of a particle in a determinate direction, then here, if we bear in mind the ultramicroscopic picture, God does play dice. But this is not Spinoza's God, not *natura naturans*. The fundamental laws of such motions, which can be observed, are statistical regularities, and this, of course, relates to each particle, individual, self-identical taken separately. Nonstatistical laws of modal motion in the ultramicroscopic aspect are excluded; the impulses, energy and space-time localization of the particle are determinate in the general case only according to probability.

Here we come up against an idea that is fully Spinozist, one that is the most alien to the classical notion. The point is that the Spinozist 'God who does not play dice' is inseparable from Bohr's 'God who plays dice': the concept of attribute in contemporary physics is inseparable from the concept of mode.

Extended substance possesses, as an initial attribute, regenerations that are undertermined as to direction. These regenerations become isolated at 10^{-13} cm and 10^{-24} sec as minimal discrete cells, inside of which there can be no modal, metrical space — the dynamic variable of a moving particle. But the concept of regeneration makes no sense without the concept of the macroscopic world-line. As we already noted, the transformation of one type of particle into another means a shift from one potential world-line to another. This contemporary physical modification of Spinoza's idea had neither antecedents nor continuation in classical science: the concept of substance includes its principal cognizability. Without any fear of being anachronistic, we can apply the modern contemporary term and speak of the *complementarity* of Spinozist *natura naturans* and *natura naturata*.

From the notion of such complementarity we can derive the invariance of the velocity of the propagation of light, of the interaction between bodies in general, and of the distribution of fields, in other words the propagation of the dissymmetry of probabilities of ultra-microscopic regenerations. These regenerations themselves are *natura naturans*, and the concept of probability cannot be directly applied to them. It is, however, applicable as a complement, and without it regeneration is without sense; if there is no notion of macroscopic motion, then the phrase 'one particle is transformed into another and then into a particle of the initial type' makes no sense; transmutation is the change of the potential macroscopic world-line, which is determined statistically. Thus the concepts of substantive certainty and probability are complementary concepts. Spinoza's God – *natura naturans* – does not play dice, but He does not have a physical sense without *natura naturata*, which is playing dice. The direction of the elementary shift at a distance of 10^{-13} cm is determined by a game of chance and is played out each time with dice. For this reason, the direction of the trajectory of the particle has a statistical distribution, and the macroscopic velocity of the particle has different values depending on this distribution, on the system of reference. But the very dissymmetry of probabilities of regeneration is not probable, it is certain; it takes place during each act of regeneration and is propagated without a statistical distribution, i.e., at one and the same velocity in every system of reference.

This deduction of relativity from complementarity could be stated in a more accurate and 'physical' form, but here it would perhaps be appropriate to mention this deduction in its most general, and therefore archaic, form in order to show the connection of contemporary problems with Spinoza's doctrine on substance and with the thesis of the complementarity of *natura naturans* and *natura naturata*.

The very fact that the discussion took the form that it did in 1937, the very fact that Spinoza was called on as an arbiter, is extremely indicative and characteristic of contemporary science. Probably this was the first time that such an appeal to Spinoza occurred. The ghosts of Descartes and Leibniz have been called upon more than once over the decades, even centuries. But this was the first time that Spinoza's name had rung out in this connection. Nonclassical science, as distinguished from classical science, directly and explicitly included in the circle of its interests the problem of the character and limits of that universal necessity which was presented in such an unprecedentedly exact form by the optical lens grinder of Amsterdam. Nonclassical science examines as a *physical* problem the limits of the power of Spinoza's God, that God who does 'not play dice'.

It must be stated that Spinoza's universal necessity of all that takes place in the world does not coincide, even as to the cardinal question, with that necessity and simple determination of processes which Laplace had in mind when he said that by knowing the positions and velocities of all the particles of the Universe, its future could be predicted with absolute accuracy. The difference is that for Laplace, the mechanism which subordinates particles to such a prediction consists in the interaction of particles. The particle moves in a simple manner under the influence of external effects. Laplace's determinism excludes freedom and whatever meaning this concept could have: the behavior of the particle is the result of *external* effects. Spinoza's determinism shifts the center of gravity of the problem onto *causa sui*, this going beyond the framework of Laplacean determinism. Spinoza's determinism frees the Universe from external constraints, it makes it free.

One problem posed by Spinoza's determinism that is essential to and perhaps most interesting for contemporary science concerns the shift from the free causation of *natura naturans* to external necessity in the world of modes, within the boundaries of *natura naturata*. As we have already stated, Spinoza identifies *natura naturans* with what he calls power, that is, with the totality of created objects. Here nature is not constrained; it is not forced to act in any way. Necessity merges with freedom, which Spinoza identifies with the expression of essence, with necessity flowing from essence. From the concept of *causa sui* it follows that nature is free, as the cause of itself, and necessary, as the consequence of itself. Out of such free necessity flows the *eternity* of substance, as distinguished from the transient character of all that is bound by external constraint, by external necessity.

Does the freedom and eternity of nature as a whole enter into the being of her finite objects, into *natura naturata*, into modes? Spinoza's fundamental ideas provide an answer in the affirmative. Unified self-identical substance with its infinite attributes, proved to be divided into finite modes that are linked by external necessity. But each of these modes possesses freedom and eternity insofar as it is a link of unified nature, insofar as every external effect, in the final analysis, expresses the free and eternal activity of creating nature.

Spinoza does not illustrate this conception with physical examples and constructions of natural philosophy: the mode, examined as a part of nature, a transitory and unfree object that dissolves again into the whole, acquires freedom and eternity. The flexibility of concepts, their ability to be transformed into their opposite as soon as we touch them, allows us to see in the mode its substantiality, eternity, freedom. It stops being a mode, it fuses with the whole. More exactly, it retains its individual being, its individuality, modality, finiteness, but at the same time, its link with the whole comes to the fore. Such a conception was described and developed by Spinoza in his doctrines on man, on man's fate and ethical ideals. It would seem that a comparison of contemporary physical ideas with Spinoza's philosophy should encompass the thinker's philosophical-ethical conceptions as well. When Einstein debated with Bohr in Princeton, they apparently did not limit Spinoza's role to his natural philosophy, but had in mind all parts of the *Ethics.*

Let us recall the comparisons of the ideas of Epicurus, Spinoza, and Einstein that were made in Chapter 3. Epicurus's basic emphasis was on the individualization of particles (whose behavior is not fully subordinate to the macroscopic laws of nature; it includes *clinamen*) and of man (who avoids the fatalistic 'destiny of the natural philosophers'). Spinoza's emphasis is different. The individual freedom of man and of any body in nature (they are united by the concept of mode) is empty without cosmic unity, simple necessity and the subordination of modes to this necessity.

For man, such subordination may become free, and necessity may become a free necessity if man's ideas are addressed to the whole, to nature as a whole. The individual being of man is a mode. Man is mortal, and his impressions and memories die with him. But his ideas and feelings, when addressed to substance, bring man to the free necessity of the whole and to its eternity. In just this connection Spinoza wrote the formula mentioned above: "a free man thinks of death least of all things, and his wisdom is a meditation of life, not of death" (cf. p. 92 above). This formula, to a certain extent, anticipates Schiller's '*Unsterblichkeit*': "Are you afraid of death? Do you want to live for ever and ever? Live in the Whole! It abides ..." (*ibid.*) as well as Einstein's statements on death and the fear of death.

For Spinoza, the fear of death and death itself is opposed by intellectual love, *amor intellectualis.* It creates for man the sensation of eternal happiness. Spinoza wrote of this happiness in the introduction to his unfinished treatise, *On the Improvement of the Understanding*

[*Tractatus de intellectus emendatione*], which Schopenhauer thought provided the best means for reconciling oneself with life. Einstein spoke of Spinoza's *amor intellectualis* as a necessary condition for scientific research. Indeed, the more fundamental the problems of science, the more they are emotionally colored by *amor intellectualis*. Nonclassical science, which systematically links more particular problems with more general and fundamental ones (and in this consists the most important cause of its boundless spiritual and economic effect) is permeated by *amor intellectualis*.

But it is not only on an emotional level that nonclassical science returns to Spinoza's doctrine of the link between mode and substance, the finite and the infinite. Nonclassical science, with its avant-garde generalizations, originates not only from the emptiness of the macroscopic picture, which has been filled neither microscopically nor ultramicroscopically; nonclassical science not only fills the world-lines with the contemporary equivalent of the Epicurean *clinamen*: it also originates from the physical emptiness of 'pure' atomism, ascribing to atoms an independence from their universal interaction; it attributes to them not only behavior but also existence itself, which is dependent on the effect of the macrocosm.

Now we ought to return to the problem of death and the fear of death in Spinoza's philosophy and Einstein's *Weltanschauung* as they are connected to *amor intellectualis*. This is love for the infinite which includes the finite being of man in infinite being. If we use the formula $\alpha \in A$, where α is finite, and A is infinite being, then the symbol $\in$ (the operation of inclusion) here means man becoming associated with immortal substance, and his freedom: for him, causal necessity becomes internal necessity. But this is a 'telescopic' conception. Another side of the question, namely the uniqueness of individual being, the known autonomy of α which is so well expressed in the Epicurean conception of *clinamen*, is not as clearly expressed in Spinoza's work as the first part of the question. For Spinoza, there is no *amor individualis*, no love for the α 'in revolt', for the individual in his unique distinction from the whole world, which is, of course, a relative distinction. For Spinoza the pathos of *amor intellectualis* is connected with the synthesis of the individual and the general, and not with their complementarity, that is, with the indissolubility of two mutually exclusive fields. We could say that for Spinoza, there is no *amor complementaris*, no love for man which is not only fused with, but also opposed to, the whole.

It is probably for this reason that we do not find in Spinoza's works that 'evening' sadness for the transient, the individual, the local, and that sorrow over the frailty of individual being that was mentioned in connection with Epicurus's formula.

Let us summarize several of the results that relate to the position of Spinoza's ideas in the evolution of rationalism, which brought rationalism to science. First of all, rationalism acquired a clearer ontological coloring. Rationalism already had this coloring for Descartes — the identification of space and matter had to determine substance and its properties so that they corresponded to a purely rationalist criterion of clarity. In this way, *ratio* became an objective concept, the initial criterion for construction of a physical picture of the world. But for Spinoza, reason fuses with extended substance; they become attributes of one substance: At this point there is no longer a substantive difference between rational thought and the rational extension of the world.

Further, for Spinoza, rationalism radically loses *a priori* elements. The starting point of deduction is not *a priori* — logical, empirical, or at all abstract. It itself submits to causal analysis. Rational thinking does not step here; it seeks the cause of the existence of nature as an initial concept and finds it in itself. In essence, for Spinoza there is no 'starting point', no 'beginning', as there were both before and after him.

Spinoza's rationalism is not limited to the behavior of bodies, the modes of substance, but takes being itself as an object of causal analysis. In this sense, Spinoza's rationalism, having gone beyond the Cartesian framework, becomes ultrarationalism.

The pathos of Spinoza's philosophy lies in the unity of substance, in that undifferentiated substance which is nature creating. But Spinoza, more than any of his contemporaries, understood the indivisibility of creating and created nature, the identity and unity of substance on the one hand, and of nature, which is differentiated, made of finite modes, on the other.

NOTES

1 "Sed hoc idem est, ac si dicerent, quòd Deus potest efficere, ut ex naturâ trianguli non sequatur, ejus tres angulos aequales esse duobus rectis . . ." (Spinoza 1925, **2**, Pars I, Prop. 17, Scolium, p. 61).

2 "Sanè sicut lux seipsam, & tenebras manifestat, sic veritas norma sui, & falsi est" (Spinoza 1925, **2**, Pars II, Prop. 43, Scholium, p. 124).

[3] "De naturâ Rationis est res sub quâdam aeternitatis specie percipere" (Spinoza 1925, **2**, Pars II, Prop. 44, Cor. 2, p. 126).
[4] "Spatium absolutum, natura sua sine relatione ad externum quodvis, semper manet similare & immobile . . ." (Newton 1972, **1**, Def. 8, Scholium, p. 46).
[5] "Per attributum intelligo id, quod intellectus de substantiâ percipit, tanquam ejusdem essentiam constituens" (Spinoza 1925, **2**, Pars I, Def. 4, p. 45).
[6] ". . . la découverte de la géométrie analytique . . . a servi de fondement à la théorie spinoziste de la connaissance . . ." (Brunschvicg 1949, p. 173).
[7] See *Ethics*, Part I, Axiom 7; Prop. 25, Cor.; Part II, Prop. 31, Cor.
[8] "Dei potentia est ipsa ipsius essentia" (Spinoza 1925, **2**, Pars I, Prop. 34, p. 76).
[9] [Kuznetsov refers here to Spinoza's letter of July 29, 1675 to G. H. Schuller. See Spinoza 1982, Letter 64, pp. 306—308; Latin original in Spinoza 1802, Letter 66, pp. 673—675. Schuller had conveyed Tschirnhaus's questions to Spinoza — Eds.]
[10] The following pages (to the end of the Chapter) include excerpts from 'Spinoza et Einstein' (Kuznetsov 1967) which were also included in the second edition of Kuznetsov's *Etiudy ob Einshtein* (1970b).
[11] "Wenn man gar nicht gegen die Vernunft sündigt, kommt man überhaupt zu nichts . . ." "Tant que l'on ne péchera pas contre la raison, on n'aboutira à rien" (Einstein 1956, p. 128 and 129).

7. THE GENESIS OF CLASSICAL SCIENCE AND THE PROBLEM OF NONIDENTITY

1. NONIDENTITY AND MATHEMATICS

Having come to know the highest point of seventeenth-century classical rationalism — the philosophy of Spinoza — we can now characterize the influence of the classical science of this century on the nature of rationalist thought.

Classical science is distinguished by differential representation, which we mentioned in analyzing its prehistory in the Renaissance. We also noted the experimental character of its reflections on nature. For the history of rationalism, it is experimental *thought* that is essential. Experimental thought is linked with the differential representation of motion and is, in essence, based on a new logic.

The distinction between the differential and the Aristotelian representations of motion was expressed by Kepler in an extremely clear form:

> And so, just as Aristotle said that there is direct opposition between one thing and another, without anything in between them, so I, considering geometry philosophically, find that there is indeed direct opposition, but with something in the middle. Thus, what was for Aristotle one term, '*other*', we may divide into two terms, 'more' and 'less' (Kepler 1941, p. 15).[1]

Kepler's sentence goes to the very heart of the new representation of the world and the new type of thinking about the world, i.e., the new rationalism. The Aristotelian theory of motion answered the basic question concerning the cause of motion by posing in its place another question about two points of motion — its beginning and its end. The question was, what is the natural place of a body? Is it at the initial, starting point of motion? The answer was 'no'. Is it at the end of motion? The answer was 'yes'. This antithesis explained motion without a series of local events on the path of motion, without describing motion from point to point and from moment to moment. Logic, with two values for each statement — 'true' and 'false' (bivalent logic) — corresponded to such a conception of motion.

The conception of Kepler and of the whole new science with the intermediate metrics 'more' and 'less' is a conception of continuous motion. It corresponds to infinite-bivalent logic, possessing an infinite number of positive values, 'true', for questions concerning the existence of a moving body at each point on its path, and an infinite number of [negative values], 'false', for questions about the existence of a body at other points in space. Such a conception operates *with an infinite diversity of predicates* — in this case, with a diversity of positions for a moving body, each position differing from the adjacent ones in an infinitely small way.

Such a conception changed rationalism in many respects that are related to its treatment of identity and nonidentity. Rationalism acquired an explicit empirical accompaniment and became dynamic. Classical science demonstrated the necessity for both the empirical cognition of the world as an integral part of rational cognition as well as the analysis of difference, nonidentity, and change without which theoretical, rational analysis became impossible. Classical science replaced the static scheme of universal harmony (the Aristotelian Universe with an immovable center, boundaries, and natural positions of bodies) with a dynamic harmony. Identity remains the basis for the rational comprehension of the world, but now it is not the coordinates of bodies but their velocities or accelerations — derivatives of coordinates in time — which seem identical. In this way, the concepts of nonidentity, change and diversity enter into the concept of identity. Identity ceases to be trivial; it now becomes a question of the self-identity of a body, which possesses, at different moments, different spatial predicates, different positions. Science radically changes Parmenides's conception: substance is now understood to be the subject of motion, retaining several predicates of motion. These are predicates of motion, differential predicates — velocity, acceleration and, further, the integrals of motion.

We shall return shortly to these new forms of the old clash between identity and difference. Before that we shall discuss the form of identity that is linked with mathematics and mathematical science.

Nowhere does the identifying function of reason acquire such a distinct form as in mathematics. Nowhere is its inadequacy as clearly revealed as in mathematical science, in that process of the ontological development of mathematics, which is one of the most fundamental attributes of the genesis of classical science. In the seventeenth century, mathematics was a spirit longing for physical, empirical, and, above all, experimental embodiment. The search for such embodiment clarified the nontautological nature of mathematics — its ability to find new, and moreover, paradoxical, ontological content, new information about being. Generally speaking, mathematics, as Russell very astutely remarked, "may be defined as the subject in which we never know what we are talking about, nor whether what we are saying is true" (Russell 1901, p. 84). Such independence from ontological meaning is the basis for the colossal successes of mathematics and, in the final analysis, of its ontological embodiment, when it finally recognizes what it has been talking about and whether it is true.

The ontological embodiment of mathematics is a demonstration of the basic feature of classical rationalism — the unification of the identifying function of reason with its individualizing function. The inseparability of these functions was systematically studied by Émile Meyerson who based his work on a great deal of historico-scientific material. Let us note that this inseparability can be systematically studied particularly in the history of science: the momentary 'three-dimensional' photography of science hides the constant individualizing activity of reason.

Meyerson's basic thought is very well expressed by de Broglie in his introduction to the posthumous edition of Meyerson's *Essais* (Meyerson 1936, pp. vi–ix). De Broglie says that a rational understanding of the world becomes possible because we find objects which are so close to each other that it becomes possible to introduce general concepts, or, in Meyerson's words, "fibers of reality". They permit science to exist despite the unattainability of its ideal. De Broglie cites the words of Paul Valéry which, in his opinion, are permeated by the works of Meyerson: "The human spirit is mad because it seeks, it is great because it finds (Meyerson 1936, p. ix)."[2] "But" continues de Broglie, "insofar as the Universe cannot, in the final analysis, be reduced to a vast tautology, we shall inevitably come up against 'irrational' elements in our scientific description of nature which resist our attempts at identification."[3]

Our previous discussion of the evolution of classical rationalism and Spinoza's philosophy differs from Meyerson's position: It is not irrationalism that is opposed to the identifying side of rationalism, but the individualizing function of reason itself, as its calculations transform rationalism into ultrarationalism, into a rational explanation not only of behavior but also of the very being of physical objects. The phrase de

Broglie quoted from Valéry can be understood as a statement of the biunique function of reason. The 'grandeur' of the spirit is measured by what is found — general concepts, general laws, a general identifying scheme of being. The 'madness' of the spirit leads to a breakdown of this scheme, to the ascertainment of phenomena that contradict it, destroy identity and, being individual, do not form part of the general scheme; it leads to the search for a new scheme. Without these breaches, without searching and nonidentity, reason loses its ontological value. 'Grandeur' is impossible without 'madness'.

Let us note a characteristic difference. Conceptions stemming from classical science tend to consider the ascertainment of identity as the basic function of reason, and the ascertainment of non-identity as a deviation, albeit an unavoidable one, from the rationalist path. Epistemological conceptions stemming from nonclassical science tend, on the contrary, to consider the individualizing function of reason as the basic rationalist virtue, and identification as an inevitable breach of this virtue. Einstein, we remember, wrote to Solovine that the process of measurement changes the magnitude of what is measured, but unless this effect is ignored, it is impossible to find out anything at all. "If we do not sin at all against reason, we will never get anywhere" (see above, p. 210, n. 11). 'Reason' here is the ascertainment of the nonidentity of a measured body to itself. 'The sin against reason' is the classic presumption of its identity. This inversion of virtue and sinfulness in relation to reason shows only that nonclassical science made explicit what in classical science had remained relatively hidden.

Let us return to Meyerson's ideas. Mathematics begins with a statement of a certain logical ambiguity of the equal sign. In the first volume of *Du cheminement de la pensée*, Meyerson expounds on Herbart's remarks concerning the contradictory quality of logical deduction (Meyerson 1931, **I**, pp. 47—48). How is it possible to acquire a new principle in the process of thought and the development of science, if such a principle is completely defined by the principle from which it is derived? Further on in the second volume of *Du cheminement* Meyerson arrives at the mathematical form of this fundamental *aporia*. Statements which seem elementary hide a real *aporia* that we constantly confront. When Frege says that two bodies equal in volume but not identical in form represent different bodies ("If I should consider my neighbor's house, equal to mine in volume, as my own house, the difference will be quickly explained to me"),[4] when Meyerson derives from this the partial character of every quality, there is always an implicit question: what nonidentical predicates are eliminated in a judgment in which there is an equal sign?

It is particularly these eliminated predicates, this baggage of travelling reason that is left at the side of the road (perhaps, more accurately, in the left luggage office) that turns out to be the source of reason's adventures *en route*. This baggage contains the reserves of nonidentity, which, in the final analysis, guarantee the nontautological quality of each logical deduction. Expressions united by the equal sign, judgments united by whatever forms of implication, words of the type 'hence it follows', etc., are distinguished from each other by what they do not contain. The links of deduction, like the links of mathematical analysis, remind us of the anecdotal question: "Without which jam would you like your tea?" This strange situation is constantly occurring in the history of science. The baggage left behind is preserved, and from it unexpected clashes appear. Nonidentical predicates, which have been driven out in the course of identification, suddenly interrupt the chain of logical deduction, and force us to turn back and review the initial premises, the more general assumptions. This is the 'flight from the odd', the 'flight from paradox' that we already know; this is the shift to a more general paradox.

One characteristic of classical science is the *gradual* accumulation of nonidentity. The ignored predicates accumulate, and only sporadically demand a review of the initial law. In periods of growth, during similar radical reviews, an idea that is identical to itself acquires additional assumptions in order to correspond to new facts, and theory, while striving for 'external confirmation', loses more and more 'inner perfection'. These moments, when theory again acquires 'inner perfection', are turning points and represent the prototypes of nonclassical science. These prototypes develop so quickly that a return to the initial, more general link in the deduction and the revision of such a link (or, at least, the sensation of the inevitability of such a return) accompanies each new identification, each new link of logical or mathematical analysis.

The result of great logical efforts and a far-reaching compromise is hidden under each 'consequently', and even more so, with each 'obviously' that is written in a mathematical treatise. The crown prince who studied geometry and had not seen why the sum of the angles of a triangle is equal to two right angles was not so naive. Let us look at this historical (or quasi-historical) anecdote. The crown prince's teacher was tired of the prince's doubts concerning the obviousness of geometry and assured him on his honor that the sum of the angles of a triangle is equal to two right angles. The prince understood geometry badly, but knew that it was impossible to doubt the word of a nobleman. "Why did you not say so at once?" he asked his tutor.[5]

If we remember that there exists a just-as-obvious geometry in which the sum of the angles of a triangle is not equal to two right angles, it becomes clear that the simple soul of the crown prince was not so far from the profound and general conflicts of mathematics. The obviousness of geometric conclusions is founded on the colossal number of nonidentical predicates of geometric objects which have disappeared, drowned in the stream of identifying thought. The very generality of obvious assertions, their axiomatic character, points out the breadth of that reserve of nonidentity that was accumulated during the birth of the Euclidean foundation of science.

When summarizing the evolution of mathematical science in his book, *De l'explication dans les sciences*, Meyerson speaks of the geometrization of the physical picture of the world, of the dissolution of physical processes in the rational, geometrized representation of the moving parts of space. Can we believe that science consists of *this*?

> Yes, [answers Meyerson] science is indeed *this*: but this is not all there is to it. In conforming with the human spirit's explanatory tendency, science certainly appears to want to reduce everything to the rational. But, on the other hand, in being infinitely careful to preserve contact with nature, it acknowledges the limits nature imposes on its efforts. This gives rise to a concept whose nature and significance has perhaps not always been clearly recognized, but has nevertheless exercised a profound influence on the evolution of science. This is the concept of the *irrational*, that is, that which, among the elements that science is led to make use of, seems as though it must, by its very essence, resist any final reduction to purely rational elements.[6]

2. A THOUGHT EXPERIMENT

What are the objective roots for the treatment of the rational as identical and the nonidentical as irrational? The source of this conception is by no means classical science as traditionally defined, that is, the science of the seventeenth–nineteenth centuries. It is, rather, the impressions of the first stage of nonclassical science. Later on, we shall encounter several times a certain difference in the style of scientific thought and its relation to classical science in the first half and in the second half of our century. For the moment, let us limit ourselves to a few brief remarks.

The theory of relativity and quantum mechanics, in the form they assumed in the twenties, showed that identifying rationalist thought comes up against experimental paradoxes and must change its initial positions. But this change consisted in the shift to another identification, another identity, another homogeneity, geometrization, another metric. In the theory of relativity, four-dimensional space-time is homogeneous. Its metrics are, for the most part, non-Euclidean. Geometrization leads to a four-dimensional scheme of world-lines, i.e., to a certain *geometrical* scheme that explains the course of events in the world. The intervention of the nonidentical is present. The identification of time, the concept of simultaneity, can take place under *experimentally* registered conditions. But this intervention changes the rational scheme of being and for the time being does not claim to include nonidentity among the initial definitions of rationalism itself. In quantum mechanics the intervention of the nonidentical is even more energetic. The indeterminacy of dynamic variables signifies an experimental correcting of rational categories. The demand for experimental comprehensibility, for experimental registrability of the constructions of reason, is even clearer here than in the theory of relativity. But in both cases, this intervention *limits* the identifying function of reason (and herein lies a basic transformation of scientific thought), but does not include the nonidentical and empirically comprehensible in the definition of the rational. Thinkers, proceeding from results in nonclassical physics that occurred in the first half of this century, saw in this physics only the limits of the identifying function of reason. Thinkers whose consciousness was occupied not so much by results, as by new and unsolved problems, saw further. Einstein, as we already know, no longer saw reason in classical identification, but the sin against reason.

In the second half of our century, the search for a synthesis of the theory of relativity and quantum mechanics, and the development of the theory of elementary particles, made a complex problem out of the identity of physical objects. When science uses the concept of a particle identical to itself and of its continuous world-line, it knows that it is sinning against reason and that here 'reason' hides nonidentity, the revolt against identity. Now science pushes epistemology toward the inclusion of nonidentity in the very definition of the rational, comprehended by reason. No matter how the shift from the concept of transmutation to the concept of self-identity occurs, no matter how the transmuting particle is transformed into one identical to itself, the deed is already done — the concept of identity has lost its absolute character.

Contemporary physics gives a very clear scheme of 'irrational' processes which resist the transformation into 'rational'. There are quotation marks here because the processes that resist rational comprehension also resist empirical comprehension. Non-relativistic quantum mechanics uses the concepts of the position and impulse of a microparticle. By themselves, the simultaneous values of both of these variables are inaccessible to experiment — not only to real experiment, but also to ideal, thought experiment. It is not a question of technical obstacles preventing the registration of the position and impulse of a particle in one experiment, but rather a question of the nonrepresentability of these concepts without macroscopic apparatus. The exact position and the exact impulse of a particle cannot be registered in one real experiment, nor can they be thought of as a single definition because rational thought, regardless of what it has become in contemporary science, considers impulse and position the result of an interaction. Such an interaction makes sense if there is a body of the interaction that possesses a constant calibration which does not change in the process of measurement — in other words, if there is a classical object, a classical apparatus. The words 'to make sense' are a rationalist criterion; they mean that a concept can enter into the chain of conclusions of reason. At the same time the words 'to make sense' mean that a concept can, in principle, be embodied in some kind of experiment.

Now we shall leave these clear epistemological principles of nonclassical science and turn to their classical prototypes. We shall be persuaded that in classical science, also, the concept of thought experiment destroys the boundaries between the rational and the irrational, between the rational and the empirical. Thought experiment is empirical comprehension become rational, and rational [comprehension] become empirical without ceasing to be itself.

There is yet another side. The rational comprehension of concrete, empirically registered objects always goes beyond the boundaries of the actual being of an object into the realm of its potential being. We include the given, concrete, single object in some kind of set and identify it with other elements of the set. Now we can consider the concrete object that acts on the sense organs as potential: if it appears, it will possess certain properties. "Thought is, in short, the activity which causes what does not exist to come alive in us," says Valéry (Valéry 1958, p. 73).[7] But when thought operates with thought experiments, it considers each concrete object as potentially changing. Here, there is no longer the simple explanation of belonging to a class: the given object belongs to a class of objects possessing a given predicate, therefore the object possesses the predicate. The possession of a predicate appears now in an experimental and dynamic form, it signifies that "something is changing in nature." Magnitudes are predicates which can be registered in an experiment: the position, velocity, acceleration of a body.

These concepts appeared in mechanics, but very quickly transformed the notion of identity and nonidentity as epistemological categories. They are closely linked with what is called 'experimental rationalism', with the theoretical thinking at work in thought experiments, and also with the results and schemes of real experiments. If each link of logical analysis represents some change, then rationalist thought unifies (by means of logical or mathematical deductions) the *states* of a given nontrivial object that is self-identical. Experiment, which in the seventeenth century brought about a new style of rationalist thought, is the change of states of self-identical objects (the best known experiment in the seventeenth century was the change of the states of motion), change that is called forth by some cause, and therefore enters into the logical explanation of facts in the rationalist (in the strict sense) fabric of scientific explanation.

The meeting of the empirical comprehension of the world with the logical and, above all, with the logico-mathematical comprehension (which is the beginning of classical science) occurred as a result of the evolution of logic and mathematics on the one hand and of experiment on the other. As we have already noted, the logic of Galileo and of all seventeenth-century and later science is the logic of continuous predicates of diversity, which draws rationalist thought close to the differential calculus. Here we must emphasize the connection of the new logic of infinite diversity with experimental thought, with thought experiment, as the basic element of theoretical deduction.

We explain any concrete phenomenon by including it in a class of analogous, and thus identified, phenomena. But for such inclusion, we must distinguish between classes whose elements are undifferentiated and classes whose elements are ordered by some kind of differentiation. Let us look at two types of explanations for the falling of a given concrete body. One explanation — "This stone falls because it is heavy, and all heavy bodies fall" — consists of the simple inclusion of a stone in the category of undifferentiated bodies and ignores distinctions: *any* heavy body falls, and a concrete body is not differentiated in this sense from any other body. A second explanation — "A stone falls at the present moment with observable velocity because it fell at the preceding moment with another velocity and possesses a constant acceleration" — assumes: (1) that the moments and points characterizing the position of the stone are essentially distinguishable; (2) that individual phenomena, entering into an ordered set, are not identical, but the second derivative of the coordinates, [constant] acceleration, is; (3) that we can determine experimentally the velocity of a falling body at every moment.

This last condition is no less essential than the others. The individualizing function of observation invades the holy of holies of logical deduction. In the set of identified objects, each individual object is now numbered. We see this distinction in the result of a particular logico-mathematical operation: we examine the future, possible behavior of a given object in the course of a certain period, and then compress this period into a moment and obtain as a limiting meaning the momentary characteristic of behavior. Of course, this is a logical, or, more accurately, mathematical operation; it is a question of obtaining a derivative in time, which becomes a distinction of one point and moment from other points and moments. But this operation is the logical equivalent (logical in the new sense of an infinite-bivalent logic of continuous diversity) of an experimental operation, of a thought experiment. The object of this operation is a self-identical body, nontrivially identical, with changing spatial predicates, and moving. Of course, differential [analysis] and the calculus can also be applied outside of mechanics, but in any case, the link with thought experiment is preserved. We impart acceleration to a body, at each moment we obtain a measurable velocity, and, integrating the derivatives, we find the measured positions of bodies. All these influences and measurements on a body represent, in their totality, not passive observation, but the observation of the results of active interference in the course of processes in nature. Thought remains thought, logical deduction remains logical deduction, but thought experiments become the content of thought, the elements which deduction links together: *thought* — hence science does not leave the soil of rationalism; *experiments* — hence rationalism itself reveals its link with the empirical, a link which had never disappeared but was sometimes invisible. Seventeenth-century science becomes experimental not only because it uses real experiment, but also because its rational logical-mathematical fabric, its thought, turn out to be experimental.

By itself, the objective *ratio* of the world now consists of events that are, in principle, observable and reproducible in experiments. The *ratio* of the world is local events occurring here and now; it is momentary impulses, velocities and accelerations; it is the differential laws of being. In this sense, the evolution of the concept of absolute space is very characteristic. Its ancient and medieval equivalent was the system, with an immovable center, boundaries, and natural locations, of Aristotelian cosmology. For Aristotle, there was no concept of *empty* absolute space, but there existed the prototype of the privileged, absolutely immovable system of reference, which was [later], in essence, superimposed on the [Aristotelian] framework of the center, borders, and natural locations of the Universe. Absolute motion here could only be a secondary concept; its absolute character was guaranteed by the absolute immobility of the *integral* framework, by the absolute character of space in which a body moves.

For Newton absolute space and absolute motion change places. The initial concept is now absolute motion, and absolute space turns out to be a derivative, secondary concept. In Newtonian mechanics, motion (accelerated!) figures as an absolute, because it calls forth the forces of inertia — the well-known push backwards during acceleration and push forwards during deceleration. The absolute character of space — the basis of universal harmony in Newtonian mechanics — is shown by means of a thought experiment (as Newton demonstrated it). The bases of universal harmony appear before man as *local*, lawlike regularities which are realized here and now, regularities which can be comprehended by observing local processes. These processes can be observed in the Venetian arsenal (as Galileo did) and, most important, they can be reproduced in real experiments. This is not enough. One can reflect on the local, differential regularities of being, and it is necessary to reflect on them by sketching ideal experiments like Galileo's, in which a ball rolls without stopping on a horizontal plane (demonstrating inertia), or like Newton's pail, suspended by a rope (demonstrating the forces of inertia under absolute acceleration), and analogous constructions. The locality of the universal *ratio* and the differential representation of motion (considered from point to point and from moment to moment) are connected with experiment, both real as well as ideal, that is, they penetrate thought, modifying the ideals of rationalism.

It is difficult to define, but completely impossible to deny, the profound influence of this active, experimental, dynamic and local character of classical science on culture, on the style of thought, on man's psychology. We shall confine ourselves to two observations. V. Muratov once wrote that in *Les liaisons dangereuses* of Choderlos de Laclos, one feels the Newtonian simplicity that is characteristic of eighteenth-century thought. Thinking over his moves, Valmont does not doubt the simple, absolutely determined result of each move. But there is another side to the matter. For medieval psychology, the normal course of events is determined not by the most immediate local impulses but by cosmic providence, by a stable system encompassing the whole universe and established before the beginning of time. In the Aristotelian Universe, man was left to attain truth and moral principles from Revelation, or to think about the world without attempting to change the course of events in the meantime. The integral harmony of the universe did not lead man to its experimental study and to thinking about the world by constructing models of local effects on events. Now all this has changed. The harmony of the world is the chain of local events. It includes states in infinitely small intervals; these states can be described, comprehended, and included in rational thought if one visualizes how events will develop, how the states of objects will change during one or another local action.

Probably nowhere is this dynamic and active relationship to the world expressed as deeply and as clearly as in the tragedies of Shakespeare, in this world of men and women who, with the utmost passion, meddle with the course of events, thinking least of all about their peripatetic 'natural places', as they trespass norms, break static harmony, and expect an immediate result here and now after each local action. For Hamlet, the ideal man, embodied in his father's image (. . . "he was a man, take him for all in all . . ."), is a man who not only thinks about the world but also embodies his thought in action. The striving to embody thoughts about the world in action, the craving for a real world harmony, the striving to unify "the time [that] is out of joint", is the basis of Hamlet's tragedy.

But this is by the by. The fundamental question concerns the transformation of the basic clashes in rationalism, the development of its new peculiarities and contradictions together with the genesis of classical science. We shall now turn to this problem. First of all we must address the modification of the concepts of being and nonbeing and then of identity and nonidentity under the influence of classical science. Let us examine from this point of view the peculiarities of classical science mentioned above and linked together: the differential representation of motion, the idea of continuous predicative diversity, and thought experiment as a form of scientific thought.

3. NONIDENTITY AND BEING

The differential representation of motion and of differentiation itself, if examined from the ontological point of view (and it is from this point of view that it must be examined), represents a widening of Democritean 'being', its emergence into 'non-being'. Each definition of being appeals to non-being, to *nothing* – this idea, as developed from Parmenides to Spinoza, has become clear enough. But in the seventeenth century, the transition from being to nonbeing acquired a new, positive and dynamic sense. Nonbeing not only limits the element of being and allows the individualization of this element (empty space, surrounding the atom, permits the separation of the atom from the surrounding milieu) but also plays the role of potential being; emptiness is regarded as the totality of the atom's potential positions. This construction derives from antiquity. What is proper to the seventeenth century and, in part to the three preceding centuries (i.e. to the Renaissance), is the definition of being in terms of nonbeing. This is a positive definition: what distinguishes being from non-being is expressed in the potential behavior of an object – in essence, in what is absent *here* and *now*. In fact, a particle is distinguished from space by its dynamic properties: the ability to act on other particles, and the ability to perceive the effects, mass, and charge of a particle. But these properties are now expressed in an experimental form: if one applies force to a particle, it will behave in a certain way. At a given moment the question is one of potential behavior.

The determination of mass and charge originates from the determination of the acceleration of a particle, from its behavior after a given moment. In this sense, the differential calculus begins with potentiality as a quality of being. But it also includes the opposite operation: potential behavior, the predicted world-line that does not exist at a given moment, is compressed into a point and into a moment, into *here* and *now*. This transition from actual to potential being and from potential to actual being, is the source of the *aporia* of motion, which becomes insoluble as soon as reason takes into account only one of two poles of being. Such a transition also serves as the source for an analysis of the *aporia* of the infinitely small.

The experimentalism of scientific thought caused a long-lasting divergence between science and philosophy, which existed for almost three centuries. Scientific, logical analysis consisted of a series of thought experiments. Even in mathematical deduction, when this analysis uses differential equations, the ontological essence of these equations is reduced to statements: this is the course processes will follow here and now, in infinitely small intervals of time and in infinitely small regions of space when the factors that determine the course of these processes change. Philosophy searched for the truth along another path, one which was semantic rather than experimental (I repeat, this is a thought experiment). It came from the more or less established meaning of words and changed this meaning by bringing concepts closer together, and then arrived at new concepts which sometimes received a new name, and often an old, but generalized, clarified, and rationalized one. But these paths did not always lead to one and the same truth. The simple grouping of concepts led to truths of the type: 'stone' is a word belonging to a broader set of words meaning a 'heavy body'. From here it was easy to shift to the traditional scheme of the universe, the 'sympathies' and 'antipathies' of Aristotelian physics. Even in the case when these ordered constructions were fantastic they were stable, they classified individual processes and generally more concrete images along the shelves of the already existing and strengthened language schemes. Such a style of philosophic thought aroused opposition on the part of science, and, in a country with a particularly sharp empirical tradition (in England, where, at that time, 'philosophical instrument', for example, meant thermometer), the motto of the Royal Society became *nullius in verba* – 'take *no man's word*'. This did not prevent Newton from being first and foremost a theorist and from remaining in the stream of rationalism, albeit modified.

The local ('here and now') character of the elements of scientific thought signified a very profound change in the fundamental logical operation — the inclusion of a given element in a more general class, the inclusion of the object, predicate, separate impression, and concrete concept in a certain set. This operation, signified by the symbol $\in$ in mathematics, ceased to be purely verbal and became a genuine operation, though an intellectual one. The inclusion of a certain concrete concept in a more general class depends on the result of a thought experiment. For example, the inclusion of the rotation of one of the more general classes of relative and absolute motion depends on whether water will rise up the sides of a rotating pail. Let us note that we can find an example of the opposite verbal-systematizing style not only in medieval but also in later discussions about the inclusion of the concept of existence in the more general class of attributes of God. For Newton himself, in the *Principia's* conclusion, which is directed against deism, the concept of dominion [господство] over nature is included in a purely semantic way among its attributes ('Lord' [осподь], says Newton, comes from 'master' [господин].[8]

The transformation of $\in$ into an operation, the transformation of logical deductions into thought experiments, to some extent excludes *a priori* constructions. But it also excludes purely empirical conclusions. Continuous predicate diversity, with its local here-now elements, signifies the experimental character of each conclusion. In the transition from nth element to another, $(n + 1)$, the result of the experiment demonstrates not only identity, but also difference. Elementary induction is already impossible here; elementary empiricism will not do. More accurately, it becomes clear that elementary induction and elementary empiricism are, strictly speaking, generally impossible.

The point is that *identity* is continually being established and destroyed. As early as antiquity, science arrived at a very important result — the notion of the natural immanent harmony of being — having assumed that certain elements of the universe (the center of the world, its boundaries, and the natural locations of bodies) were fixed. Then this presumption of static harmony was rejected, and science advanced to a new stage, ascribing immutability no longer to position, but to a derivative in time — the velocity of an isolated body. Thus the kinetic harmony of the universe was established. It limited its role to the abstract world of rectilinear and uniform motions. Generally speaking, the velocities of bodies are affected by fields of force.

Constant acceleration became the basis of the theory of gravity. Newton's dynamics was based on uniform acceleration which was unaffected by given forces (in other words, acceleration was the constancy of the mass of moving bodies) and which, along with the conservation of impulse and energy, formed the basic premise of eighteenth- and nineteenth-century science.

In relativistic physics, a more complex expression, in which impulse and energy are fused together, becomes the invariant. Generally speaking, the mass of a body changes according to its velocity. Such change can be determined on the basis of the conservation, identity, and immutability of mass at rest, which are peculiar to each type of particle. Particles lose their identity in the course of their transmutation, disintegration and birth.

In all of these cases, we meet up with a characteristic statement of science concerning the identity of the meanings of increasingly complex magnitudes. Science cannot observe, explain, or foresee changes, processes, or motions, without such identifications, but the identifications lose their physical meaning without statements about change. In other words, what is conserved is change — a change of place in a statement about the conservation of velocity, a change of velocity in a statement about the constancy of acceleration. Without such complementary nonidentity the concept of identity loses its meaning just as the concept of nonidentity loses its meaning without identity.

Thus we return to what we said in the Introduction: identity and nonidentity are complementary concepts. Each loses its meaning without the other.

What, then, is this? Bohr's complementarity? No. So far we can only say that these concepts are meaningful only when both are stated simultaneously. But they do not exclude each other. They relate to different predicates: acceleration is identical, velocity is nonidentical. Bohr's nonclassical complementarity, which figures in quantum mechanics, unites mutually exclusive concepts. Corpuscular and wave representation are incompatible, and, at the same time, inseparable and lose their physical meaning without each other. Probability without certainty loses its meaning, waves of probability without an indication of which event is probable (and this event is corpuscular in that it consists of the presence of the particle in a world-point) loses its meaning. The wave picture makes dynamically variable particles indeterminate, and the registration of each of these variables affects the conjugate variable.

What, then, is the *classical* complementarity that is characteristic of the differential representation of motion, of the genesis of classical science?

In answering this question, we must remember the historical review of classical concepts that was linked with the theory of relativity. In 1905, Einstein showed that not only mechanical processes, but also all physical processes flow uniformly in systems that move without acceleration relative to each other. Therefore motion without acceleration cannot be registered in an absolute manner (in the course of internal processes) without registering changes of distances between a given body and reference bodies. Before the appearance of Einstein's theory of relativity, no one had called the classical principle of relativity, or the Newtonian-Galilean principle of relativity, an affirmation of the invariability of the internal mechanical motions in a moving system, of their independence from motion if the latter occurs without acceleration. For slightly more than two centuries the contents of Galileo's *Dialogue* were repeated: the cabin of a moving ship where drops falling into the narrow opening of a vessel, the flight of butterflies, and the trajectory of objects thrown in different directions remain the same as on an unmoving ship. Repeated even more often was the theorem of Newton's *Principia* that concerns the impossibility of discovering motion without the acceleration of the behavior of bodies entering a moving system. But none of those who repeated these ideas of Galileo and Newton knew that they were dealing with the classical principle of relativity, just as Jourdain did not know that he was speaking in prose. But if Molière's hero had lived before the beginning of poetry, who could have reproached him for the absence of a special term for ordinary, non-poetic speech? Obviously, in the case of the classical principle of relativity, the matter was not reduced to a new term; there was an essential rethinking of the classical conception.

4. CLASSICAL REDUCTION

When we turn back from contemporary Einsteinian relativity to classical Galilean-Newtonian relativity, we are shifting to a comparatively limited principle. This is the *classical reduction of relativity*. In the framework of classical science, 'relativity' means that we cannot speak of spatial and temporal predicates (of position, velocity and so on) without indicating other, nonidentical predicates — bodies distinguished from space, masses distributed in some way or another in space. What we are saying in essence is that the characteristics of Democritean *nonbeing*, geometric characteristics, lose their physical meaning without the characteristics of *being*, without the filling of some points of space by something distinguished from space. At the same time, the basic characteristic of Democritean nonbeing — spatial distance — must be considered as the distance between bodies or as the distance between two positions of one and the same body, i.e., as its trajectory. Such classical relativity (position and motions in space are related not to space, but to bodies) demands that geometrical characteristics be linked with physical characteristics, with real or thought experiments, with something that allows empirical proof. But within the limits of classical science such a proof does not change the geometrical characteristics, it does not change the spatial distances through which a body must pass in the course of proof, nor does it change intervals of time. It does not change the predicates of 'being', the masses which take part in the motion of bodies.

In Einstein's theory, the nonclassical theory of relativity, the experimental determination of positions (coordinates) and of distances, as well as the determination of temporal intervals, changes them. Space-time scales change with motion, depending on the velocity and, correspondingly, on the reference systems.

A somewhat analogous relationship exists between nonclassical, quantum, Bohrian complementarity and its classical reduction. In the framework of classical science, the macroscopic picture loses physical meaning without microscopic filling. Macroscopic thermodynamics cannot be a physical theory without kinetic theory, without a representation of molecules. But the determinations of thermodynamic magnitudes do not change molecular models, and the models do not change the course of thermodynamic processes. In nonclassical, quantum physics the determination of each of the complementary poles changes the other pole.

The above-mentioned relation between the classical reduction of contemporary concepts and the concepts themselves in their nonclassical specifics can be seen in a more general form, by comparing the concept of continuous predicative diversity with its contemporary nonclassical form. The concept of continuous diversity is the result of a contemporary historical perspective similar to classical relativism and classical complementarity. The differential representation of motion is representation of motion as a process that goes from point to point and from moment to moment. It leads to an infinite-bivalent logic, to an infinite number of 'yes' answers to the question of the existence of a given particle on a real world-line, and to an infinite number of 'no' answers to the same question as it relates to others not filled by world-lines that appear in other problems. This has already been mentioned. The concepts we have just listed lead to the limitation of both the real world-line and a purely geometrical, four-dimensional image. What is the difference between the particle and the world-point, what is the difference between the filled, real world-line and the non-filled? This is the initial question that contemporary science raises and to which we constantly return when analyzing the past.

Classical science ascribes dynamic properties to the particle. The particle, as distinguished from a spatial point and moment, reacts with other particles. Now in examining this interaction as a fundamental peculiarity of matter (Democritean 'being'), we see that already within the boundaries of classical science the isolated 'single' particle was an abstraction that went beyond the boundaries of 'being'. But in classical science, interaction could be episodic; it was sufficient to ascribe to the particle an original impulse and the possibility for interaction with other particles, i.e. mass and charge, in order to distinguish the particle from its space-time localization and to consider the world-line of the particle filled – in other words, to consider it a physical, not a geometrical concept.

In quantum physics a particle is a particle, it has corpuscular being [but] only in the process of interaction, and of interaction with a classical body. In particular, such interaction allows us to ascribe to the particle space-time localization as well as impulse and energy. The world-line becomes a physical concept, it becomes filled when the particle interacts with a classical body. But given such interaction, the world-line in essence disappears as something determined. It pays for its physical meaning with the loss of its determinateness; in essence it ceases to be a world-line. Without a world-line, the concept of interaction also loses its physical meaning, for it is precisely within the deterioration of the world-line, in its erosion, that interaction is expressed. Once more we become convinced that nonclassical complementary concepts exclude each other, and are, at the same time, inseparable; when isolated they lose physical meaning, physical content.

The classical concept of a moving body signifies that the object possesses a varied, nonidentical position in space at various nonidentical moments. Moreover the body remains itself, identical to itself, it preserves its substantive properties; if we are dealing with a particle, then the particle preserves its mass, charge, and so on. Here identity and non-identity, as we already noted, relate to different objects. The predicates of 'non-being' — spatial relations between bodies — *are changed*, the predicates of 'being' *are preserved.*

In nonclassical science the relationship is different. The mass of the moving point changes. But this relativistic effect is only the beginning. The very existence of the particle depends on its motion (for example, in the case where the particle is a photon). The change of mass is a change of the potential world-line of the particle. The classical notion — that the position of the particle changes, and its dynamic properties (mass, charge, and existence) are preserved — is no longer adequate. Now, together with position in space (and in time), the dynamic properties also change, that is, those properties which distinguish one type of particle from another, and, as a whole, distinguish matter from space, particles from their space-time dislocation, from world-points. The classical formula of the inclusion of the individual statement ('here' and 'now') in the diversity of predicates is a declaration: before us is a momentary state of a particle identical to itself. In other words, there exists a set of states of an object identical to itself (A) and a state of a particle at a given moment (α), which enters into this indicated set. Such a formula can be written using a symbol of inclusion: $\alpha \in A$.

As we already noted, in nonclassical science the operation $\in$ ceases to be neutral. Inclusion changes both α and A. In order to determine whether a particle is to be found on a world-line A, we must find its space-time coordinates α. But the search for such a localization changes the energy and the impulse of the particle (Heisenberg's uncertainty), i.e., changes the world-line of A. And vice versa, the determination of the energy and the impulse of a particle, i.e., the determination of the direction and curvature of the world-line A makes the existence of α on this world-line problematic. Even sharper is the distinction between the nonclassical operation $\in$ and the classical one in high-energy physics. Given a highly detailed determination of dislocation, penetrating into a region significantly smaller than 10^{-13} cm or 10^{-24} sec, the substantial properties of a particle change very radically: the particle is transmuted.

This is no longer a simple existence (*here* and *now*) of the local being of the particle and of its possible being (*further* and *in the future*). We are now dealing with their mutual intervention, the effect of possible being on the actual, and vice versa. We know that the 'possibilism' of classical science, the taking stock of possible being, is expressed in 'differentialism', in the calculation of the differential predicates of the particle — its velocity, acceleration, and so on. But in nonclassical science, the determination of differential predicates (their velocity and all magnitudes depending on velocity and its change, i.e., impulse, energy, etc.) and the determination of the position of the particle and the moment of time of its observation are connected by mutual actions.

Thus the basic determinations of the differential representation of motion find their contemporary nonclassical equivalents. This could be stated in another way: contemporary, highly paradoxical quantum-relativistic ideas find their reduced classical prototypes. This is related to the transition from the integral criteria of Aristotelian physics to the local, differential criteria of universal harmony. It is related to the experimental character of classical science, as distinguished from medieval science. In relation to experiment — real and thought — we can limit ourselves to the following remarks concerning the distinction between nonclassical experiment and classical experiment.

The nonclassical experiment is a *cruel* experiment. The expression, 'cruel experiment', has been used metaphorically outside of physics: it has been applied to the characteristic method of Dostoevsky's poetics, as he placed his heroes in unbearably difficult conditions, bordering on madness or crime (and sometimes beyond that border) and then revealed deep, unexpected, paradoxical contradictions of consciousness and being. In physics itself, the expression 'cruel' experiment in the literal but ironic sense, must mean that in the experiment, in principle, the paradoxical regularities of being are revealed.[9] Thus the primary definition is related to the results of the experiment and not to its methods. But the next step in the definition of 'cruel experiment' already indicates the characteristic methods, the character of the applied agents which call forth the paradoxical processes that are observed.

But the real experiments that created classical science — the experimental natural science of the seventeenth—nineteenth centuries, and the thought experiments that were the basic form of theoretical natural science in the seventeenth century — operated with customary relationships, processes, and observations. In thought experiments this is very clearly visible. Galileo's ball, which rolls unceasingly, and Newton's rotating pail, seemed to be almost *a priori* conclusive; here, as always in usual observation, the *a priori* obvious is fused with the empirically obvious. On the other hand, in nonclassical experiments there is that 'reserve of nonidentity' which we already mentioned. In nonclassical experiment, things that science has packed away in a suitcase, things that it has tried to forget, things that resist the identifying function of reason, come into play. The point is that the concept of 'usual' (and this includes usual velocities, energies and so forth) is not only a psychological but also an epistemological concept: Parameters are usual which are already worked out and classified by thought, which have become conclusions from more general principles and are already placed out on the shelves and recorded in the inventory of science.

These were the parameters of immediately observed and practically used objects which, in the course of experimental research, did not destroy those general laws which had already been established. From the beginning, these were purely mechanical processes; then, thermodynamic processes were added to them, which did not violate macroscopic approximation, which did not demand for their comprehension the individualization of separate quantified acts of thermal radiation. These were the electrodynamic processes which did not demand a rejection of the continuity of electromagnetic radiation.

The foundation of nonclassical physics turned out to be experiments in which the speed of light (not equal to infinity) and the comparable speeds of bodies played an essential role. The most nonclassical experiments were those that destroyed the presumption of the self-identity of the particle. For example, one such nonclassical experiment was the discovery of the transmutation of particles at the very high energies which the particles possess in cosmic rays or in accelerators.

One must not think, of course, that classical experiments reveal the identity of processes and objects, and nonclassical ones, their non-identity. Both kinds of experiment reveal both. The distinction is an historical one: classical experiments demonstrate the already established relations between the identical and the non-identical (for example, the identity of mass given non-identical space-time localizations of a body); and non-classical experiments demonstrate the changes of these relationships. Jumping ahead, let us note that in the generalization of non-classical science, its epistemology must lean not only on the existent 'three-dimensional' system of knowledge, but also on its historical analysis.

The following conclusion derived from the differentiation between classical and non-classical experiment is essential for the history of rationalism.

New experimental results can be neither obtained nor understood without differentiating between what corresponds to already established theoretical representations of identity and nonidentity and what does not. For a long time a crucial fundamental experiment may not find a rational explanation. But its effect on theoretical thought begins long before such an explanation, in the new fundamental theory's prehistory. Before the appearance of theoretical thought, there is an unsolved problem, an unexplained experimental observation that provokes varied research. There is reformist research, forming complementary hypotheses in a traditional spirit, and revolutionary research, trying to transform the initial principles of science. During such critical periods, classical science, while remaining classical in content, ceases to be classical in style. It changes not only the initial physical principles, but also the mathematical apparatus and the very logic of scientific thought. The generalizing of such critical periods creates new philosophical directions. In that way rationalism comes closer to experimental science. It is farther removed from it in organic periods, and closer to it in critical ones. In the perspective of one or even more generations, the initial principles, the mathematical apparatus and the logic of science do not undergo changes, and science, to a certain extent, justifiably uses these initial principles, apparatus and logic as something immutable. It is from such an approach that the illusion of *a priori* knowledge grows. In critical periods, science demonstrates the dependence of its fundamental bases on experiment. Of course, in the history of classical science, critical and organic periods are not divided by clear chronological boundaries. The impulses for critical revision act for a long time, as do the impulses for the systematic exploitation of immovable ideas. Therefore, in the history of rationalism, these impulses are interwoven and their effect can be seen in the works of one and the same thinker, their content in one and the same theory — perhaps even in one and the same formula.

We turn now to the analysis of these tendencies and their effect on classical rationalist thought. It is now no longer a question of that philosophy officially called classical rationalism. We are now speaking about German classical philosophy and the transition from classical rationalism to it.

5. DIFFERENTIAL REPRESENTATION, LOCAL REFLECTION AND THE PROBLEM OF IMMORTALITY

Before moving on to the philosophy of Leibniz, Kant and Hegel, we should turn back to a problem we considered in the chapters on Epicurus and on Spinoza, the problem of immortality. The synthesis of the mathematics of variable magnitudes with the experimental understanding of being allows us to represent more concretely for ourselves the link between man's active work and his local sensation of immortality — 'the goblet of Oberon' — to imagine more concretely that free man who, in Spinoza's words, thinks not of death but of life.

As we have already stated, differential representation saves the local event and local object from being transformed into zero when compared with infinite nature, infinite space and time. Differential representation ascribes to the particle infinitely small predicates, whose relations (for example, the relation of an infinitely small shift to an infinitely small interval of time) are determined by law, encompassing infinite nature, the infinite number of local events and objects.

But these local events and objects are by no means absolutely subordinate to macroscopic laws. It is particularly for this reason that experiment enters into cognition as a necessary test of the application of the law and, as it later turned out, of its accuracy. Experiment and industry, which repeats it, serve as the local confirmation of the macroscopic, general law. And perhaps not as confirmation, but as the abolition, limitation, modification, clarification or generalization of the law. Experiment signifies that the sensual component of cognition must approve of the result of rationalist thought. Experiment can be the starting point for a new law, a new rationalist construction. A local event, a local object, not only embodies the infinite but also transforms it.

Of course, as we said above, the epistemological effect of experimental natural science is clearly visible only from a nonclassical historical perspective. But it is unquestionable. As early as the seventeenth century it *implicitly* changed man's notion of the relation of the local and finite to the general and infinite.

First of all, let us examine the notion of the relation of the local, limited existence of man to infinite nature. As far as rationalism is concerned, man is *homo sapiens*, the thinking man of Descartes. For sensualism, he is *homo sentiens*, man possessing sense organs. In real rationalism, in real sensualism and, above all, in experimental natural science, man is *homo faber*, man transforming the world. This is a local object, and his existence is a local event, but in them is embodied more than the 'world soul' of Giordano Bruno, or the eternal and absolute laws of nature (this is the way the new science decoded the constructions of the past). They also include the virtual change of eternal laws or, at least, the change of their sphere of application. Experiment is the correction of the eternal tablets, and by doing this, experiment secularizes them, depriving them of the halo of eternity and, in any case, of any *a priori*.

Let us move forward a few centuries and cross the boundaries of the genres of human culture. These transitions and leaps as well as their historico-cultural invariants are considered in the last chapter of this book in connection with an analysis of the fundamental problems of classical and non-classical science. In that chapter we shall introduce for comparison some images from nineteenth-century Russian literature: the work of Dostoevsky will be compared with the ideas of Einstein. We shall also mention another Russian writer, Anton Chekhov, who once remarked in his notebooks that the mortal in man is what he perceives through the sense organs, and all that is contained in consciousness above this is preserved.

Can these lines be understood to mean that only man's ideas are immortal, and all else is mortal? No matter what Chekhov may have in mind, Dostoevsky's poetics cannot possibly agree with such an interpretation of these lines. Chekhov, unlike Dostoevsky and Tolstoy, was not a writer who could be called a 'thinker' or a 'prophet', but this in no way diminishes the depth and originality of his thoughts about the world, man, life and death, and what could be called the philosophical 'subtext' of his artistic creation.

For Chekhov, the basis of such a 'subtext' was introducing the sensual elements of man's existence to immortality. Chekhov said that a moonlit night must be depicted through the shine of a bottle fragment lying on the ground. Such a fragment, gleaming in the moonlight, is more than poetics. Chekhov's work is deliverance from oblivion, from local limitation, from the transformation into non-being of all the fleeting splashes of being, of all sensual impressions and local situations. Through every-day existence, through detail, cosmic problems are no longer visible; the details themselves become part of the whole, unique, but also heterogeneous, detailed and infinitely complex immortal flow of being. This is a very democratic tendency: it is not only the great thinkers, the great ascetics, and the great sinners of Dostoevsky's unbelievably difficult 'cruel experiments' who create an immortal substratum of human being. No, the sum total of feelings, thoughts and acts (including the most modest, common and unnoticeable ones) enter into the immortal being of humanity; all are part of infinity.

The feeling of such participation is not only that of the passive local embodiment of infinity, but also of the active effect of an individual, limited being on the infinite being of the world; it grows with time, urged on by non-classical science. Here the epistemological effect of non-classical science is interwoven with the direct psychological effect of the application of science. Nonclassical in both its original principles and overall schemes, the productive and cognitive activity of man is formed from acts, each of which, in principle, can lead to a very radical transformation of the world, and of notions concerning the world.

NOTES

[1] "Itaque quam ARISTOTELES dixit primam contrarietatem sine medio, inter Idem et Aliud; eam ego in Geometricis, philosophicè consideratis, invenio esse primam quidem contrarietatem, sed cum medio, sic quidem, ut quod ARISTOTELI fuit ALIVD, unus terminus, eum nos in PLVS et MINVS, duos terminos dirimamus" (Kepler 1941, p. 15).

[2] [What Valéry wrote was: "J'ai écrit: *l'homme est absurde par ce qu'il cherche; grand par ce qu'il trouve*" (Valéry 1968—70, **II**, p. 520); "I once wrote: 'Man is absurd in what he seeks, great through what he finds'" (Valéry 1970, p. 62). This aphorism occurs in a collection intitled 'Moralités' (1930), which was published along with three other such collections in a volume called *Tel quel* in 1941. Kuznetsov has mistranslated the French 'par ce que' (in or through what) as 'parce que' (because). De Broglie's citation of Valéry is accurate.]

[3] "Mais comme en définitive l'univers ne peut pas se réduire à une vaste tautologie, nous devons forcément nous heurter cà et là dans notre description scientifique de la nature à des éléments 'irrationnels' qui résistent à nos tentatives d'identification . . ." (de Broglie in Meyerson 1936, p. ix).

[4] "Freilich sind Körper von gleichem Volumen nicht identisch, aber sie haben dasselbe Volumen" (Frege 1962, **II**, §58, n. 1, p. 71); "Wenn ich von der Verschiedenheit meines Hauses von dem meines Nachbarn abstrahirend beide als dasselbe betrachten und daraufhin im fremden Hause wie im eignen schalten wollte, würde man mir die Fehlerhaftigkeit meiner Abstraction bald klar machen" (*ibid.*, p. 107).

[5] Meyerson 1931, **II**, p. 420. There exists an analogous story about Kosh, who taught the Duke Bordovsky, as well as several other variations of this same anecdote.

[6] "Se peut-il que la science, cette maîtresse incontestable de la vie de notre époque, celle dont l'homme moderne s'enorgueillit de suivre les inspirations, tende, en fin de compte, à tout expliquer par l'Espace, lequel nous apparaît cependant d'autre part

comme quelque chose d'inerte, comme une forme vide de contenu? Est-il croyable que la science soit *cela*?

"Une restriction importante s'impose tout d'abord. Oui, la science est bien *cela*: mais ce n'est pas là toute la science. La science, obéissant à la tendance explicative qui est le propre de l'esprit humain, a bien l'air de vouloir tout réduire au rationnel. Mais, d'autre part, infiniment soucieuse de garder le contact avec la nature, elle reconnaît les limites que celle-ci impose à ses efforts. C'est ainsi que naît un concept dont la nature et l'importance n'ont peut-être pas toujours été nettement reconnues, mais qui n'en a pas moins exercé sur l'évolution de la science une influence profonde; c'est le concept de l'*irrationnel*, c'est-à-dire de ce qui, parmi les éléments dont la science est amenée à se servir, apparaît comme devant, par son essence même, résister à toute réduction ultérieure en éléments purement rationnels" (Meyerson 1921, **I**, p. 181).

[7] "La pensée est, en somme, le travail qui fait vivre en nous ce qui n'existe pas . . ." (from 'Poésie et pensée abstraite' in Valéry 1968—70, **I**, p. 1333).

[8] In the concluding General Scholium to the *Principia*, Newton writes:

> The word God* usually signifies *Lord* . . . (Newton 1934, p. 544). Vox deus passim significat dominum (Newton 1972, **II**, p. 761).

Newton's footnote, indicated by the *, reads, in part:

> * Dr. *Pocock* derives the Latin word *Deus* from the *Arabic du* . . . which signifies *Lord* (Newton 1934, p. 544).

[9] [With the expression, 'cruel experiment', Kuznetsov seems to allude to two things:

1. the *experimentum crucis* (crucial experiment) of classical, Newtonian physics, which, by deciding which of several hypotheses is correct, *resolves* conflicts and paradoxes; the 'cruel experiment' of nonclassical science *reveals* them.
2. an essay called 'A Cruel Talent' (ЖЕСТОКИЙ ТАЛАНТ) by the Russian critic and thinker, Nikolay Konstantinovich Mikhailovsky (1842—1904), published in 1882, a year after Dostoevsky's death (see Mikhailovsky, 1882, 1978). In this essay Mikhailovsky accused Dostoevsky of "perversely tormenting his readers and his characters" (Hingley 1978, pp. 195—196) because he heaped on his characters "an incredible mountain of misfortunes, calamities and insults (*ibid*., p. 50).

   In juxtaposing nonclassical science's 'cruel experiment' with Dostoevsky's 'cruel talent', Kuznetsov may be recalling that in the seventeenth century Boyle had written that the experimenter *tortures* Mother Nature. There is, for Kuznetsov, perhaps a further parallel between certain characteristics of nonclassical physics and those of the Dostoevskian hero, the 'Underground Man' who stands at the crux of Dostoevsky's poetics, the object of his 'cruelty'.

In the early twenties Bakhtin wrote:

> The Underground Man . . . has . . . no fixed definitions, . . . he figures not as a person taken from life but rather as the subject of consciousness and dream. And for the author as well he is not a carrier of traits and qualities that could have been neutral toward his self-consciousness and could have

finalized him; no, what the author visualizes is precisely the hero's self-consciousness and the inescapable open-endedness, the vicious circle of that self-consciousness. . . . Racine's hero is all objective existence, stable and fixed, like plastic sculpture. Dostoevsky's hero is all self-consciousness. Racine's hero is an immobile and finite substance; Dostoevsky's hero is infinite function. . . . The hero from the underground . . . seeks at whatever cost to retain for himself this final word about himself, the word of his self-consciousness, in order to become in it that which he is not. His consciousness of self lives by its unfinalizability, by its unclosedness and its indeterminacy. . . . The special sort of moral torture that Dostoevsky inflicts upon his heroes, in order to force out of them that ultimate word of self-consciousness pushed to its extreme limits, permits him to take all that is merely material, merely an object, all that is fixed and unchanging, all that is external and neutral in the representation of a person, and dissolve it in the realm of the hero's self-consciousness and self-utterance (Bakhtin 1984a, pp. 51, 53; 54) – Eds.]

8. DYNAMISM AND THE CRITIQUE OF STATIONARY BEING

1. THE CONCEPT OF FORCE

Modern science has shifted the center of gravity of the study of nature to local events and differential laws. But what was this new physical concept that allowed local events, with their integral causes and consequences, to be linked with the integral harmony of the universe? How did this concept modify philosophy, how did it influence the style and content of rationalism?

At the end of the seventeenth century, the curve of cognition included a new universal concept — force. This concept was an answer to the question of how local regularities of being are unified in the general, integral scheme of the universe. At a given point at a given moment, a moving particle undergoes acceleration. The cause of this acceleration is force. When analyzed further, we find that force already acts as a result, as the consequence of a spatial distribution of bodies. Newton's program — the explanation of nature in terms of the distribution and interaction of bodies — has these two tasks: first of all, the determination of the motions and positions of bodies by force; and second, the determination of force by the motion and position of bodies. The first task is the basis for mechanics, the second is the prototype of physics (or, more accurately, of field theory).

Newton worked out the first part of his program in a system that was unprecedented in its accuracy and rigor. The second part, the truly physical one — the theory of gravity — addressed the future. The law of gravity explained the motion of the planets, but did not explain why bodies possess weight, why they fall, or why they attract each other. Here the question 'why' was replaced by the question 'how', the causal mode was replaced by the phenomenological conception. It would be incorrect to ascribe Newton's phenomenological theory of gravity — as it was called against the background of Cartesian models — to Newton's 'left hand'. His law of gravity was the only theory of gravity possible in the seventeenth—eighteenth centuries and was the creation of the 'right hand' of a great thinker. Nevertheless the kinetic temperament of the seventeenth, eighteenth and nineteenth centuries could not be reconciled with the unexplained force of gravity and the unexplained concept of force as a whole. Force, as the abstract symbol of undiscovered causes of acceleration, aroused opposition from scholars and philosophers.

The opposition took different forms. Researchers in the natural sciences strove to reduce gravity to motion and impulse, a rather hopeless struggle between the departing Cartesian kinetics and dynamism, which had just appeared. The philosophers Leibniz and Kant modified the doctrines concerning substance in such a way as to include force.

The second part of Newton's program – the determination of force by the positions of bodies – was the program of field theory. But this was only a program, only a question addressed to the future. This lack of an answer was an essential circumstance that determined the effect that the concept of force in the seventeenth—eighteenth centuries had on philosophy. In field theory, force is fused with space and becomes spatial, but given a variable, or a field propagated with finite velocity, force becomes a spatio-temporal representation. So far there are no differential laws that determine the voltage of a field of force at a given point at a given moment, one that is independent of the presence or absence of an experimental body. Force remains behind the scenes of a space-time representation; it determines what occurs on stage, but does not itself appear there. The law of gravity talks about the space-time picture of the accelerations of gravitational bodies, accelerations that depend on the distances between bodies. Without these bodies the law of gravity makes no sense. The Newtonian force that figures in the basic law of mechanics in the law of forces is applied force. The force that still makes sense without being applied to a body, force which is calculated for each point of space, reflects a transition to the concept of field. But there was no physical concept of field in the seventeenth—eighteenth centuries.

The claim of force to the role of an original, primary concept posed a threat to the kinetic conception of universal harmony. Galileo's universe, in which the *ratio* of the universe was embodied in the scheme of motion, had sunk into the past. There now existed a new scheme of universal harmony based on the forces that acted between bodies and determined their acceleration. The inert component of motion, i.e., rectilinear and uniform motion, which is curved and accelerated by forces, also depends on force, the initial impulse that determines the velocity and direction of motion that is maintained by a body when left to its own resources.

Newton ascribed this impulse to God. To its credit, eighteenth-century philosophy did not accept this explanation. Kant, in particular, in his search for the active impulse which forces planets to revolve around the Sun, for the impulse which explains the tangential component of this motion, arrived at the hypothesis of the primary nebula. But the basic philosophical problem consisted in the generalization of the concept of force, in its inclusion in the determination of its place among the basic philosophical categories.

Force did more than change the meaning of the basic categories. It distributed these categories and concepts between empirical and logical thought in a new way; it changed the relations between the sensual and rational understandings of the world. Force was a category which brought these poles closer together. In this sense, dynamism was an important stage in the basic evolution of rationalism, in its transformation into an ontological conception, in the doctrine of being. Since force is a rational concept, it also enters into the new, experimentally comprehensible world. Here not only the registration of bodies, but also the measurement of their effect on other bodies, is required. Bodies must become comparable, that is, homogeneous not only in their substance but also in their effect and their reaction to effects. Otherwise they do not acquire the right to enter into the experimentally comprehensible world. Beginning with Newton, the basis of quantitative experiment became the determination of the mass of bodies as proportional to their weight.

Mass had already been represented as one condition for the effect of bodies on other bodies and, consequently, for the possibility of their sensual apprehension. An analogous evolution occurred in the character of the *rational* paths of cognition. Dynamic properties allow bodies to be separated out, individualized, and given a meaningful character as regards thinking about the world. Now dynamic properties — the weight of bodies and the mass proportional to them — became the basis for a new form of rational comprehension of the world, for mathematical deduction. The concept of force transformed both the empirical and the logico-mathematical paths of cognition of the world.

At this time it was necessary to transfer the dynamic categories to a higher stage of cognition. Philosophy had to take into account empirical knowledge and mathematics in a new way, one that was different from the seventeenth century, when philosophy not only generalized but also, in essence, absorbed the weakly differentiated branches of knowledge, which did not possess as wide an experimental base as in the eighteenth century. Positive science existed in the lap of metaphysics. In the period between the seventeenth-century English revolution and the eighteenth-century French revolution, science (which, together with philosophy, had earlier freed itself from the spiritual authority of the church) separated itself even from philosophy. Philosophy could not separate itself from science, but neither could philosophy subordinate itself to science. Philosophy, which could have acquired sources of knowledge independent of science, could also have false notions about the world. Without sources other than the empirical data of science, philosophy would have lost its meaning. An apriorism that divided philosophy from science would not have been of the least use here; the same is true of an empiricism that had been raised to the ranks of an, in principle, unique path of cognition and had transformed philosophy into a passive reflex of differentiated science. The problem lies in the generalization of science, but not in the simple introduction of empirical facts under the *a priori* constructions of reason. Made absolute and isolated, the identifying and classifying tendency, the presumption of the full independence of partial statements from inclusion in sets, from the formation of more general concepts, and also the presumption of the independence of sets themselves from their elements, no longer led to new results.

In the search for a way out, eighteenth-century philosophy undertook a rather radical review of its basic categories, filling them with new empirical content, and at the same time, it radically rethought this empirical content. The inclusion of this empirical content in the traditional categories gave eighteenth-century science what Einstein called 'inner perfection'. Insofar as the categories themselves were verified in this process, the inner perfection acquired external confirmation.

Such relations between inner perfection and external confirmation are characteristic of the radical turning points of science, which call forth radical turning points in philosophy and in the notion of being as a whole. For contemporary science and contemporary philosophy, such relations between inner perfection and external confirmation are no longer sporadically created for radical turning points, but in practice exist all of the time. It is easy to see that Einstein's criteria correspond to a "deepening of reason within itself" and to "the progress of reason" — the widely known concepts of Laplace.

The link between experiment and philosophy, now closer and more immediate than previously, the dependence of cosmological and epistemological generalizations on new experimental data, and the more intensive dependence of the interpretation of experimental data on the change of epistemological positions did not mean that philosophy and the natural sciences were fused into one. Such a fusion was finished off as decisively as other seventeenth-century metaphysical traditions. Natural science ceased to be a passive reflex of metaphysical speculations, and philosophy did not become a passive reflex of the empirical comprehension of the world. The era of fusion was over; the era of the interaction of philosophy and science, the era of the 'deepening of reason within itself' along with its 'progress', had begun. This was the era of nonlinear interaction. Perceiving impulses in science, perceiving new experimental data and their original generalizations in the framework of differentiated branches of science, philosophy now has a free rein; it arrives at what was not contained in the empirical data and even contradicts them. Philosophy reserves the right to paradox. The experimental study of nature as a whole, which receives impulses from philosophy, in its turn also has free rein. Experimental study goes further: It arrives at what is new, unexpected, paradoxical, in principle. But with the interaction of philosophy and natural science, the notion of the *a priori* constructions of reason and of the purely empirical nature of scientific data disappears or, at least, it ought to disappear.

Now the historico-philosophical and historico-scientific riddles begin. In the period when philosophy was engrossed in mastering a series of new empirical data, which required the review of its basic categories, Leibniz transferred the metaphysics of the seventeenth century into the eighteenth century; Kant's philosophy, which closed the eighteenth century, seemed to rehabilitate metaphysics by declaring space and time to be *a priori* forms of cognition. This rehabilitated metaphysics brought down on science a series of *a priori* judgments and systems of *Naturphilosophie*. On the other hand, in a period when theoretical thinking in the form of mathematical analysis and qualitative-logical constructions led to great results, empiricism began to develop in science and, in its sensualist surge, was included to deny completely all that lay beyond experiment.

How did this happen? In this contradictory situation, what role was played by science and, above all, by its most important new link, the concept of force, the dynamism of Newton?

Let us return to the concept of force. In mechanics force is a pseudonym, an undecoded cause of acceleration. In the theory of gravity, when bodies are considered not as the passive victims of gravity but rather as its active sources, then force acts not so much as a coded concept than as a formal mathematical correlation which expects physical embodiment.

Now we must somewhat clarify the concepts 'formal-mathematical correlation' and 'physical embodiment'. The question obviously concerns fundamental epistemological concepts, the basic problem of being, its bases and the inalienable components of classical rationalism.

'Formal-mathematical correlation', in the traditional sense, means a series of magnitudes that are connected by the mathematical signs of multiplication, division, differentiation, equality, etc. and moreover, no visual models whatever correspond to these magnitudes. In this traditional notion, 'visual' models are models of displacement and collision of impenetrable solid bodies. A similar Cartesian variant of 'visibility' and 'formality' in the eighteenth century was no longer appropriate. Newton's mechanics and dynamism consisted, in effect, in the irreversible transition to the more general concepts of visibility and formality. These more general concepts are easily seen from a nonclassical historical perspective.

Strictly speaking, it is not the motions of solid bodies but the oscillations of fields, the oscillations of forces, and electromagnetic vibrations which are graphically clear in a direct physiological sense. But we are interested not in the physiological but in the epistemological meaning of 'visibility'. Viewed from a nonclassical historical perspective, the epistemological meaning of 'visibility' is as follows: A model can be said to be 'visible' if (1) it has physical content; (2) when viewed macroscopically, we can imagine the processes that constitute it as parts of universal harmony; (3) when viewed microscopically or ultramicroscopically, we can see the differential mechanism governing these processes, track these processes from point to point and from moment to moment or register the boundaries of such tracking and show some kind of ultramicroscopic quantification of the spatio-temporal representation of the processes. If these conditions are present (for example, if we can imagine the world-line of a particle and the ultramicroscopic events which give the world-line physical meaning), then the motion of the particle has a visible character. The word 'visible' keeps its original link with observation, as the process for which we found the macroscopic world-line, and the ultramicroscopic filling-in of the macroscopic world-line can figure in experiment. Experiment, a concept that has been refined by quantum mechanics, represents the sum total of macroscopic processes, but these macroscopic processes are, in essence, physical processes, not geometrical concepts, because they are filled in by strictly or intuitively represented ultramicroscopic events.

From this point of view, the concept of force does not seem to be a departure by Newton from the true, physical, visual path (as it was regarded by the Cartesians in the seventeenth–eighteenth centuries, and by the imitators of Cartesianism in later times) but rather a transition to a more accurate notion. But from this perspective the concept of force has a different character in the theory of motion and in the field theory of the eighteenth century. In the theory of motion, in the law of forces and in the analytical mechanics which grew out of it, we know both the macroscopic skeleton of the process (for example, the initial and final conditions for the determination of an actual trajectory according to the law of least action), and the differential mechanism at each point and each moment. The question of the determination of force as the result of the distribution of sources of gravity is completely different. Field theory as a physical theory is a child of the nineteenth century when it played the role of the stepson rather than the child of classical science. In any case, in the eighteenth century there did not exist a concept of the field as a positive answer, a solution that was in some sense stable, 'classical' in its style.

In the history of science and philosophy, the cause for a *too radical* rejection is often the impossibility or incapability of formulating a *sufficiently radical* rejection. At the turn of this century the inability of science, working within the framework of traditional philosophy and of traditional philosophic notions, to radically reject the old notions concerning matter, led to a 'radical' rejection of any objective notion about matter, to a rejection of the epistemological value of science as a whole, i.e., to a very traditional agnosticism. In the eighteenth century, the impossibility of rejecting the traditional narrow notion concerning extended substance led to a rejection of the inclusion of force in the spatial world, the representation of force as something nonextended, the transfer of extension itself into the subjective world, and a rejection of the objectivity of space and time. All these are examples of the absolutization of the spirals of cognition. In what did the spiral itself consist in the given case, upon the appearance of the concept of force?

In the history of science certain positive steps can be defined only negatively. In the first variant of positron theory, Dirac spoke of 'holes', free places where there were no electrons, as if they were particles, physical objects. Eighteenth-century mechanics possessed such a 'hole; in the picture of the world there was no *spatial* image of force, no concept of field as an independent substance distributed in space. It is not a matter of the absence of physical models that fill space with specifically 'gravitational matter', or with gravitational ether. There were enough such models. There was no theory which would have allowed the determination of a field of force independent of the motion of experimental bodies. *Therefore*, and not as a result of the absence of mechanical models, there existed a vacuum in the development of science. The vacuum was only filled by Maxwell, for whom the electromagnetic field became an independent reality, distributed in space and in time.

As long as there was no physical conception of field, its absence could not be discovered. The existence of a vacuum in eighteenth-century science can only be seen now, in retrospect. In the eighteenth century, any philosophical conclusions from dynamism without the concept of field could be only very indirect. When it came to the impulse, which determined the inert component of the motion of heavenly bodies, philosophy could not basically alter its position, and Kant, without transforming basic ontological and epistemological principles, found in the nebular hypothesis a rational explanation of the notorious first impulse. But a field of force could become substance only by means of a radical transformation of the very concept of substance. The bridge had to be built from Democritean 'nonbeing' to Democritean 'being', not simply to unite the new form to the existing forms of being, but also to change the definition of being. In atomistic notions, 'being' was a synonym for something homogeneous and unchangeable. A field, as being, was something heterogeneous and movable. A field is a function of space and time, which, generally speaking, has different meanings at different points and different times. This is 'being' in diversity and in motion. Therefore, the field is the basic natural-science precondition for the transformation of rationalism, for its dynamic transformation.

In the eighteenth century there were not yet data for this. There were attempts to use the old 'being', the mechanical body, to substantialize the field. But science, if we are speaking of its basic course, could not fill 'non-being' with a hypothetical mechanical bearer of force and thus return to the Cartesian all-filling 'being'. Such efforts were headed in a direction opposite to the main course of science. Philosophy, in the person of Leibniz, refused to place force in space. It placed force 'behind the scenes' of space, as its cause, its base.

Moreover, substance ceased to be unified. Force does not lie behind all extension, behind the whole Universe; it is an interaction that presupposes the existence of another subject, which separates the given subject from another, is opposed to, resists and acts on the other subject. We shall now turn to this problem of the individualization of bodies and the plurality of substances.

For a modern understanding of the role of force in Leibniz's philosophy, which corresponds to modern science, we must take into account the modern tendencies of field theory. Field theory tries now to explain not only the behavior of bodies under the influence of fields of force, but also the existence of the very sources of these fields. In other words, it attempts to become a unified theory that not only explains the totality of the manifestations of nature in terms of the existence of fields, but also deduces from the field itself the existence of its sources. Modern tendencies in field theory bring this theory close to Spinoza's *causa sui*, the modern ideal of field theory. But this is no longer only a philosophical idea; it is rather the physical embodiment of the idea of heterogeneous being in which stationary elements — bodies, sources of fields — are created by the field itself, by the heterogeneous, moving, changing being. It is particularly this tendency that we will keep in mind when speaking of the field character of the conception of force and the 'pre-field' character that is opposed to it.

2. THE DYNAMISM OF LEIBNIZ AND THE PROBLEM OF INDIVIDUALIZATION

From the multiplicity of intellectual sources of Leibniz's monadology and from the varied impulses determining its appearance, let us try to single out one source and one impulse — the effect of the dynamic *Weltanschauung* and the concept of force. In other words, let us try to separate out the line Newton-Leibniz. In Leibniz's *Weltanschauung*, the criterion of individualization, the requirement that the individual possess clear borders dividing it from its environment, is most closely connected with physical sources. Such a requirement forces Leibniz to turn away from the Cartesian identification of extended substance with space itself. Leibniz goes further. He rejects any unified substance which encompasses the entire world. The 'world' consists not of one substance, but of substances. Spinoza replaced the homogeneous extended substance of Descartes with a heterogeneous unified substance consisting of various modes, with innumerable attributes. For Leibniz, the plurality of modes is transformed into the plurality of the substances themselves. Each substance possesses extension, but this extension is the result of *force*.

Thus the concept of force moves from physics into metaphysics. Newton's phenomenological force is transformed into Leibniz's metaphysical force — the basis of the individualization of the monad.

In reality it individualized the monad. Force is what really, physically distinguishes one body from another; it is what makes a body an actual, real, separate body, what gives it passive individuality (allowing it to resist influences) and active individuality (allowing it to act on other bodies). Thus force is the prototype of the individualization of bodies as well as of their connection i.e., of universal harmony.

Leibniz's thought goes from a critique of the Cartesian indistinguishability of bodies to the concept of force as the basis of their individualization. In his letter "On whether the essence of a body consists in extension", Leibniz says:

> I still agree that naturally every body is extended and that there is no extension without body. None the less we must not confound the notions of place, space, or of pure extension with the notion of substance which, besides extension, includes resistance, that is to say, action and passivity (Leibniz 1951, p. 102).[1]

The metaphysical force of Leibniz is the same as the mechanical force of Newton, except that it includes absolute stability, primordiality, which for Newton was conditional (in that there was the question addressed to the future concerning the origin of forces, a question that could be answered only in field theory), and which for Leibniz became metaphysical and *a priori*. Force, eternal and indivisible, is found in all things; it represents the basis of the thing's being, its substance. Thus individual substance — the monad — appears. The monad is distinguished from a mathematical unit in that the latter, according to Leibniz, can be split. "If the same were true of the arithmeticians' unit, one could dispense with fractions, which would please school children very much."[2] The monad is also distinguished from a geometrical point, which does not really exist, and from a physical point, which is extended and consequently divisible. In particular, the indivisibility of force, i.e. that property of force which is characteristic of 'pre-field' conceptions, transforms Newton's force into Leibniz's force, into an individualized, separate, eternal basis of being for each real object.

The shift from the physical to the metaphysical concept of force stems from the fact that in the physical world it is impossible to deduce the *being* of force. The *being* of force always explains the behavior of physical objects, but outside of field theory, this behavior does not explain the existence of force. Therefore, Leibniz places force in the metaphysical world, behind the scenes of the physically comprehensible world of moving, extended objects. The monad turns out to be non-extended substance. Monadology was the philosophical equivalent of Newton's motion of forces; but it was not only an equivalent of a positive, enforced, stable, Newtonian conception; to a certain extent it also reflected the hidden contradictions in Newton's mechanics, the 'vacuum', searches and approaches to the concept of a field of force – in other words the predictive component of dynamism.

Force as substance separates the being of a physical object from the rest of the world. But it does not remain isolated. A monad 'represents the whole Universe', 'contains individuality in itself as infinity in an embryo', 'each body touches all that occurs in the Universe'. These theses of Leibniz have an immediate relation to natural science, for matter in Leibniz's philosophy is the linking of monads. The unity of the world is material unity. In transferring force as unextended substance into the microworld, Leibniz opened the way for new notions about force. In the extended world, these forces act as material forces, establishing the unity of the universe, its material unity. But Leibniz did not follow this path.

Leibniz focusses on the unreproducible, the specific, the non-identical. Elements of the universe are by no means non-elementary. Feuerbach compares the philosophy of Spinoza to a telescope and the philosophy of Leibniz to a microscope.

"Spinoza's world", he writes,

> is a plain mirror of the godhead, a medium through which we see nothing but the colorless heavenly world of unified substance; Leibniz's world is a multi-faceted crystal, a diamond, which, through its unique essence, reproduces the simple light of substance in an infinitely diverse richness of colors and obscures it.[3]

In Leibniz's philosophy, as we see it today in historical perspective, the link between two ideas is noteworthy: the inexhaustibility and the harmony of the Cosmos. The combination of infinite inexhaustibility, originality, unreproducibility, and the complexity of micro-objects, with the idea of the harmony of the world, allows us to define somewhat more accurately the traditional contrast between Spinoza and Leibniz.

The substance of Spinoza was unified but heterogeneous. It included modes, among them a certain microstructure of modes. Leibniz's notion of the inexhaustibility of this microstructure is, to a certain extent, a continuation of Spinoza's achievements, a continuation of the shift from the homogeneous Universe of Descartes to a unified but diverse Universe that was studied in the eighteenth century with the telescope and microscope. What is new in Leibniz's work (as compared with Spinoza's) is the *infinite* complexity of the microworld as the prototype of differential representation, of the idea of the infinitely small.

3. PLURALISM AND ECLECTICISM

The so-called Leibniz-Wolff philosophy presents one of the clearest examples of the relationship between the ideas of a truly creative person and those of an epigone. Christian Wolff's pedestrian interpretation of Leibniz does preserve a link with the creator's intention, but this is the link between a monument and its shadow on the city street — distorted, flat, a pale reflection of the sculptor's masterpiece. In such a reflection, monadology is deprived of its element of inquiry, of all that looks forward, that is addressed to the future. It becomes extremely ordered; it cannot arouse any doubts, it is intended to win unanimous recognition. But this calculation was incorrect; as often happens, creations intended for unaminous acceptance are soon unanimously rejected.

The basic question which Leibniz took over from Descartes, modified and passed on like a baton of scientific thought, is the question of the relationship between extended substance and those predicates of substance which are not reduced to extension. Leibniz affirmed that extension is the manifestation of nonextended dynamic substance. The nonextended nature of force was a departure from the main course of science and, at the same time, it was objectively a question, a search, an expectation. It was the expectation of a new conception in which extension acquires dynamic properties and force becomes an extended field.

Wolff cut off this inquiring tendency of monadology. He took from Leibniz the elementary substances, the monads, but for Wolff, the monads turn out to be simple dynamic centers, physical but nonextended atoms. Here Leibnizianism and Cartesianism are united in a single conception. Cartesianism is represented by a spatial picture: all that occurs in the world occurs in space. Monadology is represented by a nonextended physical substance: nonextended objects move in space and, moreover, do not acquire extension. The bridge between nonextended force and extended particles does not enter into Wolff's natural philosophy.

The idea of a pre-established harmony is also very colorlessly reflected in Wolff's works.

> The highest general idea to which this natural science attained was that of the purposiveness of the arrangements of nature, the shallow teleology of Wolff, according to which cats were created to eat mice, mice to be eaten by cats, and the whole of nature to testify to the wisdom of the creator (Engels 1954, p. 36).

Actually, both philosophy and natural science are responsible for this kind of thinking. It marked the unification of a simplified version of Leibniz's ideas with the empirical statements of natural science — with what had already received a finished form in natural science, and not with its tendencies and prognoses. In a strictly philosophical sense, this unification did not change, modify or concretize anything either in philosophy or in natural science. It resulted from a selection of both those philosophical ideas which corresponded to already discovered and proven scientific ideas and those scientific results that could easily be fitted into simplified philosophical constructions.

Wolff's teleological harmony was the basis for an equally dull representation of 'this best of all possible worlds'. Such Panglossism is diametrically opposed to Leibniz's active and nontrivial optimism, which is based on overcoming and reconciling opposing tendencies; for Wolff, optimism consists of ignoring their opposition, of an eclectic, purely external unification of opposing forces.

Wolff's eclecticism was a pale reflection of Leibniz's *pluralism*. Leibniz tried to unite within his philosophy all that was positive in the philosophy of the past. But as a result of this unification, ideas of the past were radically changed. Leibniz imparted to them new content, and a more general and, at the same time, more concrete appearance. This was not a dialectic in our sense — synthesis did not hide in clear form contradictions, incompatibility, the struggle of synthesized concepts. But this also was not eclecticism — an external unification of ideas which remain unchangeable and, in essence, alien. In effect, this was pluralism, the representation of the co-existence of regularities in nature, which differ in principle, and of the reflection of such regularities in theories, which differ in content. From here, thought can go further along the dialectical path, if the synthesis of philosophical concepts is subordinated to the objective logic of objective processes. Such a path leads to a subsequent dialectical *Weltanschauung*, if thought penetrates into the depths of this objective logic, if it generalizes the objective contradiction of the world, infinitely opening itself up to science; if it discovers the contradictory, moving harmony of the world, its spontaneous motion, to which the cognitive activity of the human spirit is subordinated in its searching. Moreover, thought creates its own categories, including those categories which it encounters in philosophy itself and in its history — flexible categories, which thought does not break up or eliminate in the manner of nihilistic irrationalism, nor put together as in an eclectic sum. Thought modifies categories, changes the direction and content of philosophic conceptions; it reveals an increasingly exact, general and, at the same time, concrete sense of these conceptions when they lie on the main road of cognition. Writing about Hegel's *Science of Logic*, Lenin says:

> The flexibility, applied subjectively = eclecticism and sophistry. Flexibility, applied objectively, i.e., reflecting the all-sidedness of the material process and its unity, is dialectics, is the correct reflection of the eternal development of the world.[4]

Leibniz's pluralism stopped at a crossroads. One road led to an arbitrary, subjective and purely external compromise of ideas. Leibniz could not go down such a road. The feeling of the unity of being, of the Spinozist 'telescope', the thought of the real link between opposed, nonidentical, diverse definitions of the world too strongly pierced the feeling of the heterogeneity of being, of the diversity and the individuality of the elements of being, through what Feuerbach associated with the microscopic. Leibniz could not shake off this real link between opposed definitions, in other words, their objective contradiction; he could not construct his system in an external, subjective, arbitrary manner as Wolff had subsequently done. But he also did not reach an objective dialectic of the basic definitions of being; the essence of being was drawn out beyond the borders of the living modal diversity of the *extended* world. The concept of substance turned out to be outside of the reconstructing effect of the concepts which develop and modify the picture of the universe during the study of what occurs in space and time.

Leibniz's pluralism, like essentially all of his philosophy, was connected with dynamism, with the concept of force. Force is the pseudonym for the cause of motion, which obscures the nature of force. Perhaps the cause of motion is the motion of another body. Perhaps it is motion in some kind of more general form, including change of a non-mechanical character. Pehaps it is mass movement, and as it is being analyzed, individual shifts of particles can be ignored. These possibilities force the thinker to allow the existence of a series of manifestations differing in their regularities as well as the coexistence of various types of theories. To the purely mechanical conceptions of Leibniz's own time are somehow joined the qualitative-logical constructions of ancient and medieval thought. Leibniz often returned to such older material and perhaps he more willingly looked there for material to use in his own construction than in sources closer to his own time. Leibniz suspected that the cause of motion, whose nature is hidden by the concept of force, could be different from, not similar to, those processes which we encounter and study in mechanics. In other words, outside of purely mechanical processes, there extends a living force, when we register that there is none. In Leibniz's explanations of the seeming disappearance of forces after an inelastic blow, the principle of conservation is formulated by means of the following analogy:

The forces are not destroyed but scattered among the small parts. The bodies do not lose their forces, but the case here is the same as when men change great money into small (Leibniz 1969, *To Section 38*, §99, p. 713).

Leibniz understood that the laws dealing with 'small money' could differ from the laws dealing with 'large money', that force as the cause of motion could be the name of processes of a different nature. The theory of the conservation of forces included for Leibniz the concept of 'dead force', which is not motion. Is 'dead force' a spatial category? Leibniz often considers rest as infinitely small motion. On the whole, Leibniz's scientific program is the exchange of large money for increasingly smaller, infinitely small money. But it always remains money. Space does not disappear. Each of the smallest spatial areas is a complex structure. The well-known image of a garden or a pond remains a symbol of the complex structure of being, no matter how small an area we take. "Each portion of matter", writes Leibniz,

> may be conceived as like a garden full of plants and like a pond full of fishes. But each branch of every plant, each member of every animal, each drop of its liquid parts is also some such garden or pond.
>
> And though the earth and the air which are between the plants of the garden, or the water which is between the fish of the pond, be neither plant nor fish; yet they also contain plants and fishes, but mostly so minute as to be imperceptible to us.
>
> Thus there is nothing fallow, nothing sterile, nothing dead in the universe, no chaos, no confusion, save in appearance, somewhat as it might appear to be in a pond at a distance, in which one would see a confused movement and, as it were, a swarming of fish in the pond, without separately distinguishing the fish themselves (Leibniz 1898, §§67–69, pp. 256–257).

Nature does not repeat itself in the shift from the larger to the smaller. The air between plants is not made up of plants, and the water between fish is not made up of fish. The picture drawn by Leibniz is a far cry from the scheme of repeating structures which has survived until our time ("perhaps these electrons are worlds in which there are five continents" . . . wrote Valerii Briusov [Symbolist poet, 1873–1924] at the beginning of the century). Even in Leibniz's preformist theory there is no question of the infinite recurrence of the same constantly shrinking organism. Nature is infinite, given the decrease in the scale of the hierarchy of complex structures, which are continually shrinking, but preserve their complexity and are by no means of one type. Therefore, the reduction of the *ratio* of nature to any simple law is alien to Leibniz.

The representation of such a hierarchy of complex spatial structures is linked with Leibniz's version of the infinitely small, which is in turn linked with the idea of diversity in the very hierarchy of the parts of matter; laws which are essential for one level of this hierarchy are not essential for another. Leibniz considered a grain of sand as an example of the infinitely small: it is a body of finite measurements, but both the measurements and the very existence of a separate grain of sand are immaterial to the fate of mountains and mountain ranges.

For a thinker who sought diversity, nonidentity and individuality in being, but without forgetting the unity of the world, the idea of a modification of the unique laws of nature was natural. But in the seventeenth century, in the beginning of the eighteenth century and up until the second quarter of the nineteenth century, the basic tendency of science was the search for unique laws whose effect was the same in the world of atoms, molecules, macroscopic bodies and the cosmos.

His thinking about the difference and the plurality of the laws of nature defined Leibniz's relation to philosophical schools. Both the rigid traditionalism of the scholastics and the radical negation of old science were alien to him. Leibniz was called the genius of compromise, but the word 'compromise' should not be understood too literally. In Leibniz's *Weltanschauung*, the ideas of the pre-Socratics, Plato, Aristotle, the Scholastics, Descartes, Newton, and Spinoza acquired a second life, but Leibniz used their ideas selectively and, most importantly, he radically modified accepted views.

The boundary between Leibniz's philosophy and Wolff's demonstrates a characteristic peculiarity of eighteenth-century philosophy. Earlier, for example, in the Cartesian conception, a philosophical trend could become a philosophical school in the sense that ideas expressed by a teacher would be applied without essential modification to new series of phenomena and would then receive increasingly newer illustration and confirmation. In the eighteenth century, philosophers, considerably earlier than scientific researchers, understood that the dissemination of general principles in new areas in general leads to a modification of these principles. Here the transforming operative function of the formula of inclusion in a set, $a \in A$, was apparent before its relation to the particular laws of natural science was seen. The inclusion of new particular regularities in general philosophical principles modified these principles. In essence, the entire transition from the metaphysics of Descartes, Malebranche and Spinoza, through Leibniz, to the philosophy of the eighteenth century, was stimulated by the inclusion of the dynamic interaction of bodies in a number of areas, which illustrated universal and uniformly behaving general laws and principles. There was still a long way to go to the era when each great scientific discovery transformed the form of a philosophical trend; there were as yet no philosophical trends that numbered independence among their first principles. But there were already sufficient sporadic effects of the 'progress of reason' and the 'deepening of reason within itself' so that rationalism changed, so that it was no longer opposed to experiment. Only experiment could now decide which of the laws, and which modification of the laws, was applicable to a new series of phenomena.

There is another side of the matter which is interesting for the historian of philosophy and science. If the 'progress of reason' is accompanied by the 'deepening of reason within itself', then cognition depends on the depth of reason, on the level of generalizing thought. Now the 'z-axis' is already essential, that is, the level of treatment and exposition of problems; this level affects other axes, it affects the direction and content of philosophical opinions.

4. CRITICAL PHILOSOPHY: THE REJECTION OF RATIONALISM AND THE RETURN TO IT

One hundred years ago, Kuno Fischer originated the now traditional comparison between Kant and Copernicus. As Copernicus pointed out that the picture of the world depended on the existence of man on the Earth and thus revealed the subjective character of geocentrism; so Kant transferred the center of gravity of philosophical analysis to the very perception of the world, to the conditions and possibilities for cognition. This analogy is based on a somewhat simplified evaluation of what Copernicus and Kant had done. The fundamental aspect of Copernicus's theory is not an analysis of our perception of the starry sky from the Earth and the proof that it is an illusion. 'Observers' are not really necessary for the proof of geocentrism and the relativity of the Sun's motion. Copernicus had to show that objective processes on Earth do not prove its absolute rest. As regards the observation of the firmament from the Earth and the picture of the Sun moving about the firmament, Copernicus was presented with a positive task: to deduce this picture from the motion of the Earth, to reveal a series of dynamic processes and not the mechanism of subjective perception. Copernicus destroyed the anthropomorphic representation of the Solar system, but this very destruction was not anthropomorphic. The explanation of the Sun's apparent motion around the Earth is based not on the subjective mechanism of perception, but on the fully objective fact that the Sun really does move in the Earth's coordinate system. But this side of the question was secondary; what is most important is the positive statement of the Earth's movement.

Kant's task was also the substantiation of a positive statement: man comprehends the world in a space-time representation, because space and time precede experience as *a priori* forms of contemplation. If Kant had done what the traditional comparison ascribes to him, i.e., if he had freed the picture of the world from subjective illusions, then he should have obtained a world that was free from space and time. But for Kant, space and time could not be eliminated by subjective illusions, but they could be eliminated by the subjective assumptions of adequate knowledge. In this sense, Kant made the vision of the world subjective — the opposite of what Copernicus had done.

The real link between the ideas of Copernicus and Kant is a far cry from the analogy that usually joins them together; a link exists, but it is an historical one: heliocentrism and the classical principle of relativity raised philosophical questions which all the great philosophers of the modern period have answered in one way or another. Copernicus's ideas touched the basic problems of 'being' (in Democritus's sense, i.e., of the totality of quantified bodies moving in emptiness) and of 'non-being' (empty space).

'Being' means bodies which are sensually comprehensible. The image of bodies is associated with immediate perceptions; the perception of bodies can be called a sensual component of cognition, if such a component were really divisible from the rational component.

'Nonbeing', emptiness, is indirectly linked with sensual comprehension. The idea of empty space is born during the mental elimination of qualitative differences, that is, of places occupied by bodies; during the ideal elimination of separate, differentiated, non-identical objects from the physical world; or during the ideal extension of a body, in part, during the tracking of its possible motion to all sides. The major function of the understanding – the identifying function – allows the extraction of the concept of empty space from experience.

What comes into play next is not identifying common sense, but the original hero of the rationalist comprehension of the world: reason. Reason is convinced that thinking without non-identical objects has no content, that 'non-being' has content only given the presence of 'being', that empty space is the totality of the distances between bodies. The distances themselves turn out to be the tracks of moving bodies or their possible trajectories.

This is already a relativist idea: every distance is distance from body to body, every position is the totality of distances from a given body to a reference body; without them the concepts of position and motion lose their meaning.

Some people try to make emptiness, 'nonbeing', independent of 'being', of bodies. Gassendi says that if God had removed all bodies to another place, space would have continued to exist. But Descartes completely deprives 'nonbeing' and 'being' of a separate existence, space and matter are identified, and moreover, matter receives from 'nonbeing', which has fused with matter, the clarity and certainty of pure *cogito*, which is deprived of content. Thus the problem of cognition is solved: thought and nature are cleansed of qualitative content to the same degree.

This solution to the problem leaves cognition devoid of content (thought cannot single out separate bodies in the world of its objects). Spinoza links extension with other predicates as inalienable attributes of substance. Newton takes a decisive step towards the separation of 'nonbeing' from 'being' and toward the emancipation of 'nonbeing'. In Newton's mechanics, empty space becomes absolute; motion, which is not related to other bodies, is related to empty space and, in the case of acceleration, gives rise to the forces of inertia. The concept of force makes absolute even that which is empty and independent of the events of time. The momentary transmission of force signifies that the momentary state of the Universe has a physical meaning, that one and the same moment occurs everywhere, and that this stream of simultaneous moments forms absolute time. Science returns to the *a priori*, which precedes experience, creates a background for the sensual world, for extension, and returns to immobile, absolute space.

This *a priori* but objective space is the embodiment of a rationalism that is shut off from the sensual world. Philosophy does not accept such a gift from science. Leibniz starts from the concept of force, but he does not agree with Newtonian absolutism nor with the idea of space that is independent of bodies. For Leibniz, force is not found in space; it serves as the substance of a body, a multiple substance of individual being. Leibniz goes further: he is not satisfied with such a role for force; he considers it a non-extended basis of extension, the basis of the very existence of space.

We have briefly repeated an evaluation given earlier in this book in order to bring into sharper relief Kant's place in the evolution of the problem of 'being' and 'nonbeing'. *The Critique of Pure Reason* was an attack on the stronghold of *a priori* objective space and objective time that is dependent on 'being'. Like Leibniz, Kant proceeds from dynamism, from the concept of force, from Newtonian mechanics. The space that figures in the *Critique*, is permeated by bodies that interact instantaneously; this is no longer a system of distances, but a dynamic system, a system of interactions. These interactions determine the positions of bodies, the structure of 'being', which in its turn determines the character of interactions, the structure of 'nonbeing'. But this second problem of Newton's program was not realized either in physics or in philosophy. 'Nonbeing', space, the system of interactions acting through space — all of this did not meet with sympathy. Attempts at the philosophical generalization of a new, dynamic picture of nature included hypothetical schemes for avoiding *a priori*, immobile, empty space — 'nonbeing'. Kant's idea was a radical evasion: space is not a pre-existent, *a priori* background for real events. It is an *a priori* category, but it is not objectively *a priori*; it is subjectively *a priori*. Space (like time) is not an experiential but an *a priori* form of the contemplation of the world.

Kant shows the error of *a priori* objective Newtonian space by a series of antinomies. Reason wished to be freed of its objective partner, the infinite Universe that was dynamically fastened to it by 'nonbeing'. It ascribes to the Universe the quality of limitedness in time and space and gets back an unlimited Universe. Then it ascribes to it the quality of unlimitedness and gets back a Universe that is limited in space and time. Hence the conclusion that 'non-being' — space as such — is not found in the world as a whole. Reason returns from the world, having taken space and time with it as the subjective forms of the comprehension of the world.

Kant and his students considered this shift from the cosmological problem ("what is the Universe like?") to the epistemological problem ("how, by what path, shall we find out what the Universe is like?") to be *critical* philosophy; they opposed it to the *dogmatic* philosophy of Descartes, Spinoza, Leibniz, and, in general, all of the earlier thinkers who aspired to a comprehension of the substance of the world. Sometimes they compared critical philosophy with kinetic and evolutionary opinions in the natural sciences. Nineteenth-century physics turned from the question "what is heat?" to the question "how does heat arise and how does it move in bodies?" Evolutionary biology turned from the question "what is life?" to the question of its origin and development. In reality, neither kinetic physics nor evolutionary biology was phenomenological. They searched for the substance of heat and life and found it not in the traditional, *motionless* and unchanging body giving off heat, not in the traditional and unchanging 'life force', but in motion — thus shifting from the mechanics of molecules to irreversible statistical processes – as well as in the specific interaction of highly organized matter, of organic molecules, cells, organisms, with an external environment. Substance did not disappear from science; motion itself became an inalienable attribute of substance. And in philosophy the problem of being did not disappear, but included the problem of motion.

In ordinary usage, 'critique' and 'criticism' do not at all mean the repudiation of a question; rather they are associated with dissatisfaction with the existing answer, a search for a new answer, with the discovery of contradictions in the old answer. In some sense, the new answer may broaden, concretize or modify the given question; science is characterized by answers that are broader than the questions raised. In general, the critique of a given answer consists in the shift from a certain regularity, correctness, invariance, constancy, to another regularity, another constancy. Here is an example that is classical in many ways: the critique of Aristotelian cosmology. This cosmology is based on the constancy of the position of the center of the earth and the natural places of bodies. When it turned out that such constancy, such identity, does not correspond to reality, that the position of the Earth in one moment is not identical to its position in another, human reason transferred its identifying function elsewhere, to the velocities of bodies at different moments. For a body left undisturbed, the values of velocities at different moments are identical: velocity is constant.

A second example, [really] a continuation of the first, is the shift from Galileo's kinetic picture to Newton's dynamic one. When it turned out that the velocities of planets that revolve around the Sun are non-identical, that the velocities change, that it is not motion as a result of inertia but motion accelerated under the influence of forces that constitutes the basis of cosmic harmony, the identifying activity of reason turned its attention towards acceleration; and so science came to include the concept of uniformly accelerated motion as well as violations of this uniformity, the concept of the identical relation of force and acceleration, and the idea of the constancy of mass.

Logically and historically, the next example illustrates the shift to relativistic mechanics. If the identity of the observed mass of a body is violated, the identity of mass at rest becomes the basis of the world picture. Then this picture becomes suspect, and science searches inside quantum-relativistic theory in order to find constancy, correctness, and conservation in the processes of the transmutation of particles, the losses, gains and changes of rest mass.

Thus the critique of a scientific law in general consists of a statement of some nonidentity that violates the traditional notion and a statement of a new identity, a new law. It is precisely such critical acts that constitute the history of science. And it is precisely this type of criticism that characterizes the evolution of science.

Similar critical acts form those turning points of the curve of cognition, which are generalized by the theory of cognition, and they are often made absolute, are transformed from elements of the curve into rigid straight lines.

Kant's official criticism was just such an absolutized straight line. The real curve of cognition from which such absolutization grew is real criticism, the statement of the inadequacy of some kind of identifying scheme. In the case of space, the target of the real critique, which underlay official criticism and the inclusion of space in the subjective forms of cognition, was absolute space, which was traditionally unchangeable, equal to itself, identical everywhere, independent of physical processes. This absolute space did not correspond to new dynamic notions of space, if we bear in mind the *tendency* of these notions, their future, the prognosis contained in them. This prognosis consisted of a picture of space that was permeated by fields of force, which were no longer self-identical, no longer dependent on physical processes. But Kant did not build such a scheme, in fact he could not have built it. He made space a subjective condition of cognition, and it became subjectively absolute, subjectively independent of physical processes, subjectively *a priori*.

Hegel radically changed this. He made nonidentity, change, and contradiction the basic characteristics of reason itself. Now reason could make far-reaching claims. Hegel's system is dynamic rationalism, which examines not only being, but also being in motion, in spontaneous motion growing out of contradictions. But here we find the illusion of panlogism: the motion of being is considered the motion of reason. Overcoming this illusion should have changed the posing of the question of space. Dependence on objective physical processes leads to changeability and to the *visible*, non-*a priori* quality of space. 'Nonbeing' becomes inseparable from 'being'.

These brief and fragmentary comments concerning the place of Kant's critical philosophy in the evolution of the problem of being and of space will help us see the dependence of his philosophy on the concept of force, dynamism, and the genesis of classical mechanics.

The first stage of the evolution of Kant's ideas was the rejection of rationalism, the lack of belief in reason's ability to comprehend reality in its deductions, and the shift to the positions of empiricism. The second stage was the return to reformed rationalism. After rather lengthy efforts to reconcile Newton's mechanics, dynamism, and the new post-Newtonian style of natural science with the traditional rationalist metaphysics of the seventeenth century, Kant came close to Hume's positions. The second stage in the evolution of Kant's ideas was crucial to philosophy. Space and time became subjective forms, forms of contemplation. After this, reason again acquires the possibility for adequate cognition of the world. The object of cognition is sensually comprehensible objects. They are space-time; in other words, they represent the result of active contemplation, the placing of empirical facts in *a priori* forms, which characterize reflection, in space and time. Kant later elaborated this thought in his *Metaphysical Foundations of Natural Science*, bringing himself to the concept of purely rational science. Science studies the spatio-temporal world; as long as it limits itself to space-time relations, which are *a priori* by nature, it remains a purely rational science. All its deductions can be arrived at by means of mathematics, and rational science remains rational, that is, coinciding with mathematics: as Kant wrote in the *Metaphysical Foundations of Natural Science*, "... a doctrine of nature will contain only so much science proper as there is applied mathematics in it" (Kant 1970, §470, p. 7). The fusion of mathematics and natural science is based on the fact that not only the method and the initial axioms of mathematics, but also the object of science — the world as space and time — turn out to be *a priori*.

But, continues Kant, the world is not only space and time. It contains one definition which we can deduce from *a priori* forms: the *being* of the objects of the world which transforms these objects from geometric figures, possessing temporal duration, into real objects. Now we can say that it transforms the four-dimensional, spatio-temporal world into a material, *filled* world, filled with matter which is not identical with space-time.

Thus Kant arrives at the most important and pervasive problem of the doctrine of being: the filling of the spatio-temporal framework. Let us remember once more that within this framework lies the most critical problem of physics, the problem of filling the macroscopic relativistic framework with ultramicroscopic models.

Kant concludes from the *being* of the world, which is not reducible to *a priori* forms of contemplation, that there must be a metaphysical as well as a mathematical natural science, a metaphysics of nature that operates not with spatio-temporal forms of contemplation but with the categories of reason.

Here Kant has a peculiar inversion. Earlier, contemplation was the gateway to the multicolored world which could not be reduced to space; contemplation seemed to be an area of reason, identifying objects of the world, reducing them to space, making the world geometrical. For Kant, on the other hand, contemplation transforms the world into space, and reason returns to it non-Cartesian being. This inversion is caused by Kant's transcendental aesthetics, the transformation of contemplation into an activity originating from *a priori* forms.

Kant builds his metaphysics of nature. 'Metaphysics' signifies here a shift beyond the borders of contemplation, i.e., a return to rationalism with its claims to the cognition of the objective world. Contemplation ceased being empirical when it included space and time as *a priori* forms. Reason did not become empirical as a result, but it now possesses the prerogative of moving beyond the borders of contemplation, it fills the forms of contemplation with a description of being. What distinguishes the objects of the world from spatio-temporal forms is matter. It is the object of the metaphysics of nature as distinguished from the objects of mathematics.

The first problem of the metaphysics of nature is motion. At first Kant considered motion as a purely spatial process, ignoring the mass of a moving body. This is Kant's 'phoronomy' [kinematics]. Later he shifts to the concept of motion as a material and not a purely spatial process). This is *dynamics*. Here matter is considered as the cause of motion, i.e., the concept of force.

When he turns to dynamics in the *Metaphysical Foundations*, Kant differentiates between matter and space. Force is a new, essential definition of matter which separates it from space and, consequently, from objective forms of contemplation.

Comparing classical dynamism and the classical concept of force with the contemporary, nonclassical representation of a field of force, we see a profound difference in the understanding of the link between force and space. In classical science and in classical philosophy, force served to distinguish matter and space. In Kant's philosophy, this function of force led to the exclusion of matter from the sphere of subjective contemplation and to the objectivization of this concept. In nonclassical science, force does not so much fill space as transform its properties. The general theory of relativity affirms that the geometry of space, in which the gravitational field is essential, is non-Euclidean geometry. As far as other fields are concerned, contemporary science has not yet gone that far, but the conception of least action has long been depicted as a conception of curved space. Hence arises the historico-philosophical and historico-scientific thesis: contemporary science discredits not only the subjective-*a priori* understanding of force, but also the subjective-*a priori* understanding of *space* and *time*.

In classical science, physical processes are considered processes *in space*, which Weyl said resembled a 'hired barracks'. There is now a tendency to consider physical processes as changes of space itself. The non-*a priori* quality of space becomes not only a philosophical judgment but also a physical, experimentally confirmed statement.

From an historical perspective, what is the significance of Kant's *volte-face* from rationalism in the spirit of Descartes and Leibniz to empiricism in the spirit of Hume and then back again to rationalism — to the activity of reason, to the comprehension of nature as distinguished from subjective forms of contemplation?

Let us first analyze this question not within the framework of the history of philosophy as a whole, which investigates both the regularities of the twists and turns of cognition as well as their arbitrary absolutization, but rather within the framework of the history of science, which is particularly interested in these regular convolutions.

The starting point of this evolution is rationalism which identifies being, and, as a result, acquires fixed abstract categories. We saw that such rationalism, strictly speaking, never existed, that it always included a certain individualizing tendency. But the identifying function of reason was dominant.

At the end of the seventeenth century, science transcended this function. It introduced into mechanics the concept of acceleration, that is, distinctions in the behavior and state of bodies, the nonidentity of states; and it considered force as the cause of this nonidentity. Leibniz, and later also Kant, differentiated this individualizing tendency of reason from its identifying tendency. The individualizing tendency brought Leibniz to the concept of force as a substance that individualizes the monad. For Kant, the identifying tendency of reason and its fruits — space and time — led to contemplation, and reason ('the metaphysics of nature') was presented with the individualizing function.

As a whole, the leap from traditional rationalism across empiricism to a new understanding of the function of reason was a philosophical interpretation of the concept of force: it violates the rationalism of *identity* and leads to the rationalism of *nonidentity*. Jumping ahead we should note that the next shift, which took place in the nineteenth century and was linked with Hegel, was the shift to the *rationalism of contradiction*.

What is the twist in the curve of cognition, which, under absolutization, leads to a departure from the basic direction of science, to non-extended substances — to the monads of Leibniz and the transcendental idealism of Kant? Let us qualify this question; as earlier, this is not a question of all the twists and turns of science, nor does it refer to all of the historical and logical sources of monadology and of transcendental idealism. To return to the effect of *dynamism* on the evolution of rationalist thought: In this framework monadology and transcendental idealism were connected with the Newtonian phenomenological treatment of force. Newton's 'left hand', the pathos of his phenomenological *hypotheses non fingo*, his inductivism, had a definite positive meaning: the single-valued theory of motion and the single-valued theory of gravity can be developed without kinetic models. Force, as a phenomenological pseudonym, separated the sphere of single-valued science from the sphere of hypothetical constructions. It separated eighteenth-century science from seventeenth-century metaphysics, the new era from the old. This was indispensable.

But the phenomenological treatment of force divided the first part of Newton's program ("to find the positions of bodies according to force") from the second part ("to find forces according to the positions of bodies"). It divided the eighteenth century not only from the seventeenth but also from the nineteenth, from the field theory of force which united the concept of the field with the concept of space. When phenomenological force cut the eighteenth century off from the past, it turned out to be an accelerating force of scientific progress; when it divided this century from the future, it acted as a brake. The brake was necessary; it preserved science from the fantastic models of late Cartesianism, but the absolutization of the brake, the notion that the brake is the train's moving force, distorted the representation of its motion.

It is characteristic of classical science to be able to solve the problems of mechanics in linear approximation without taking into account the changes of the field, assuming that it is independent. This is a twist of cognition whose absolutization leads to the representation of principal independence, primacy, the substantiation of force. Philosophy went further along the tangent to the given curve of cognition, and arrived at the nonextended dynamic substance of Leibniz, or translated extension itself into the subjective sphere.

5. DYNAMISM AND THE IDEA OF A NATURAL ORDER

The philosophical generalization of dynamism inevitably linked it with social thought and, in the last analysis, with social struggle. Philosophical generalization is not reduced to the inclusion of the concepts of natural science into quasi-*a priori* categories that are independent of such inclusion. It is not reduced to natural philosophy in this traditional sense. It can be non-linear and dynamic: a particular natural scientific conception, when included in a more general philosophical category, changes its meaning, acquires a higher concreteness and forces the general category including it to change, to become enriched and concrete. Such a generalization of natural scientific categories may be invisible and may occur without the intervention of philosophers, through a direct shift to social ideology. But, in essence, this shift always turns out to be a philosophical generalization and, in the final analysis, takes a visible form.

Dynamism gave rationalism a local and experimental character: the objective *ratio* of the world became local; its comprehension became experimental. It gave an active and local character to the social *ratio*: the fate of the individual became the criterion for social harmony (or so, at least, the promoters of rationalist social ideas imagined this criterion), and the achievement of social harmony became a question of action.

A clear line from dynamism to social thought runs through the physiocrats and English classical political economy. Here the link is not only logical, but also truly historical: Adam Smith drew arguments in defense of economic liberalism from both Newton's *Principia* and the arsenal of the concepts of classical mechanics. We shall begin with the logical links.

Newton's dynamism opposed itself to the peripatetic, purely integral harmony of the world; in this sense it was a continuation and strengthening of the kinetic conception of the world. Newton focussed attention on the differential laws of motion (on laws which soon were given a differential form and which, essentially, were already differential for Newton), which solved one problem of the general program of the *Principia*, namely the question of the local behavior of bodies in fields of force, of their acceleration at a given moment with a given force. The second part of the program — the definition of force — forced him to return to the integral scheme, but the derivation of local conditions (the values of the tension of a field of force at each point and each moment) was a task that the eighteenth century handed down to the nineteenth along with several premises already worked out concerning the theory of a potential field. Therefore the dynamic picture of the world turned out to be somewhat ragged, and the influence of the integral *ratio* of the world on local events was hidden.

The focus of eighteenth-century social thought shifted to local events and to individual destinies and acts. Statism, the criterion of the general good, *raison d'état* as the basic criterion, Machiavelli's conceptions — all receded into the past. And conceptions that were medieval in character and close to peripateticism, in which purely integral universalia justified the power of the church, faded even further. Now concrete, local events connected with activity and with interests of the individual, with personal interests and stimuli, came to the fore. Individual stimuli play a role in society analogous to that of impulses received by separate bodies. And just as local impulses guarantee universal harmony in the picture of the universe, so the optimal integral result is ascribed in social theory to the sum of individual stimuli and individual interests.

This motif, often with direct references to the harmony of nature, is already found in the works of the physiocrats. Let us remember, by the way, that the term 'physiocracy' is found in Kant's *Critique of Pure Reason*, in the exposition of the third antinomy — of necessity and freedom. In this antinomy, transcendental freedom plays the role of the thesis (the violation of universal causality), and the role of the antithesis (the transcendental omnipotence of natural causal laws) is played by physiocracy.

The idea of a *natural order*, without macroscopic integral acts, which do not enter into the causal chain, expressed, in essence, a dynamic tendency. Separate forces (they were not yet linked in some type of integral field of force) act on bodies, and this somewhat incomplete scheme explains universal harmony. Its incompleteness consists in the fact that here there is no interaction of integral conditions and individual actions, there is no *field* and its extra-physical analogies.

Interaction is one-sided: there are forces with no visible form, but this integral result does not act on forces, the non-linear link is not taken into account.

In this incomplete form the idea of natural order and the concept of force and impulse were brought into economic science.

Adam Smith's starting point was dynamic production, which is governed not by tradition but by certain developing methods of technology linked with a manufacturing division of labor. These new methods shift from one enterprise to another and bring about a fluctuation in the productivity of labor and in the volume of production in different fields. A mechanism exists which brings the distribution of labor into balance, into correspondence with demands. This mechanism has a natural character, that is, it is made up of individual wills, governed by individual impulses — the strivings of men to increase income. Individual impulses assure the balance of production in the same way that impulses acting on bodies form a balanced Universe.

"According to Adam Smith", says Brunschvicg, ". . . the economic world consists, in an orderly and peaceful fashion, of the simple play of necessary relations, in the same way that in Newtonianism the order and harmony of the physical world result from laws governing the forces of inertia and gravity".[5]

There is a general postulate for classical mechanics and for classical political economy: the free play of dynamic laws guarantees harmony, order, the real being of the world. Let us emphasize that these are *dynamic* laws. In classical science, a static scheme of natural places no longer determines the harmony of the universe. In classical economic theory, it is no longer a scheme that is peripatetic in the spirit of 'natural' positions, profits, and so on, but rather dynamic effects, local events, the motion of goods, the changes of the mass of goods that lead production to harmony and equilibrium.

Thus the concepts of equilibrium, force, and impulse, transferred to political economy by Adam Smith, kept the features that they had in classical mechanics. First of all, these concepts in both classical mechanics as well as in political economy originated from local processes that occur here and now. When economic dynamics leads to broadening production in one branch, the local, separate, individual acts of individual wills bring production into equilibrium and, in general, into a dynamic equilibrium. Impulses in Adam Smith's economic theory are the impulses of forces, which are applied to individuals in the same way that the forces of Newton's mechanics are applied to bodies. For Newton the initial concept of physics is not the peripatetic integral scheme, but forces acting here and now on a given body. For Adam Smith the initial concept is the effect of general laws on individual wills. Only by acting on individual wills do economic laws become laws of being; this is the basis of economic liberalism. The struggle of stimuli and the choice of behavior are the economic analogues of the conception of virtual displacement.

It must be stated that the dynamism of Adam Smith's scheme does not encompass production as a whole. The regulator of dynamic equilibrium guarantees necessary shifts in the structure of production as a whole, just as in Newton's system it is not the general evolution of the solar system or even of the cosmos, but the repetition of these cycles that constitutes the result of action.

Smith's dynamism is also related to the dynamism of *Das Kapital* in the same way that the dynamism of the Kantian nebular hypothesis is related to the dynamism of contemporary cosmology. Kant searched for a mechanism that would allow a shift from elementary molecular forces to some type of cosmic structure. Initial cosmic forces, events encompassing astronomic scales, did not figure in his hypothesis. The equilibrium that has been acquired is stable; one and the same cycles are repeated for an indefinitely long time; the mechanism for further cosmic transformations is not indicated.

In an analogous way, Adam Smith seeks a mechanism that would allow a shift from the 'molecular' forces governing the behavior of individuals to a macroscopic balanced structure of production. The origin of this structure, the manufacturing division of labor, and technical progress all result from the summation of individual wills and from the impulses dependent on them. The 'cosmogony' of production, the macroscopic forces which bring about general changes in individual interests, remain in the background, as does the prognosis of further transformations of production as a whole. Having reached equilibrium, production can continue the established cycles for an indefinitely long period. The symbol of fundamental and total transformation, the ancient skeleton, despite the old tradition, does not appear at the feast of social harmony.

Such an analogy corresponds to real links. Adam Smith used the idea of natural order in all his works on political economy, philosophy, morality, and law. But this idea and its natural-scientific roots are clearly visible in Smith's works on classical mechanics, astronomy, and their history.[6] The intervention of the classical concepts of force, impulse and the equilibrium of forces was obvious.

Both the social harmony growing out of quasi-autonomous individual wills and the extra-historical character of such harmony proved to be illusions. Once the illusoriness of social harmony was understood, the disappointment in Smith's economic liberalism became the starting point for a new phase in the evolution of rationalism. But we shall discuss this later.

A comparatively direct and clear line of logical and historical connection links the basis of Newton's mechanics with French materialism. This materialism was by no means a thorough-going sensual one. A rationalist streak runs through the conception of the French materialists; this is a well-known, historico-philosophical evaluation and does not need to be justified and developed here. Let us just note that without the rationalist tendency, French materialism would not have become an encyclopedic doctrine nor a fighting, active doctrine; it would not have become one of the main channels of the ideological background for the French Revolution. The expression 'encyclopedic doctrine' here has a double meaning. Besides the usual one, it contains a suggestion of the *Encyclopédie* of Diderot and d'Alembert. This publication was based on a certain rationalist conception: the sum total of human knowledge represents a unified system which can be constructed and expounded, starting with the most general laws of motion. The latter determine the behavior of separate particles of matter, and the harmony of the world grows out of these purely local processes — a unified system of the world agreeing in all its parts and details.

Science comprehends and explains this system, which occupied the minds of the Encyclopedists, particularly that of Diderot, who wanted to survey the sum total of knowledge, social relations and political institutions from the position of reason and condemn to oblivion all that did not answer to reason, that did not grow out of the natural causal harmony of being.

In the *Encyclopédie*, reason sits in judgment over being. This trial was in session in the philosophic works of the eighteenth-century French materialists and then became a real trial: the French Revolution annihilated all that did not correspond to the criteria of reason. This was the first purely practical experience of the division of being into that which possesses reason (and is therefore real) and that which does not (and is therefore non-real). When owl of Minerva began its flight, when the Germans began to think about what the French had done, Hegel's famous thesis, which had caused so much puzzlement, turned out to be the algebra of revolution, owing not only to its possible outcomes, but also to its beginnings. The revolution to which Herzen refers in his evaluation of the phrase "all that is real is rational" is not only the coming revolution (embodying dialectical philosophy) but also the revolution which gave it impetus.

But the French Revolution itself received an impetus from philosophy, in part from French materialism. Its rationalist Cartesian component (interwoven with the sensualist one that had come from England) included the following idea. Local processes, in this case, social ones, the human analogues of Newtonian dynamics (actions of individual people) are fused in social harmony. Whatever comes from elsewhere, whatever does not flow from the natural impulses which govern human deeds, violates the natural harmony, contradicts it, is at odds with the *ratio* of the world and must be destroyed. We are already convinced that the new form of rationalism that is characteristic of the seventeenth—eighteenth centuries – distinctive local dynamism – possessed a great penetrative force in the shift from classical mechanics to philosophy and social thought.

The recognition of such a penetrating force was itself part of the ideological preparation for the revolution. The isolation of philosophy from the scientific revolution of the seventeenth—eighteenth centuries (this was a revolution in that the most fundamental principles changed) signified for philosophy the defense of *a priorism*, the defense of its independence from experiential knowledge. For social thought, isolation from the 'line of Newton' signified the defense of providentialism. In all cases, such isolation signified the defense of theism against deism, the defense of a pre-established harmony *prescribed to nature from on high* against the idea of *immanent harmony flowing from local dynamic impulses.*

Thus Newton's *Mathematical Principles of Natural Philosophy*, despite its ending in defense of theism against deism, in defense of god-the-all-preserving against god-nature, became the basis for the widespread diffusion of deism, the departure from theism, a departure which did not stop with deism and led to atheism.

In order to bring these logical links somewhat closer to historical ones, let us look at the work of Voltaire, the author of *Elémens de la philosophie de Newton* (1738), a defender of deism, a participant in the *Encyclopédie.*

His work itself was encyclopedic, and, as was the case with the Encyclopedists, the content of articles and ideas of the first volumes differed from those of the last. Nevertheless, one tendency pervades all of Voltaire's phases, all the genres, and even all the contradictory judgments.

His enthusiasm for Henry IV is connected with features of the new period [found] in the psychology and actions of the hero of *La Henriade* [1728]. Henry very decisively violated the traditional, providential, *a priori* bans in the name of local necessity ('Paris is worth a mass') and thought that this activity, dictated by this necessity and by these logical impulses, would lead him if not to full social harmony, then at least to a chicken in every French peasant's pot.

His earlier enthusiasm for Shakespeare was linked to the reflection in Shakespeare's tragedies and histories of the power of local impulses. In *Richard III*, the scene around the corpse of the slain king and the effect produced by the murderer's monologue break all 'macroscopic', *a priori*, divine, and human norms in the name of local effect, and moreover, of rational effects, with simple reaction. What has already been said about Valmont [of *Les liaisons dangereuses*; see above, p. 222] could also be said about Richard III (although Valmont is an intriguer and villain of a lower order by far); he would not hesitate to react in the same way in response to an action or an argument. When Voltaire changed his attitude toward Shakespeare's work in his old age, his original impression did not waver.

Among Voltaire's ideas, which have so profoundly entered into humanity's consciousness that they seem anonymous and seem to have existed always, let us note the content of the *Treatise on Toleration* (*Traité sur la tolérance*). In actuality this was a discovery that puts the year 1763, the year the *Treatise* was published, on a par with the dates of the most important scientific discoveries. The idea of toleration was an outcome of rationalism and its new forms. For Voltaire, intolerance is the use of other than rational arguments, the use of other than logical deductions and generalized empirical statements in a conflict of opinions. It is the use of authority and of authoritarian power, of tradition — of all that falls outside of the framework of reason, which is fused with science. In other words, all this is 'non-rational'. It must be destroyed. Intolerance is personified by the church — the object of the famous '*écrasez l'infâme*'. Thus Voltaire's intolerance was not abstract. It signified the pluralism which was characteristic of the new representation of harmony. This pluralism did not put static and *a priori* limits on local impulses; it is formed from them, and the framework of universal harmony leaves *a priori* and authoritarian coercion outside its borders.

A similar idea permeates Voltaire's struggle against *Panglossism*. The name of the ill-starred character of *Candide* is probably an apt one for the *a priori optimism* he personified: "All is for the best in the best of all possible worlds" (*Dans ce meilleur des mondes possibles . . . tout est au mieux.*) Wolff's teleological optimism simplified Leibniz's pre-established harmony beyond recognition. Voltaire's optimism is not *a priori* harmony, nor is it *a priori* providentialism. His is a belief in the compatibility of individual freedom and order, a natural order moreover, the only real one: it is an order that grows out of freedom. This very freedom is the dependence of local events on local impulses.

By comparison with Voltaire, Rousseau is very far from Newton, from rationalism fused with science, from the visible natural-scientific roots of eighteenth-century social thought. But the invisible roots remain essential. Rousseau and Voltaire differ in their interests, ideas, style, their very lives — and above all, they differ in the chaotic nature of these things: sentimentalism versus reason, the apology for pre-cultural existence versus the encyclopedic synthesis of all culture and science of his time, suspicion and solitude versus an active social life. And, to take something completely personal, Thérèse Levasseur [Rousseau's widow] versus Newton's interpreter, the Marquise du Châtelet, symbolizes the difference in the two thinkers' immediate entourage.

But for Rousseau, inner contradiction is characteristic. In his famous characterization, Chuquet speaks of a man "seeking solitude and wanting to be known throughout the whole world, calling forth his century to battle and flattering it, dreaming only of peasant houses and living in palaces, involved with a servant girl and falling in love with society ladies . . ."[7] Next to each 'anti-Voltairean' characteristic is a 'Voltairean' one, bringing the ideas, moods, life and entourage of Rousseau closer to those of Voltaire.

We are interested here in the 'anti-Voltairean/Voltairean' antithesis that is most closely connected with the evolution of rationalism. This is the antithesis of reason and feeling. Are they so separate? Eighteenth-century rationalism could not enter deeply and intimately into the culture of the century without a certain emotional component, without a certain shift from reason to feeling. And Newton, science, and dynamism participated in this shift.

Bodies of nature, which compose nature, could not create it as a whole without the interactions of forces, without dynamic predicates, without, above all, the forces of attraction and repulsion. People cannot form society without the dynamic component of man's inner world; they must have not only reason, but also will and feeling. Pure thought is analogous to motion. Thought goes from one representation to another, from one judgment to another, identifying subjects and individualizing their predicates, and, at the same time, identifying predicates and individualizing subjects. For Descartes thought was a certain [i.e., reliable] analogue to motion, which became equal to thought in its certainty: motion is the shift from one position to another, and the indicated positions become the individualized, self-identical predicates of a moving object. Man as a thinking person, *homo cartesianis*, is not social man. This is an epistemological category. Real man possesses not only inner impressions but also inner impulses, and the receptor for these impulses are human feelings. Foremost among these, in Rousseau's opinion, is compassion. It makes man an active creature, the analogue to the dynamic particle which figures in the conceptions of Leibniz and Newton.

This is by no means the thrust of Rousseau's arguments nor that of his thought. The thought, argumentation and style of a philosopher change, depending on the material on which he bases his constructions. Rousseau speaks of feelings: they serve as the object of his thoughts, but here thought itself becomes different: feelings permeate it, the philosopher summons up feelings, strives to arouse them, and sentimentalism enters into philosophy. The purely rational base of sentimentalism cannot be obvious.

Therefore the dynamic statement of the link between dynamism and the philosophy of Rousseau is not the reconstruction of his thought. In general this statement is not biographical but, strictly speaking, historical. In order to find the equivalents of Newtonian mechanics with its particles, which were already not only moving but also attracting, pushing and resisting forces in human society, it was necessary to go from Descartes's man with his *cogito* as the only predicate of the soul to feeling and compassionate man.

Such a shift was hostile to theism. Dynamism makes God as a cosmological authority unnecessary. The presence of a natural, inborn compassion in man makes god unnecessary as a moral authority. Cosmic harmony is established as the sum and combination of local impulses; moral harmony turns out to be the result of the integration of individual human acts of compassion.

NOTES

[1] "Je demeure d'accord que naturellement tout corps est étendu, et qu'il n'y a point d'étendue sans corps. Il ne faut pas néanmoins confondre les notions du lieu, de l'espace, ou de l'étendue toute pure, avec la notion de la substance, qui outre l'étendue, referme la résistance, c'est-à-dire, l'action et la passion" ('Lettre sur la question, si l'essence du corps consiste dans l'étendue', *Journal des savans*, 18 juin 1691, reprinted in Leibniz 1840, p. 113).

[2] "S'il en estoit de même de l'unité des Arithmeticiens, on pourroit se passer des fractions, ce qui accommoderoit fort les écoliers" (letter, April 22, 1702, from Leibniz to Queen Sophie-Charlotte [1668–1705, wife of Frederick I, king of Prussia] in Leibniz 1877, **10**, p. 145).

[3] "Die Welt des Spinoza ist ein achromatisches Glas der Gottheit, ein Medium, durch das wir nichts erblicken als das ungefärbte Himmelslicht der einen Substanz, die Welt des Leibniz ein vieleckiger Kristall, ein Brillant, der durch sein eigentümliches Wesen das einfache Licht der Substanz in einen unendlich mannigfaltigen Farbenreichtum vervielfältiget und verdunkelt" (Feuerbach 1969, p. 40).

[4] Lenin 1961 (*E*), p. 110; 1963 (*R*), p. 99.

[5] "Suivant Adam Smith . . . l'univers économique se constitue en ordre et en paix, par le simple jeu de relations nécessaires, comme dans le newtonianisme l'ordre et l'harmonie du monde physique résultent des lois qui régissent des forces d'inertie et de gravitation" (Brunschwicg 1949, p. 28).

[6] These works are analyzed in 'A propos de quelques travaux d'Adam Smith sur l'histoire et la philosophie des sciences' (Moscovici 1956). Here the significance of Smith's historico-scientific interests for his economic ideas is pointed out.

[7] ". . . combattant et flattant son siècle, . . . recherchant la solitude et voulant être connu du monde entier, . . . ne rêvant que de chaumières et habitant des châteaux, s'acoquinant à une servante d'auberge et n'aimant que des femme de haute volée . . ." (Chuquet 1893, p. 199).

PART II

9. HETEROGENEOUS BEING

1. RATIONALISM AND IRRATIONALISM IN DIALECTICAL PHILOSOPHY

The seventeenth century is the age of classical rationalism. Within the limits of Descartes's physics, rationalism was expounded in its most complete and decisive form. Later, with Spinoza, rationalism went beyond the explanation of the behavior of bodies and embarked upon the explanation of their being without reference to absolutes that could not be comprehended by reason. It merged with science and became a doctrine not only of rational cognition, but also of the objective *ratio* of the world. In the eighteenth century, rationalism was tied to science still more tightly by a great number of diverse threads. It triumphed in differentiated branches of science, despite the fact that the degree of differentiation became so great that it was not always possible to discover a single rationalizing tendency in the various branches and in the discoveries and generalizations made in them. Nonetheless such a single rationalist tendency did exist. On the whole this amounted to the transition from the search for the static *ratio* of the world to the search for its dynamic *ratio*. But such a change of direction had to bring about a change in the methods of the rational comprehension of the world. To begin with, abstract analysis could not exclude nonidentity and variety in nature, it could not impoverish and bleach nature, but was obliged to make the image of nature more multicolored and concrete. Such a modification of the scientific method became explicit in the nineteenth century, but even in the eighteenth century it was possible to observe in science the beginnings of a new, dynamic and heterogeneous ideal in scientific explanation.

In philosophy itself, rationalism became the target of criticism. English empiricism, French sensualism and German critical philosophy attacked rationalism from different sides. After each attack the similarity of the warring camps became clear; the empiricists, the sensualists and the advocates of transcendental idealism revealed that they were imbued with a rationalist spirit and could not separate themselves from it.

But even rationalism itself emerged from these battles transformed, defeat being more effective here than victory. The criticism of rationalism led to the inclusion within its framework of new ideas connected with the dynamism of post-Newtonian science. The real historical limitation of seventeenth- and eighteenth-century rationalism lay in its inability to explain motion as an inalienable component of being. Rationalist thought could not detach itself from the empirical comprehension of the world; the construction of general identifying concepts could not take place without the simultaneous individualization of the objects of cognition, the appearance of nonidentity, the irreducibility of phenomena to more general definitions, the demonstration of heterogeneity in the world. But early in the nineteenth century, when the philosophical understanding of the scientific revolution that produced the victory of classical science was roughly complete, it became clear that the results of this revolution demanded a more profound transformation of rationalism. Rationalism had to account for more than motion, heterogeneity, the nonidentity of the objects of cognition. It had itself to become mobile and heterogeneous in its fundamental categories.

It is sometimes maintained that the relation of Hegel to rationalism consists of the inclusion of *irrationalism* in its foundations. Kroner wrote that Hegel's dialectic may be considered as much an embodiment of irrationalism as rationalism (Kroner 1921—1924, **2**, p. 271). Hyppolite expresses similar ideas. He sees the connection between rationalism and the purely logical analysis of ideas and the necessity for some other sources of cognition in order for the ideas to have content. "If there were only ideas", Hyppolite reminds us, "there would not be any ideas" (see p. 177, n. 7). But it is no longer a question of the simple need for empirical data; it is a question of the inclusion of new methods of cognition, of the reform of these methods. Therefore Hyppolite considers Hegel's achievement to be the inclusion of irrationalism, not empiricism, in the foundations of epistemology. "Here we touch upon perhaps the decisive point of Hegelianism," he continues, ". . . which makes Hegel at once the greatest irrationalist and the greatest rationalist who ever lived."[1]

Jean Wahl takes a somewhat different, yet similar position. Irrationalism is included in the activity of reason. Speaking of the Hegelian conception of reason, Wahl says:

Given that *Vernunft* is not ordinary reason, one could ask whether the philosopher of

Vernunft is a rationalist or an irrationalist. However, does not one part of Hegel's attraction lie in this sort of wave of the philosophical wand with which he introduced the irrational into reason and satisfied the need for both the irrational and the rational, the one dissolving into the other, by means of the philosopher's same magic? [2]

In the first part of this book, I expressed a somewhat different view. Rationalism does not give way to nor does it unite with irrationalism. Rationalism was always inseparable from the statement of motion, change, nonidentity; without this, rationalism would have no content. Beyond rationalism, in its turn, there is no cognition nor could there be any; rationalism develops side by side with science, side by side with being. It seeks the objective *ratio* of being, and the Hegelian reform of rationalism is the continuation of this evolution.

Such an evaluation of the reform does not negate its depth. Rationalism never turned away from contradictions. Before Hegel, attention was focussed on the contradictions of the Logos; for Hegel rationalism was the doctrine of the contradictions of the Logos and of the Cosmos as the embodiment of Logos; and after Hegel, it became the doctrine of the contradictions of the Cosmos and of the Logos as a reflection of the Cosmos. As early as Spinoza, rationalism had become a doctrine of being, of *all* being.

In the nineteenth century, rationalism became the doctrine of contradictory being — of heterogeneous, multiform, moving being.

2. 'BEING' AND 'BEGINNING'

But is not such heterogeneous and contradictory being already the first category of the analysis of being, the beginning of such analysis — Hegelian *pure being*?

"Mere being . . . is mere abstraction . . . ," says Hegel (Hegel 1975, §87, p. 127). Generally speaking, abstraction in Hegel's philosophy is the result of the analysis of being, an analysis which identifies the objects of cognition and simultaneously allows for their individualization; it approaches being as a rich accumulation to a growing number of determinations. But *pure* being appears as a single abstraction, as the result of the liberation of being from determinations as the ultimate liberation that transforms being into nothing. Hegel's pure being is not opposed to a being rich in determinations. For Hegel the wealth of determinations is the result of the evolution of 'pure being', its identification with the concept of 'nothing', the appearance of a number of other concepts including 'existence' and 'reality'. 'Pure being' is the *beginning* of analysis.

The concept of beginning in Hegel's philosophy means that specific concept which is the starting-point of this philosophy as well as of its criticism and historical evaluation. Hegel differentiates an objective beginning, i.e., the *beginning of things*, their substance, from beginning in an epistemological sense, the beginning of cognition itself. They must, says Hegel, coincide in the sense that "what is the first for thought ought also to be the first in the *process* of thinking" (Hegel 1969, pp. 67—68).

Here, in essence, is already the statement of the fundamental question which divided dialectical philosophy after Hegel. The objective principle, the objective beginning of being, what serves as the content of thought, toward which thought proceeds and is directed, must be the beginning of the *process* of thinking, i.e., its fundamental concept, which can be seen in two ways. It may appear as something *a priori*, independent, not mediated by some influences and reflections of the objective *ratio*, the objective principle of nature, the objective content of science. Then what is fundamental in cognition appears to be thought, free of content, pure idea.

The second view ascribes the fundamental role in cognition to something already having content, already mediated. Thus, in essence, the question of an absolute beginning is nullified; the beginning itself is a result; it already includes the results of prior cognition. In other words, there is nothing in cognition which could be mediated in the process of cognition itself. Then the process of cognition loses its independent character, it depends upon its object; thinking turns out to be a reflection of nature.

In analyzing the process of cognition, Hegel ascribed to each new concept the character of content and also some dependence on prior concepts, that is, to each, excluding the first, he ascribed the absolute beginning of the process of cognition, "the first in the *process* of thinking." Hegel says that "there is nothing, nothing in heaven or in nature or mind or anywhere else which does not equally contain both immediacy and mediation . . ." (Hegel 1969, p. 68). The logical beginning also possesses both immediacy and mediation. The mediation of logical analysis consists in what it represents as an abstraction of the entire beginning deduced from this, the subsequent knowledge. But this mediation cancels itself out, is transformed into immediacy if we ignore all this subsequent knowledge and take this beginning as only a beginning, if we cut off the sources of the latter. Then we are left with thinking as such, without any of its previously mediated links. We approach absolute beginning, complete immediacy, without any suppositions on which to base this beginning.

In the introduction to the first book of the *Science of Logic*, Hegel demonstrates how the logical beginning loses its mediation and content and is transformed into a contentless abstraction of pure being indistinguishable from pure nothing:

> Here the beginning is made with being which is represented as having come to be through mediation, a mediation which is also a sublating of itself; and there is presupposed pure knowing as the outcome of finite knowing, of consciousness. But if no presupposition is to be made and the beginning itself is taken *immediately*, then its only determination is that it is to be the beginning of logic, of thought as such. All that is present is simply the resolve, which can also be regarded as arbitrary, that we propose to consider thought as such. Thus the beginning must be an *absolute*, or what is synonymous here, an *abstract* beginning; and so it *may not presuppose anything*, must not be mediated by anything nor have a ground; rather it is to be itself the ground of the entire science. Consequently, it must be purely and simply *an* immediacy, or rather merely *immediacy* itself. Just as it cannot possess any determination relatively to anything else, so too it cannot contain within itself any determination, any content; for any such would be a distinguishing and an interrelationship of distinct moments, and consequently a mediation. The beginning therefore is *pure being* (Hegel 1969, pp. 69–70).

For Hegel the path toward an increasing number of concrete determinations of the world always turns out to be a return to previous determinations and concepts. They thus obtain a foundation, mediation, development. In this sense even the beginning of philosophical analysis is grounded and mediated in all the subsequent processes of thought. But this mediation differs from non-fundamental determinations. In the case of the absolute beginning, the return to it, its mediation, destroys itself; it does not add new determinations to the beginning, it does not make it concrete and does not fill it with content. It consists of the ultimate impoverishment of the beginning and leads to the blunt statement: the thinking process has begun. But this statement says nothing of the content of the thinking process once it has begun, and in essence says nothing. Pure being is therefore the same as pure *nothing*.

The cycle of cognition returns it to the initial determination which turns out to be not only the basis of the subsequent process of thought, but also its result. But in the case of the absolute beginning, the result is zero-order: we have returned to the beginning as to something completely indeterminate and without content — to *pure* being.

Hegel writes:

> Being, pure being . . . is pure indeterminateness and emptiness. There is *nothing* to be intuited in it, if one can speak here of intuiting; or, it is only this pure intuiting itself. Just as little is anything to be thought in it, or it is equally only this empty thinking. Being, the indeterminate immediate, is in fact *nothing*, and neither more nor less than *nothing* (Hegel 1969, p. 82).

Thus pure being severs the head of the snake biting its own tail. The beginning has become an *a priori* statement. The *a priori* here is zero-order; what is *a priori* is the contentless concept; it is without content because it is *a priori*. Therefore pure being cannot develop and be transferred into other concepts spontaneously; it demands an encounter with the 'nothing' equal to it, and from the transition of one of these concepts without content to another there arise concepts with content — becoming, determinate being [being-then-and-there], being-for-itself, essence, existence, reality, etc.

Thus the course of philosophical analysis as a whole and, on the other hand, the beginning of philosophical analysis — the concept of pure being — demonstrate to us two types of thought processes. One type makes the initial determinations concrete and increases their content. Thus thought moves, returning to *relative* beginnings. The second type of thought process returns to the initial determinations and thus deprives them of their content, their inner contradictions, their heterogeneity. The second path leads to the concept of pure being.

Later I shall attempt to show that these two paths correspond to the two tendencies in the generalization of classical science. Such correspondence will be the theme of one of the following chapters. Here it is necessary to emphasize the principle underlying this transition from the first path to the second, which led Hegel to the concept of pure being.

It is the path to the *a priori* version of being. It is the path to the idealistic system which came to light as a result of the dialectical method. Hegel's method is the ascent[3] to abstraction, which is, moreover, the ascent to the greatest concreteness, and includes abstractions devoid of inner contradictions, inner content, inner heterogeneity as relative and approximate representations. These abstractions block subsequent analysis, the demonstration of their spontaneous development and the search for their mediation. They are lawful, they serve as necessary elements of dialectical analysis, as long as they are not made absolute, are not frozen, as long as noncontradiction, spontaneity and, consequently, lack of content are not ascribed to them in principle.

For Hegel the concept of pure being is mediated. But *how* is it mediated? Its mediation is reached by an essentially simple pre-Hegelian path, a path of abstraction that eliminates the concrete, the heterogeneous, the nonidentical. Such mediation, as even Hegel says, exhausts, destroys itself and its ultimate result is nothing, devoid of content, empty, non-existent 'being', equivalent to 'nothing'.

This 'being-nothing' exists only as a thought of being without an object of thought, a thought that excludes its object. And it is precisely this thought that turns out to be the beginning of dialectical analysis. Thought begins its reign at Khodynka: It destroys its subjects.[4] The entire process of subsequent development bears the imprint of the beginning, which, in principle, lacks content — the result of a simple, pre-Hegelian abstraction which does not make a thing concrete, it *only* makes it abstract. Nonbeing enters philosophy as an initial concept. But this is the nonliving, contradictory 'nonbeing' of ancient atomism, which is indivisible from being. It is nonbeing that negates matter, negates its own sensualistic mediations and connections; it is an apparition claiming the role of a demiurge.

When Hegel's philosophy was turned upside down, it was not a simple kinematic transposition. It was a victory of the abstraction that fills the universe over the abstraction that empties it, a victory of the thought that improves its content by embracing the vital heterogeneity of the universe over the *purely* abstract thought that destroys it. This was the movement of the Logos, which never *began* its movement toward the Cosmos because it was always a vital content reflecting the moving, heterogeneous Cosmos, which has no beginning.

3. 'PURE BEING' AND 'PURE EXPERIENCE'

Pure thought about being — 'pure being' — is thought devoid of physical content. It finds its content by pure logic. Hegel introduces new concepts; they clash, merge and are differentiated. But this is already an advance from 'pure being' to concepts with content, as opposed to 'nothing'. It does not include the '*beginning*'. It resembles not a segment (in mathematics a line including both its ends is called a segment), but an interval (by definition the beginning and end points are not included in it). The beginning of the dialectical path here neither appears to be dialectical nor does it have content.

The other end of the segment — 'the purely physical' — likewise does not enter the dialectical line. Hegel, in attacking traditional empiricism, says that in science it is impossible to be free of metaphysics, i.e., to be free of the constructs of reason. "The only mere physicists are the animals: they alone do not think . . ." (Hegel 1975, §98, *Zusatz* 1, p. 144).

Here the term 'metaphysics' is interesting because it designates what goes beyond the framework of empiricism. Pure physics, according to Hegel's definition, is a purely sensual sphere; it is an accumulation of direct impressions. . . . What makes it an object of cognition is already beyond the limits of physics — it is metaphysics. Such a distinction does not correspond to the concept of Aristotle, who coined both terms. For Aristotle, his *Physics* was very rational, while his *Metaphysics* was quite sensual. Nor does it correspond to the Hegelian conception of the 'ideal' and the 'real' (cf. Hegel 1975, §96, pp. 141–142). Physical reality does not coincide with 'pure physics' just as it does not coincide with 'pure being', i.e., with pure thought, with ultimate abstraction.

In fact, 'pure physics', i.e., immediate perceptions, are, strictly speaking, an illusion. Cognition is not an interval and not even a segment in which the ultimate points are included; cognition as a whole is a circle where there are no ultimate points. The 'pure physics' that Hegel attributed to animals probably does not exist even for them. In any event, for people, long before science and philosophy, immediate perceptions demanded a certain amount of apperception. Let us recall once again Darwin's story in *The Voyage of the 'Beagle'* in which the aborigines on one island paid no attention to the ship anchored at the shore — they did not associate it with anything — but looked with astonishment at the sloops which were somehow associated with canoes.

Ethnography has accumulated a great deal of analogous data. The history of the language and culture of the most ancient peoples also contain such data.

Ethnographic and historical data support Hegel's idea of the absence of 'pure physics' in man. These data are related to observations of the development of perception in the child; together with data on the psychology and physiology of the sense organs and with the equally important sources of dialectical epistemology,[5] they eliminate 'pure physics'.

The physiology of vision created by Helmholtz is based on a certain apperception without which the eyes and brain could not receive visual sensations so as to create images of the observed world. But beyond this elementary apperception, perception always separates the elements of the extended world, likening and identifying them with others. Helmholtz says that in childhood he *saw* some beautiful dolls at the top of a tower standing in a square and asked his mother to reach out and take one of the dolls. "After that I often looked at the landing of this tower when there were people on it but they no longer wanted to be transformed into dolls for my refined eye."

The initial associations were replaced by others, and Helmholtz no longer saw dolls, but people. There were two fundamental operations – the identification of visual impressions and their individualization. They are in fact the operations which are transferred into logic; they are prelogical but not illogical operations. They illustrate the transition from sensation to thought; they already contain the direct identification of objects according to their predicates and the individualization of objects, the analysis of the concrete with the aid of general definitions. "Not only is the transition from matter to consciousness dialectical, but also that from sensation to thought, etc.," wrote Lenin in his conspectus of Hegel's *Lectures on the History of Philosophy*.[6]

During the transition from one concrete impression to another, the entire system of general concepts, which permit the isolation and individualization of the initial impression are affected and altered to some degree. In the typical case, each concrete impression is surrounded by a field of general concepts without which it does not exist as such, it is not isolated from the uninterrupted chaos of impressions. In its turn, the transition from one concept to another cannot, in the typical case, avoid touching the clouds of concrete images, observations and associations which are united in this concept. The connection between a concept and the concrete impressions sometimes allows purely logical transitions without rethinking and altering concrete associations. But this is a special case of weak, unclear or remote connection; it is an approximate representation. In the typical case, even a purely mathematical correlation suggests a law which is founded and varied on the basis of precedents. Such a law changes its meaning, its scope, the level of its severity from case to case, from one interpretation and legal decision to another. In mathematics such a change at times becomes explicit and radical. Such, for example, was the transition to non-Euclidean correlations. As a whole, as was already mentioned, logic, mathematics, speculative thought can never be separated from matter as the source of sensory impressions. Therefore, in dialectical philosophy there are no *a priori* concepts in an absolute sense. There are no concepts in it which would be independent of the flow of sensory perceptions and of their content. The content of the concepts of dialectical philosophy does not consist in their being filled with some content which does not affect them; they are not like a bottle containing various liquids which does not alter its form when the liquids are replaced.

Let us refer again to the example of space in modern physics, which depends on physical, comprehensible processes that occur in space and affect the sense organs sensually. Space in which there is a gravitational field becomes non-Euclidean. This example is by no means a special case: space in Cartesian physics represented being, while in atomist conceptions it represented nonbeing. Hegel's abstract logical schema — 'pure being' which turns out to be equivalent to 'pure nothing' — possesses an entirely concrete and an entirely historical equivalent. Descartes's matter — all-pervasive and homogeneous — has no predicates other than extension, which itself is not (within the limits of Cartesian physics!) being and only being. But such being, absolutely self-identical, absolutely homogeneous, is not distinguished from space — 'nonbeing'. Seventeenth-century physics fought against this annihilating function of identity; Descartes wanted to isolate the separate impenetrable bodies in matter-space, in being equivalent to nonbeing and return heterogeneity and multiformity to extended substance. But for this it is necessary to endow portions of space with dynamic properties. Then the bodies can interact, possess 'becoming', 'being', 'existence', 'reality', the entire hierarchy of heterogeneous being not equivalent to 'nothing'.

Hegel's dialectic as a whole reflects and generalizes this evolution of classical science; indeed it does more than that. Hegel's thought, dispensing with language and its sources, and with impressions of social processes, embraced a much wider range of ideological tendencies than the evolution of the science of the seventeenth and eighteenth centuries.

Even the logical connection is apparent here only in retrospect. But it allows us to see much in both the science of the seventeenth and eighteenth centuries and the philosophy of Hegel.

In particular, we can see the profound content of the categories and concepts in the Hegelian doctrine of being, the content of all except 'pure being'. This concept does not have a physical equivalent. On the contrary, it expresses in fact the absence of a physical equivalent, a physical lack of content. The physical content of a concept demands that it be mediated, that it contain in itself a certain heterogeneity, a certain ability for spontaneous transition to other concepts.

What is the origin of this concept which does not reflect the evolution of science and is represented as *a priori*? It represents, as was already stated, the result of abstraction and in this sense is mediated, but the mediation here destroys itself and leaves its result without mediation, which gives it an *a priori* character. Does such a construct have roots in science? The answer to this question will be given, or rather an attempt at an answer will be given, later, in the chapter devoted to the connection between Hegelian dialectics and classical science and its historical evaluation from a modern, nonclassical perspective. But now we must return to the concept of being which contains physical content.

4. BEING IN MOTION

The concept of being containing physical content is thereby distinguished from 'nothing'. The search for such a concept with physical content has not stopped and is not finished. And at present this search is the most important current of scientific thought. The most difficult question in the modern theory of the microcosm is the question of the ultramicroscopic processes which fill the world-line of the particle and transform it from a geometrical concept into a physical one. The struggle for the content of being is now interwoven with the struggle for the content of 'nonbeing': space as the image of 'nonbeing' became the field, a physical reality which also does not want to become the victim of annihilating identification, which levels reality and transforms it into 'nothing'.

In modern science, the *a priori* beginning — abstract 'pure being' — not only does not find an equivalent in the generalization of new tendencies, but also these tendencies urgently demand that fundamental concepts of philosophical analysis not cut the head of the snake biting its tail, not claim an *a priori* character, but embrace a wealth of determinations, and be, in this sense, concrete to a great extent. Meeting such a demand means to some extent retracing one's steps from Hegel back to Spinoza. A return to Spinoza is characteristic of each step of science and philosophy away from *a priori* conceptions to conceptions of a continuous cycle in cognition (without a beginning that can claim to be *a priori*). Spinoza's substance is *causa sui*; it represents the objective cycle of causally connected processes and, accordingly, the cognition of substance is represented as an eternal cycle; it can never arrive at the absolute beginning of a causal series. Therefore a conditional 'neo-Spinozism' will remain a constant motif woven into the development of materialistic philosophy (in a few lines it will be explained why it is conditional). In the eighteenth century Diderot (in the article 'Spinozist' in the fifteenth volume of the *Encyclopédie*) demonstrated the similarity between the materialism of his time and Spinoza's doctrine; and in our century, G. V. Plekhanov devoted a special historico-philosophical excursus to the constant return of materialist philosophy to Spinoza (cf. Plekhanov 1976, pp. 330—339).

But the return to Spinoza takes place not on a plane but in ever-rising spirals. What comprises the rise of the spiral, what is the difference between post-Hegelian 'neo-Spinozism' and Spinozism, and what is the conditionality of this term in quotation marks?

Here, in connection with the problem of being and motion, we must note the following.

Spinoza's conception of infinite substance reflects the old pre-differential representation of the world, when the concept of infinity was associated primarily with the infinitely great Universe. Physics and philosophy were also concerned with the infinitely small, but less so. Spinoza's substance possesses infinite extension as its attribute.

But what has remained obscure is the transformation of the finite mode, the finite body in the infinite set of infinitely small elements and, what is especially important here, the *reflection* of the infinitely great object, in this sense, in each of its infinitely small elements.

For a long time this idea was latent [not explicit]. The differential representation of motion and the localized criteria of being were a natural deduction from the concepts of acceleration and its dynamic conditions and effects; they were a natural physical interpretation of mathematical analysis. But neither science nor mathematics noticed that in science a representation of being, new in principle, had matured. Instead of 'To be means to be motionless and unchanged', there appeared 'To be means to move and to change'. Parmenides's criterion was replaced by Heraclitus's.

Thus the relation of the abstract to the concrete is altered in a fundamental way. The transition from the concrete to the abstract is no longer a simple grouping of concrete objects and the ascription to them of identical properties, the ignoring of the nonidentical and the replacement of individual designations by a single abstract designation, common to the entire identified set. Now individual elements are grouped in a diversity characterized by the law of motion: for example, the law which permits locating a particle if its initial position and the field of force are known. Being included in such a diversity enriches the collection of determinations which belong to the individual. The particle as the subject of motion, located at a given point, is viewed as the reflection of the infinite number of previous positions; now, velocity (the particle's future behavior is gathered into a point, is reflected in the point), acceleration, the dependence of acceleration on the field, i.e., mass, the series of new predicates, are all ascribed to it. These predicates are contradictory: motion is a contradiction; the particle must possess both location and a change in location; it is determined locally, but this determination reflects the behavior of the particle in space.

The new representation of being has remained obscure, implicit, but that has not prevented it from very energetically influencing physics, mechanics and mathematics. The idea of the dependence of localized being on infinite diversity, the inclusion of this localized being as its element, has penetrated into philosophy in a very complex manner – initially in the form of the antinomy of infinity, and generally in a critical form. Hegel made the philosophical generalization of a new type of abstract analysis positive, leading not to an abstract schema poor in determinations, but to a rich concreteness. For Hegel the Logos ceased to be a synonym for the homogenizing, levelling identification that annihilates the universe; it included the ascent to the concrete.

Hegel connects such a representation of the transition from the *particular* to the *universal*, which retains all the wealth of the particular, with the content of logic, with the generalization in logic of the content of the scientific disciplines.

In his conspectus to Hegel's *Science of Logic*, Lenin, having noted down the phrase "Not merely an abstract universal, but a universal which comprises in itself the wealth of the particular", writes:

A beautiful formula: "Not merely an abstract universal, but a universal which comprises in itself the wealth of the particular, the individual, the single" (all the wealth of the particular and single!)!! *Très bien.*[7]

But in this transition to the abstractly-universal which was "not only abstractly universal", Hegel stopped short of the highest, ultimate abstraction: 'Pure being' is the highest abstraction but by no means the highest concreteness. On the contrary, this concept expresses the utter lack of concrete determinations, the utter annihilation of the universe.

If we proceed further along Hegel's path and continue to ascend, through abstract analysis, to ever higher concreteness, then *concrete being* will be revealed to us. It embraces the entire Universe, but the Universe in its diversity and motion, heterogeneous and evolving. And this is no longer the simply heterogeneous and simply evolving Universe of Spinoza. The infinite concreteness of the Universe stands behind the series of the concepts that have physical content in Hegel's dialectics. But this infinite concreteness, this infinite wealth of determinations, includes dynamic determinations that express the influence of being as a whole on each of its elements. Hegel introduces the concept of true infinity and places it in filled space and time. True infinity is reflected in each of its elements. Being as a concrete, the most concrete, category, as the filling of space and time, is reflected in its infinitely small elements; these localized reflections represent statements of motion and change. Being as a whole is the infinite set of localized events which fill space and time. It is acceleration, interaction, change, which create variety, heterogeneity, evolution — the diversity of the world in space and time.

Thus we see in Hegel both the conception of 'pure being', in which mediation cancels itself out, the liberation of being from mediation, and the conception of concrete being which is the result of a specifically dialectical analysis moving toward increasing concreteness. The transition from the first conception to the second is a transition from a concept lacking physical content, in essence *a priori* and fixed, to a concept both fluid and containing content. The fact that the concept of 'pure being' coincides with the concept of 'nothing' while concepts containing physical content and expressing motion are not equivalent to 'nothing', signifies that real being, not degraded into 'nothing', is being in motion, heterogeneous, evolving and contradictory.

5. THE HIERARCHY OF FORMS OF MOTION

Let us look more closely at heterogeneous, evolving and contradictory being and also at the course of the rational cognition of this being.

When reason moved from the simplest laws of being — in the nineteenth century, the laws of mechanics were considered such — to more complex physical, then chemical and then biological laws, it was a transition not only, and not even so much, from the general to the particular as from the particular to the general. In the seventeenth-eighteenth centuries no one thought so; then the laws of mechanics were considered to be universal laws, and all the other scientific laws to be modifications of mechanical laws. The goal was to discover universal laws in their particular modifications, the true mechanical nature of heat, electricity, chemical reactions, biological processes, etc. But in the nineteenth century, and even more in the twentieth, the transition from mechanics to physics and beyond became a transition from particular laws to more general ones. Physics was emancipated from mechanics, chemistry from mechanics and physics, and biology from mechanics, physics and chemistry. Now the transition to what was then considered a more complex series of phenomena turned out to be not an extension of detail but an expansion of the scope of the scientific explanation of the world, a transition from *exclusively* mechanical processes to a *general* world picture.

But this transition to the general, this generalization of scientific representations, meant not their levelling, but rather a concretization, an increase in the number of determinations. The world that was revealed in such a generalization was a heterogeneous, multicolored and fluid world, and the closer cognition drew to the world, the more it had to differentiate methods and concepts, the more it had to make them concrete and flexible. The generalization of science itself could no longer be reduced to the identification of its arsenal nor could its developing and diverse methods be reduced to certain impoverished, homogeneous categories. The generalization had to become a picture of motion, a picture of the transition of science from some laws to others, a picture of the complication of its content, object and means.

Such a generalization had already been made in the second half of the nineteenth century. At around this time there appeared new social stimuli for the dialectical generalization of science, and this time the generalization was continuous and was not halted by any *a priori* and protective absolutes. Science itself was altered; the first half and the middle of the nineteenth century were a time of essential displacements in both the content and the style of science.

In the words of Engels, Hegel's goal was to generalize the natural science "of the *old* Newton-Linnaeus school" (Engels 1954, p. 331), in other words, of that period in the development of natural science when the fundamental goal was a systematization that would identify the inclusion of isolated phenomena within more general groups and classes. In the second half of the nineteenth century, after such strides in science as the discovery of the principle of the conservation of energy, the kinetic theory of gases and the theory of evolution, scientific research changed.

Engels writes:

But since the general evolutionary connection in nature has now been demonstrated, an external side by side arrangement is as inadequate as Hegel's artificially constructed dialectical transitions. The transitions must make themselves, they must be natural. Just as one form of motion develops out of another, so their reflections, the various sciences, must arise necessarily out of one another (Engels 1954, p. 331).

Therefore the new dynamic generalization of science had to contain a new classification of science, a natural, genetic classification demonstrating the origin of each science.

Classification of the sciences, each of which analyses a single form of motion, or a series of forms of motion that belong together and pass into one another, is therefore the classification, the arrangement, of these forms of motion themselves according to their inherent sequence, and herein lies its importance (Engels 1954, p. 330).

Such a classification of science is, in essence, its dialectical substratum; it includes the genesis of the branches of science and the explanation of the transition from one branch to another. Here is the path for the transition from the natural *classification* to the *periodization* of science, just as in biology the natural classification of appearances was the transition to the representation of their evolution.

An example of the new philosophical generalization is the treatment of the principle of the conservation of energy in the *Dialectics of Nature*. Engels shifts the focus of the problem from quantitative *conservation* to qualitative *transformation*, i.e., from identity to nonidentity, or rather, a combination of both. Thus the law of the conservation of energy becomes an historical law.

From our historical perspective, we can more concretely define the historical meaning of the ideas of the hierarchy of forms of motion, its connection with the pervasive *aporias* of rationalism and its predictive aspect, its predictive function, its connection with modern nonclassical science.

The clash between irreducibility and reducibility is the old contradiction between identity, immobility and invariance, on the one hand, and nonidentity, motion and transformation on the other. One of the fragments of the *Dialectics of Nature* demonstrates a characteristic peculiarity of nineteenth-century science as compared with seventeenth- and eighteenth-century science, and allows us to see how far the nineteenth century had come. If, in Newton's time, concepts were distinguished as being reducible to mechanics or not, then reducible concepts were designated as the main ones and irreducible ones as subsidiary. Engels, in generalizing nineteenth-century science, considered nonidentity the main form. He called the main form of motion the accumulation of the natural laws, the phenomena, the processes that were not reducible to mechanics, that were not reducible to more general and simpler forms of motion; those which were reducible he called subsidiary forms of motion (Engels 1954, p. 328). The twentieth century went even further. At present we see not only the irreducibility of the statistical, irreversible processes of thermodynamics to the mechanics of molecules, the irreducibility of chemical reactions and the regrouping of atoms to more simple laws, the irreducibility of the phenomena of life to general physico-chemical schemata; we see not only the specificity of each series of phenomena and not only their dialectical, inclusive irreducibility, nonidentity, universal connection: In modern nonclassical science we also see that the fundamental, most general laws of the universe are the laws of the motion, interaction, generation and destruction of elementary particles. But it is precisely these laws that are now represented as the least stable; it is precisely here that almost continuous radical alterations of fundamental views take place, and it is these alterations that threaten the established conceptions in all spheres of knowledge and thus promise a general modification of the world-picture. Nineteenth-century scientific discoveries brought with them a certain autonomy for the isolated disciplines, a certain liberation from the former all-absorbing power of mechanics. At present, when compared with the modern upheavals embracing the entire foundation of science, they seem to be reforms. But these reforms led to the present revolution. Their generalization — the dialectical substratum of classical science — is represented at present as the forerunner of the nonclassical revolution.

The modern, dynamic, rapidly changing schema of the hierarchy of the forms of motion was by no means hidden in a preformative aspect in the *Dialectics of Nature*. In the history of thought preformism does not explain real evolution as it does in biology. In the history of thought a logical connection consists not of identity, but of transformation; but the concept of transformation does not have real meaning without the concept of identity, without the subject of transformation, in this case, without a pervasive, self-identical, but modifying idea.

This idea is inseparable from science; without it science ceases to have an object. It consists of the heterogeneity of being. Being is heterogeneous in space: one spatial point is distinguished from another point, one accumulation of points is distinguished from another accumulation. Being is heterogeneous in time: one moment is distinguished from another, one temporal interval from another temporal interval. But both spatial and temporal distinctions lose meaning without identity, homogeneity, symmetry, conservation.

What is true of the universe is true of its reflection. The idea of the heterogeneous world forms a heterogeneous line in the history of thought; from Heraclitus to nonclassical science the concept of non-identity was different at different stages. And at the same time it was identical. The idea of the irreducibility of complex forms of motion to the simple forms growing from classical science turned out to be an introduction to the modern conception of the irreducibility of each new stage of science to the previous, simpler conception of the ceaseless renewal of the fundamental principles of science.

NOTES

[1] "Nous touchons peut-être ici au point décisif de l'Hégélianisme, . . . à ce qui fait de Hegel à la fois le plus grand irrationaliste et le plus grand rationaliste qui ait existé" (Hyppolite 1953, p. 131).

[2] ". . . on pourrait se demander si, la *Vernunft* n'etant pas la raison ordinaire, le philosophe de la *Vernunft* est un rationaliste ou irrationaliste. Une partie d'ailleurs de l'attrait de Hegel ne vient-il pas de cette sorte de coup de force philosophique par lequel il a fait rentrer l'irrationnel dans la raison, et a donné satisfaction à la fois au besoin d'irrationnel, et au besoin de rationnel, l'un venant se fondre dans l'autre, par la magie même du philosophe?" (Wahl 1931, p. 451).

[3] Hegel's term is *Aufhebung*, "the dialectical transition in which a lower stage is both annulled and preserved in a higher one" (Taylor 1975, p. 119).

[4] [Khodynka or Khodynskoe Polye was a vast military parade ground in pre-revolutionary Moscow (it is now the site of a sports stadium) where, by the dawn of 30 May 1896 (N.S.) 500,000 people had gathered to celebrate the coronation of Tsar Nicholas II, emperor since 1894, and his empress Alexandra, both of whom planned to distribute commemorative mugs and free beer to their subjects at the traditional fête.

The crush was great: as people ran forward to claim their gifts, over one thousand were trampled to death and hundreds were injured because of inadequate security arrangements.

The idea of annihilating reason is mentioned again by Kuznetsov on p. 288 and p. 289 below. — Eds.]

[5] Cf. Lenin 1963 (*R*), p. 314; Lenin 1961 (*E*), pp. 352–353.

[6] Lenin 1963 (*R*), p. 256; Lenin 1961 (*E*), p. 283.

[7] Lenin 1963 (*R*), p. 90; Lenin 1961 (*E*), p. 99. The passage is from the introduction to the *Science of Logic*:

> It is only after profounder acquaintance with the other sciences that logic ceases to be for subjective spirit a merely abstract universal and reveals itself as the universal which embraces within itself the wealth of the particular . . . (Hegel 1969, p. 58).

10. EXISTENCE AND ACTUALITY

1. BECOMING

In the whole series of concepts containing content and not equivalent to 'nothing' in Hegel's *Science of Logic*, we encounter a clash between two tendencies – the 'tranquil', stable tendency that seeks and discovers identity, and the 'unstable' tendency that reveals nonidentity and provokes the search for new identity. For all their contradiction, they lose their meaning without each other and in this sense are complementary. Complementarity does not mean here that, besides its identifying tendency, reason seeks multiformity, change, motion in a complementary way. It means that complementary concepts lose their content without each other, as in the case of the wave and corpuscular aspects of the microcosm. Modern nonclassical representations make Hegel's abstract constructs very concrete and physically perceptible.

The first in the series of concepts with content in the *Logic* is *Becoming*. Hegel leaves pure abstractions behind and moves on to concrete categories.

Becoming is the first concrete thought, and therefore the first notion: whereas Being and Nought are empty abstractions. The notion of Being, therefore, of which we sometimes speak, must mean Becoming; not the mere point of being, which is empty Nothing, any more than Nothing, which is empty Being. In Being then we have Nothing, and in Nothing Being; but this Being which does not lose itself in Nothing is Becoming. Nor must we omit the distinction, while we emphasize the unity of Becoming: without that distinction we should once more return to abstract Being (Hegel 1975, §88, *Zusatz* to Para. 5, p. 132).

Thus being, which does not lose itself in nothing, which does not disappear, does not lose its content (in other words, concrete being) is becoming. In the concept of becoming, the emphasis is placed not on the *conservation* of being in 'nothing', but on its change, its nonidentity, on the fact that conservation has a place within the limits of *nothing* where "being does not lose itself". The emphasis here is on the synthesis of the concepts of *being* and *nothing*; the mobility of being, its motion, its nonconservation in the given form are emphasized.

'To become' is the true expression for the resultant of 'to be' and 'not to be'; it is the unity of the two; but not only is it the unity, it is also inherent unrest — the unity, which is no mere reference-to-self and therefore without movement, but which, through the diversity of Being and Nothing that is in it, is at war within itself (Hegel 1975, §88, Para. 4, p. 131).

Hegel contrasts becoming with conservation. The traditional thesis "from nothing comes nothing" is at odds with the inclusion of 'nothing' in the definition of being, with "being which does not lose itself in *nothing*", with the nonidentity of being.

The ancients saw plainly that the maxim, 'From nothing comes nothing, from something something', really abolishes Becoming: for what it comes from and what it becomes are one and the same. Thus explained, the proposition is the maxim of abstract identity as upheld by the understanding (Hegel 1975, §88, Para. 5, p. 131).

Further, Hegel finds it strange that in the nineteenth century the position that "from nothing comes nothing, from something something" was accepted without reservation. Apparently he considers nineteenth-century science to be a demonstration of becoming. In fact, the concept of becoming and the unification of being and nonbeing in this concept generalize nineteenth-century science — the highest stage of classical science. Such a generalization allows, above all, a more precise and broader evaluation of the historical sources of the dialectical conception of being. Hegel says that it is Heraclitus's system, as opposed to that of the Eleatics, that serves as the historical antecedent of becoming.

"Everything flows" means that becoming is the basic definition of the real.

Becoming for Hegel is more than the basis of an historico-philosophical examination of the past.

Becoming, in his words, "taken at its best on its own ground, is an extremely poor term: it needs to grow in depth and weight of meaning" (Hegel 1975, §88, *Zusatz* to Para. 5, p. 132).

Now we can define the subsequent evolution of the concept of becoming in the science of the middle and second half of the nineteenth century and its most recent evolution within the framework of non-classical science. But it may be most clearly defined in the light of the purely logical evolution of becoming which was discovered by Hegel in his *Science of Logic*.

The above-mentioned logical evolution begins with a shift of emphasis from the negative side, with the merging of being and nothing and their disappearance in such a merging, to the result of becoming, to a being no longer devoid of content, but determinate, possessing several predicates. Becoming includes coming-to-be and ceasing-to-be — such are its *moments* (Hegel 1969, p. 105). If we begin from *nothing* and see how it passes into *being*, we encounter coming-to-be; when we begin from being and see how it passes into nothing, this is ceasing-to-be. But such a definition makes no sense without a representation of something coming to be. Thought ascribes to what has come to be the predicate of coming-to-be and, further — bearing in mind the second moment of becoming — the predicates of ceasing-to-be, transience, disappearance, being doomed, the transition into something different. These predicates belong to something stable, determinate, that has come to be; they limit and regulate its stability, its determinateness. When we speak of the subject of coming-to-be and ceasing-to-be, we bear in mind that nothing was transformed into being and being into nothing and that in their function of rejecting each other they have disappeared.

Being and nothing are in this unity [becoming] only as vanishing moments; yet becoming as such *is* only through their distinguishedness. Their vanishing, therefore, is the vanishing of becoming or the vanishing of the vanishing itself. Becoming is an unstable unrest which settles into a stable result (Hegel 1969, p. 106).

Such a relatively firm and stable result of coming-to-be and ceasing-to-be is deduced by Hegel purely logically. There must be a subject of these unstable predicates which itself is stable. Coming-to-be makes no sense without the representation of what has come to be, of a result, of something present. But behind these logical transitions stands all of classical science. And as for prediction, there stands, no longer *behind* logical transitions, but ahead of them, as their affirmation, modification, concretization — there stands nonclassical science.

The Hegelian, tranquil, stable result is not an analogue to the absolutes in classical science, absolutes 'classical' in the strict sense that claim an *a priori* and fixed character. It is an analogue to the stable and balanced concepts which are a necessary link in the demonstration of unrest and imbalance. It may be most clearly illustrated by means of nonclassical models.

Let us take as such a model the determination of the dynamic variables of the particle by means of a macroscopic classical device. At the moment of measurement, these variables are completely in flux, they are unstable. Generally speaking, they change because of the influence of the measuring apparatus. But such a change, the coming-to-be and ceasing-to-be of the variables, their indeterminacy, represent predicates of something that has come to be, something determined and stable. The apparatus changes some *given* value of the dynamic variable and replaces it with another *given* value after passage through the apparatus. In the first case we ascribe ceasing-to-be to these given values and in the second, coming-to-be. The condition of physical meaning is the device itself to which we ascribe stable values of location and impulse at each point, for example, at the opening of a screen through which the particles pass. The concepts of the unstable quantum object and the stable classical apparatus are complementary.

2. DETERMINATE BEING

The result of becoming — that which has *become* — Hegel calls determinate being. It is itself the stable aspect of being, but it is by no means a rejection of motion, multiformity, heterogeneity, nor is it a return from becoming to pure being, or an equation with 'nothing', or an annihilation of concrete being. It is the starting-point and the result of becoming, the prediction of subsequent nonbeing and the 'recollection' of previous nonbeing. Coming-to-be and ceasing-to-be enter the determination of concrete being. Here the rational, mathematical and natural scientific equivalents of the comparatively complex and abstract passages of Hegel are quite precisely apparent. When abstract concepts contain physical content, their concrete deciphering is not difficult as the history of science increasingly demonstrates. In the case given, the connection between determinate being and a differential representation of motion is apparent. In classical science, motion is examined from point to point and from moment to moment. But *what* is measured in such an examination, *what* distinguishes each point and each moment from the others? It is differential quantities: a particle at this point undergoes acceleration. It allows us to speak of the qualitative determinateness of the point and moment. The measurement of the behavior of a particle before and after its presence at a given point is typical of classical science. After this the characteristics of behavior are gathered together, and we can obtain a qualitative determination of the given point and distinguish it from others.

Let us transfer this method to the language of philosophical categories, to the language of a doctrine of being.

We must refuse to ascribe any determinateness to the presence of a particle at a given point if we do not introduce its nonbeing into the analysis, i.e., its coming-to-be and its ceasing-to-be — in other words, its becoming. For the determination of the being of a particle at a given point, we examine this being as a result of its coming-to-be here, its transition to this point. Later we represent its subsequent disappearance, its ceasing-to-be at another point. For this we also need the category of the result of becoming — in other words, the determinate being of the particle at the examined point. We ascribe a transient character to precisely this result.

There are many such situations in classical science; their character illustrates not only the content of natural scientific representations and goals, but also the style of classical science and its influence on thought, psychology, language. After Galileo, Descartes and Newton, people began to consider being, concrete being, as a cessation of motion. The tirades of Salviati and Sagredo against the Aristotelian identification of perfection with immobility; the Cartesian system that explained the world in terms of the motion of bodies; the Newtonian system that explained the harmony of the universe in terms of inertia and gravitational acceleration; and the entire sum of new knowledge, new experience, new methods, prepared the soil for the dynamic representation of being, for the idea of being without content and without becoming, for determinate being with content as a result of becoming. The abstract and difficult periods of Hegel's [thought] were hard to understand but they did not contradict people's style of thought, they did not provoke inner protest.

For Hegel the concept of determinate being was a *logical* deduction from the concept of becoming. But for Hegel the word 'logical' has a different meaning than for Kant or for the other philosophers of pre-Hegelian times. It signifies not only the subordination to certain logical norms, but also the demand for physical content. Hegel's logic is forbidden to speak of nothing (i.e., to eliminate subjects from its deduction) and is forbidden to say nothing (i.e., to eliminate the predicates).

Let us cite an excerpt from Hegel's *Logic* that contains the definition of determinate being.

Even our ordinary conception of Becoming implies that somewhat comes out of it, and that Becoming therefore has a result. But this conception gives rise to the question, how Becoming does not remain mere Becoming, but has a result? The answer to this question follows from what Becoming has already shown itself to be. Becoming always contains Being and Nothing in such a way, that these two are always changing into each other, and reciprocally cancelling each other. Thus Becoming stands before us in utter restlessness — unable however to maintain itself in this abstract restlessness: for since Being and Nothing vanish in Becoming (and that is the very notion of Becoming), the latter must vanish also. Becoming is as it were a fire, which dies out in itself, when it consumes its material. The result of this process however is not an empty Nothing, but Being identical with the negation — what we call Being Determinate (being then and there): the primary import of which evidently is that it *has become* (Hegel 1975, §89, *Zusatz*, p. 134).

This tale of adventure about unrestrained motion and a dying out like fire is by no means a game that the concept plays with itself nor is it an explanation of the relations between concept and object; it is rather a description of real processes. To the degree to which logic for Hegel is a science with content, the logical passages hide ontological assertions. And they are indeed *hidden*, but not so carefully that it is impossible to reveal them. From this point of view the secret of (present) being has a precise physical interpretation.

In classical science it consisted of the conception of localized processes. The simple spatio-temporal localization of the particle does not signify any determinate being if the particle does not possess some non-Cartesian predicates, if we cannot say anything about it other than that it *is* at the given point. We cannot even assert in this that it is at the *given* point; we are not able to distinguish the given point from others if nothing is occurring at these points.

The assertion that the particle is, period, is an assertion without physical content. It is an analogue to pure being, equal to pure nothing: after all, if the particle possesses no properties other than coordinates, it is not distinguishable from the empty spatial point.

Now let us represent the particle as moving. We hope to save its being from identification with *nothing*. In addition to coordinates, it possesses their derivatives, velocity and acceleration. It enters the given point and leaves it. This is an analogue to becoming, the simplest and most elementary analogue; it is limited (as it seemed in the time of classical science) by an elementary form of motion — displacement.

Now we may determine the position of the particle differently. It is a point which the particle reaches and from which it departs. The arrival of the particle and its departure are moments of becoming, which is like a fire that "dies out in itself, when it consumes its material".

When 'disappearance disappears', we arrive at the position of the particle as something achieved. "The result of this process however is not an empty Nothing, but Being identical with the negation — what we call Being Determinate (being then and there)". The position that results from motion is already a concept that contains physical content.

Thus the classical physical analogue to determinate being is the position of the particle as well as the spatial position of the system of particles in the framework of a dynamic representation of the world.

Nonclassical science has not made this analogue entirely satisfactory. From the point of view of the theory of relativity, the motion of the particle is geometrized in the form of the four-dimensional world-line, and again we cannot distinguish it from the geometrical image, this time four-dimensional. But here quantum mechanics and, most recently, relativistic quantum mechanics, come to our assistance. They provide a transition from Hegel's formula to modern physical language that is the most complete as well as the closest to the original. Modern representations of particle motion (we speak of these representations in the plural because they are extremely diverse) include a picture of the radiations and disintegration of various particles, of the interaction of moving particles with a vacuum and of a schema of the quite complex correlations between the self-identity of the moving particle and its transmutation. At a given moment the particle possesses determinate being as a result of the processes of birth and annihilation which could just as well be called coming-to-be and (with the reservation connected with a neologism) ceasing-to-be.

But when we seek analogues to Hegelian concepts in nonclassical science, we are immediately forced to move on to more contentful and concrete concepts of existence and reality. Before we turn to them, we must remember that the search for analogues has nothing (or almost nothing) in common with the *Natur*-philosophical behavior of physical categories under *a priori* philosophical schemata. There exists another tradition about which we must speak later in greater detail. It is the comparison of Hegelian categories with nature-scientific conceptions that modifies both of them.

3. ESSENCE

Unfortunately, it is impossible here to compare the entire series of concepts that form Hegel's logic with the categories of classical and nonclassical science. It is necessary to select only a few of them, and, strictly speaking, this is inadequate: after all, the content of these concepts is inseparable from the transition to other concepts that immediately follow. Therefore, the goal of this section will be limited to several extracts from the dialogue between modern science and the dialectics of Hegel, somewhat analogous to the dialogue between Einstein and Spinoza in the first part of this book.

In the doctrine of being, the dialogue is limited to concepts of pure being, becoming and determinate being. In the doctrine of essence it is limited to concepts of existence and reality.

The basis of existence is *essence*. The object of thought goes beyond the limits of determinate being; it is viewed as self-identical, and essential predicates are retained in it. Abstract analysis selects the predicates present; it itself frees them from differences and makes them *be*, abiding and conserving, and leads to the self-identical object. "This identity becomes an Identity, in form only, or of the understanding, if it be held hard and fast, quite aloof from difference" (Hegel 1975, § 115, p. 166). Thus thought is raised to its universal laws, or rather to what are called the universal laws of thought.

Hegel limits and makes relative the identifying function of reason. As soon as this function is separated from the statement of difference, the universal laws that identify and free the objects of thought from differences turn out to be the laws of *abstract common sense*. Such is the law of identity — the assertion that everything is identical with itself. Hegel says that abstract identity leads to tautologies: "Utterances after the fashion of this pretended law (A planet is a planet; Magnetism is magnetism; Mind is mind) are, as they deserve to be, reputed silly. That is certainly matter of general experience" (Hegel 1975, § 115, p. 167).

Everything that Hegel says of identity is directed against such abstract identity. "It is important to come to a proper understanding on the true meaning of Identity: and, for that purpose, we must especially guard against taking it as abstract Identity, to the exclusion of all difference" (Hegel 1975, § 115, *Zusatz*, p. 167).

In order to see the equivalents of the Hegelian conception of identity in classical and, later, in nonclassical science, we shall mention the connection of this conception with concepts of absolute space and absolute time.

The principle of relativity identifies the behavior of a body moving in space. Space is homogeneous; it does not interfere with the behavior of a moving body, and a body retains certain predicates while in motion. They are *invariant* predicates, invariants of motion. The essence of a moving body is its invariant predicates. The essence of a solid body is the invariant distances between its points. The essence of a system of interacting bodies is the invariant interactions that do not change their character during the motion of the system.

The invariance of the indicated quantities does not permit the measurement of motion other than through change in the distances between the moving body, or the system of bodies, and other bodies.

Absolute space and absolute time are also the result of identifying abstraction. Hegel says that the absolute is self-identical and that the concept of the absolute can make sense depending on *what* self-identity lies at the basis of this concept (Hegel 1975, § 115, pp. 166—167).

Hegel says further that absolute space is abstract space, and that absolute time is abstract time (*ibid.*). What kind of abstraction is this? Is it the true abstraction which is inseparable from the concrete, or is it the abstraction which ignores concreteness and leads to the absolutes cited?

We will note that it has only been since Einstein's theory of relativity that the concepts of absolute space and absolute time could become the object of historico-logical analysis and that we could interpret, in a precise and *physical* form, the Hegelian thesis that equates absolute space and time with abstract space and time. When the absolute space and the absolute time of Newton are no longer viewed as the objective background of real processes, it is possible to pose the question of the logical roots of this illusory category.

Abstract space and abstract time claim real meaning by remaining purely abstract, purely identical, by not including nonidentity. Let us compare two tendencies in the abstract construction of space. One tendency is true identity, which retains differences, and an abstraction which does not destroy the concrete. In this case a body, i.e., something sensorily comprehensible and distinguishable from space, serves as the fundamental concept. We mentally extend this body on all sides and ascribe to every other body the ability to come into contact with the extended body, which can be called a reference system or relative space where the location, the coordinates of each point, are relative to some body. If the concept of motion assumes such a construct, it is relative motion.

The second tendency — a picture of space in which we do not see differences — renders its parts identical and thus removes material bodies from it. Such an abstraction eliminates the multiformity of being; it is bleaching and devastating abstraction. Its result is empty space to which several invariant properties are ascribed, and it is considered responsible for phenomena, for the appearance of the forces of inertia during acceleration.

The appearance of absolute space in Newton's system illustrates the mechanism of the coming-to-be of this absolute. Initially, while Newton speaks of the motion of a system without acceleration, this motion has a concrete character. It is related to other bodies, it makes sense when it takes place against the background of the heterogeneous Universe where discrete bodies are interspersed with spatial intervals. But in the transition to accelerated motion, Newton replaces this heterogeneous Universe with the absolutely self-identical, abstract, devastated background devoid of sensorily comprehensible interspersions. This Newtonian absolute space is the result of devastating abstraction, the result of purely abstract identification.

But the fundamental principle of classical mechanics remains the principle of relativity. It is limited only by the absolutism of Newton, and in seventeenth- and eighteenth-century science, the relativist opposition to such limitation did not cease. Motion itself is the physical equivalent of true self-identity, which includes difference. The essence of motion is the self-identity of the moving body, but it includes differences, nonidentity — it is not trivial. Motion in the sense of dislocation is the simplest equivalent of nontrivial identity and, accordingly, of the doctrine of essence in classical science. The most general equivalent is the concept of predicative multiformity with which we are already familiar.

4. EXISTENCE

The evolution of the principle of relativity allows us to understand more precisely the concrete character of such an abstract category of Hegelian logic as *existence*, namely the concept of essence. The object of thought is identical to a set of other objects but it is also distinct from them. Therefore the identification with other objects does not lead to tautology, to the destruction of the multiformity of the world in identical being that lacks content and is equivalent to 'nothing'. In uniting the given object with others, in discovering their identity and thus finding the *basis* for the given object, we do not destroy its individuality, we do not dissolve it in identity, we do not identify it with the essence lying behind it. The essence must be realized in some new object of thought that is distinguished from it. "Essence must appear" (Hegel 1969, p. 479). This thesis of Hegel signifies that true thought, i.e., thought containing the truth, thought with content, unites identification with individualization.

This is one of the most characteristic features of the new rationalism, which had drawn closer to science, to the experimental comprehension of the world in its becoming, motion and evolution. Thought with content is thought about something that appears and is sensorily comprehensible.

For abstract rationalism appearance seems to be immediate, not having logical mediacy and therefore foreign. But not for the new rationalism, which embraces the motion and connection of objects. For dialectics an appearance is connected with others, is substantiated, has a ground and only in this case is it real.

For Hegel there is a very complex and especially abstract-logical construction which expresses the most concrete tendency of his philosophy — the transition from thought to its content as to something independent of thought.

Appearance is something immediate. The concept of the 'immediate' means that the appearing object is not a logical conclusion but rather something independent of the logical conclusion. Appearance is the appearance of *essence*, it is substantiated, it has a ground. But the ground has already appeared, it has already moved into the appearance. Hegel says that the ground "has *fallen to the ground*" (Hegel 1969, p. 478). Here is the resolution of the *aporia* of which Émile Meyerson wrote so much in the works mentioned in the first part of this book: if the ground and the effect are identical, then everything is compressed into a localized act which is self-identical in time and space; the world loses its multiformity, is compressed into a point and moment, and thus disappears. In accordance with ground and appearance, Hegel emerges from such an *aporia* by means of what appears at first glance to be a linguistic sleight of hand. He says that once the ground has completed its business, it disappears, and grounding subsequently acquires immediacy.

Hegel writes: "Ground, therefore, does not remain behind as something distinct from the grounded . . ." (Hegel 1969, p. 478). But it does not merge with it. What has been grounded turns out to be no longer grounded but absolute, not mediated but immediate. Meyerson's *aporia* of ground and result, cause and effect, is removed.

"Consequently", Hegel continues,

> the fact is not only the *unconditioned* but also the *groundless*, and it emerges from ground only in so far as ground has '*fallen to the ground*' and ceased to be ground . . .
>
> This immediacy that is mediated by ground and condition and is self-identical through the sublating of mediation, is *Existence* (Hegel 1969, p. 478).

Existence is immediacy. Hegel repeats, develops and considers this thesis in its various aspects. The first assertion of the doctrine of being (of *concrete being*!) is "being is *essence*" (Hegel 1969, p. 479). In other words, it is mediated, relative, contains content. The first assertion of the doctrine of essence is "*Essence is being*" (*ibid.*). It appears. In other words, it becomes immediate. "The truth of being is to be, not a first immediate, but essence that has emerged into immediacy" (Hegel 1969, p. 481).

Consequently, the truth of being is the removed, i.e., to some degree retained, ground and conditionality, and, to the degree to which they are removed, immediacy and unconditionality.

> But when further it was also said, '*Whatever exists has a ground and is conditioned*', then equally it must also be said that it has no ground and is unconditioned. For Existence is the immediacy that has emerged from the sublating of the mediation by which ground and condition are related, and in emerging it sublates this emergence itself (Hegel 1969, p. 481).

The science of the seventeenth and especially of the eighteenth centuries was driven toward immediate knowledge. It was experimental science, and the data in the experiment seemed independent of the traditional apparatus of the mediation of scientific truths and their deduction from fundamental principles in the spirit of the old metaphysics. But, as was already mentioned in the first part of this book, real experiment was the continuation of thought experiment — of the form of logical mediacy and the deduction of adequate representations of nature. Logic led science to experiment, grounded it and 'fell to the ground', when experiment found and demonstrated the ground of appearance — directly observed appearance. But these appearances were "essence that has emerged into immediacy" (Hegel 1969, p. 481).

Hegel's doctrine of existence was a reflection of the vital dynamics of classical science, a very logical reflection, but one that both hid and expressed the departure of science from the purely logical mainstream. More precisely it expressed the invariable irreducibility of the real development of science to the purely logical mainstream.

The theory of relativity was a demonstration of the transition from essence to appearance inherent in science. It was not the first demonstration but it was much clearer, more open, and revealed its epistemological meaning. The theory of relativity is in this sense a theory of existence. Like every physical theory, it operates on grounds which have passed into appearances, on essences which have appeared. The questions that Einstein's theory answered were questions of *existence* — the existence of motion in relation to the ether, the existence of the ether itself, the existence of a single simultaneity for various reference systems, the existence of velocities that exceed the velocity of light, etc. In each case, existence signified the possibility, in principle, of the experimental proof of existence, i.e., of proof of the transition of essence into appearance, "the submergence of essence", the appearance of immediate criteria of being. In other words, we are dealing with the criterion of external confirmation, which was the heuristic criterion that led Einstein to the theory of relativity. The essences listed above (motion in relation to the ether, the ether itself, absolute simultaneity, etc.) did not pass the test of existence; in principle they could not be revealed experimentally, they could not appear.

Is there not a certain *Natur*-philosophical tendency in the inclusion of Einstein's theory in the more general concepts of Hegelian dialectics? No, there is no such tendency here, because such an inclusion changes the meaning of both the theory of relativity and the epistemological concepts of the *Science of Logic*.

The assumption of such a change in meaning of unusual general philosophical concepts, on the one hand, and of the physical theory supported by experiment, on the other, may seem not only a relapse into *Natur-philosophie* but also an aggressive *Natur*-philosophical intrusion into an alien sphere; it may seem an intrusion of experimental data into the sphere of pure reason. The next chapter will be devoted to a critique of such fears. At present, while not raising the question of the

Monroe Doctrine for physics ('physics for the physicists') and for philosophy ('philosophy for the philosophers'), we must try to examine how the concept of existence is reinterpreted under the influence of science and how several tendencies of the theory of relativity change along the same lines in which Hegelian 'existence' is evolving and becoming more complex and concrete.

5. ACTUALITY

If it is considered as one of the latest equivalents of the constructs of Hegel's dialectics, the genesis and evolution of the theory of relativity is the genesis and evolution of a theory of *actuality*. It suffices to recall, with no need to prove, that in this case, 'equivalent' signifies not so much a resemblance to as a contrast between question and answer, between a philosophical question and a physical answer, between a question raised by the nineteenth century – and an answer that reverberated in the twentieth.

"Actuality", says Hegel, "is the unity, become immediate, of essence with existence . . ." (Hegel 1975, §142, p. 200). After some fleeting acquaintance with these concepts – essence and existence – the meaning of this phrase is not difficult to understand. Existence is something separate from essence, grounded by essence, but grounded in such a way that essence "has fallen to the ground". Thus essence no longer figures in existence. On the contrary, what is actual is existence taken as an expression of essence; it is something indivisible, where essence and existence have united.

Such a unification eliminates the extremely prevalent and the extremely trivial contrast between thought and actuality. This contrast is the target for Hegel's poisoned arrows, the occasion for the transition from very difficult logical constructs to a burst of polemical temperament. Hegel says that the representation of thought as something removed from actuality and of actuality as something removed from thought is the representation of people who have not understood, as they should, either the nature of thought or the nature of actuality: "Thought in such a case is, on one hand, the synonym for a subjective conception, plan, intention, or the like, just as actuality, on the other, is made synonymous with external and sensible existence" (Hegel 1975, §142, *Zusatz*, p. 201).

For the philosopher, actuality is illusory if it does not embody thought, just as thought is illusory if it is not directed toward actuality. Actuality is cut off from thought and unreasonable only for those who "have not properly apprehended the nature either of thought or of actuality" (Hegel 1975, § 142, *Zusatz*, p. 201). The thinker does not turn away from actuality; he sees thought in it. But Hegel finds alien the conception of actuality as an emanation of thought, just as he finds alien a conception of thought which descends to actuality and is degraded, as in the Neoplatonists or which is seen as the single true substance, as in Plato himself. In expounding his conception of actuality, Hegel returns to the sources of the antithesis between thought and the sensory perception of the world. When we project into the past the treatment of thought as a synonym for a "subjective conception" and actuality as a synonym for "external and sensible existence", we see Aristotle as the founder of empiricism. Hegel's position is different. He sees that the Aristotelian philosophy of actuality was a philosophy of thought inseparable from actuality. From a true historical perspective, Aristotle is seen, correctly, as the founder of actual rationalism, which is not limited by speculation, but perceives the world in its sensorily comprehensible concreteness. Hegel says that, unlike Plato, who saw in the idea only the moving force of the world (*δύναμις*), Aristotle viewed the idea as 'energy' (*ἐνέργεια*),"as the inward which is quite to the fore, or as the unity of inner and outer . . ." (Hegel 1975, § 142, *Zusatz*, p. 202).

Here we approach the well-known "what is rational [reasonable] is actual and what is actual is rational [reasonable]" (Hegel 1967, p. 10; 1975, § 6, p. 9).

This phrase, included in the preface to *Philosophy of Right* and in the introduction to the *Logic*, expresses one of Hegel's fundamental epistemological and ontological ideas. It found many advocates who saw in it a reconciliation with surrounding actuality, even more opponents who also saw such a reconciliation, and who protested against it, and a very few people, like Heine and Herzen, who understood its real meaning. When we examine eighteenth- and nineteenth-century classical rationalism and classical philosophy in this book in the light of nonclassical science, we must limit ourselves to the ontological meaning of "what is actual is rational".

Philosophical literature threw less light on the ontological than on the sociological and moral conclusions drawn from Hegel's formula. Apparently classical science yielded neither enough impulses nor enough data to connect concrete conceptions with the general problem of the reality of the world and its rationality, the existence of the objective *ratio* of the macroscopic ordering of the Universe. Nevertheless, several physical conceptions had already raised the issue in the nineteenth century. At present it has become an actual, specifically physical problem, about which we will speak in the next section.

In the introduction to the *Logic* (Part I of the *Encyklopädie der philosophischen Wissenschaften*), Hegel returned to the phrase "What is reasonable [rational] is actual and what is actual is reasonable [rational]" and explained what it means.

Hegel says that the discontinuous and transient elements of actuality, as the word is understood in everyday speech, do not enter into the strict and non-contradictory concept of actuality. "But even our ordinary feelings are enough to forbid a casual (fortuitous) existence getting the emphatic name of an actual; for by fortuitous we mean an existence which has no greater value than that of something possible, which may as well not be as be" (Hegel 1975, §6, p. 9). It is necessary to dwell on this characterization of the contingent and the *possible.*

One's first impression of the passage quoted above is immediately associated with the statistical lawlikeness of classical science and with the distinctive, nonclassical, statistical natural laws of quantum physics. Do such natural laws contradict the tendency expressed by Hegel — the extension of the contingent beyond the limits of the actual?

There is no hint here at the subordination of concrete natural laws to *a priori* schemata. The widespread authority of such natural laws under the concept of Hegel's dialectics ("oats grow according to Hegel") contradicts, first and foremost, this very concept. If "oats grow according to Hegel", it means that they grow in their own way; the essence of dialectics lies in the irreducibility of the concrete to the general. But such irreducibility becomes precise when compared with general categories and, what is especially important, it itself forces the modification of these categories. What happens with statistics — classical and nonclassical — and what happens with the general category of the contingent as an element that is included or not included in actuality when these poles are compared?

The Hegelian conception of the contingent does not lead it beyond the limits of the objective; it does not view it as something purely subjective. Chance is only one moment of actuality but it is found in the world of objects.

On the surface of Nature, so to speak, Chance ranges unchecked, and that contingency must simply be recognized, without the pretension sometimes erroneously ascribed to philosophy, of seeking to find in it a could-only-be-so-and-not-otherwise (Hegel 1975, §145, *Zusatz*, p. 206).

Hegel's philosophy in general and the conception of chance in particular are by no means a generalization of Laplacean determinism. But they contain only the vaguest hints at the transformation of the contingent into the necessary and at the statistical path from essence, law, natural law to existence, appearance and the synthesis of existence and essence — actuality. The 'operative', transforming function of the comparison to the Hegelian conception of chance consists in the fact that the designated conception becomes different, it loses its linguistically logical character and acquires a very concrete aspect. Of what does the transforming role of comparison consist in relation to the physical concept of statistical lawfulness?

Philosophy and its history, in this, as in the majority of other cases, cannot indicate the direction in which the elaboration of the concepts of chance and necessity in physics will go. If we portray this direction by means of a vector, then historico-philosophical analysis cannot change this vector. It can only enlarge its module, hasten the realization of some determined tendency. However, perhaps the general direction of science will to a certain extent turn out to depend on the module of the tendencies noted at present.

Among these tendencies we now observe the expansion and deepening of statistical probability concepts. In the original, nonrelativist quantum physics, the conception of wave probabilities appeared as something that limited the Laplacean determinism of being. The correlation of indeterminacy indicated the limits of a simple and precise picture of the world; it found in it several pockets where the world point is diffused and the world-line acquires a blurred appearance.

Now in elementary particle theory and in quantum and relativistic conceptions as a whole, we are dealing with something different, namely macroscopic and cosmic processes that are essentially dependent on microscopic processes, quantum ultramicroscopic natural laws from which the macroscopic, relativistic objective *ratio* of the Cosmos grows. Here physical ideas become so broad that they find their historical antecedents not so much in classical physics as in classical philosophy. This stage in the development of science no longer corresponds to the tendency to hem in classical rationalism with corrections, and classical determinism with nonclassical reservations. The goal is to search for a new, universal, objective *ratio* of the world which will explain both Cosmos and Microcosm on the basis of single premises. Therefore the idea of the rationality of the universe ("what is actual is rational") and the idea of the objectivity, the materialism and the experimental knowability of the rational schema of the world ("what is rational is actual") are very close to modern science.

6. THE THEORY OF RELATIVITY AS A THEORY OF ACTUALITY

The negative aspect of the theory of relativity, at first glance, is placed in the framework of the conception of existence. The positive aspect – the assertion of four-dimensional space-time, the alteration of its metrics and gravitational field, the homogeneity of pseudo-Euclidean (the special theory) and curved (the general theory) space-time – includes more than the demand for the experimental proof of essence, its transition into appearance. These assertions meet the opposite demand: appearance must possess ground, must flow from more general conceptions and be identical to some degree to other appearances. Compared to the Hegelian categories, the positive aspect of the theory of relativity seems to demand the transition from existence to reality.

But in fact everything here is more complex. The negative aspect of Einstein's theory is the rejection of not only absolute simultaneity, absolute space, time and motion as they were in the eighteenth and nineteenth centuries, but also Lorentz's initial conception of contraction.

We have come so far from the Introductory essay where this conception was mentioned that perhaps we should recall it again. Lorentz explained the results of Michelson's experiment – the independence of the velocity of light from the motion of the system in which its velocity is measured – in terms of a longitudinal contraction of moving bodies and the dilation of time. The inadequacy of such an explanation lies in its artificial character. The hypothesis of longitudinal contraction was expressed *ad hoc* – namely to explain an isolated experimental fact. It was not derived from a more general principle. Because of the epistemological inadequacy of Lorentz's hypothesis, science turned to Einstein's conception in which there was no longer a 'real' length of a rod at rest and no reduced length of the longitudinally moving rod in an absolute sense. Einstein viewed these rods as having equal rights; it is possible to ascribe rest to one, and then the other rod will be shortened. This reciprocal contraction stems from very general assumptions, and the conception lacks an artificial character; it possesses inner perfection. Such an epistemological advantage becomes physical: Einstein's theory explains a set of other experimental data; it predicts new data and finds unexpected spheres of confirmation and application.

It is not difficult to see that the criterion of external confirmation moves along the same line as the criterion of inner perfection, only in the opposite direction. In the first case 'essence becomes being', it *appears*. In the second case, 'being becomes essence', existence merges with essence and becomes *reality*.

The criterion of inner perfection, which supplements the criterion of external confirmation, transforms the theory of relativity into a theory of actuality. It thus reveals a very important aspect of this category. Hegel saw in actuality something active, developing, effective. But this aspect of the matter was not precisely expressed, as well it could not be: Hegel's system not only explained the real transformation of the world but also *replaced* this transformation with a logical explanation. The merging of the active transformation of the world with its explanation is no longer Hegel, but Marx. Active dialectics is the dialectics of Hegel turned upside down, a dialectics in which the initial definition of being is already the definition of *concrete* being, i.e., not only initial but also mediated, and grounded by the real influence of material substance on the sense organs and by the active influence of man on material substance — by experiment.

The comparison of Hegelian categories with nonclassical physics pushes these categories to the fore; it likens them to the real dialectics of real being. The evolution of the theory of relativity brings relativity close to actuality as the synthesis of existence and essence, as an increasingly consistent and broad revelation that exhibits an increasing degree of external confirmation and inner perfection. We might say that it is a growing revelation of essence, which is [*appears*], i.e., passes into existence, and a revelation of existence, which seeks and finds a new essence, a new ground in exchange for what has already been realized and has "fallen to be ground". We might say this if the content of the theory of relativity did not go beyond the limits of Hegelian constructs, if it had not been the *transforming* equivalent of them.

The evolution of the theory of relativity moves in two directions, which may be distinguished to some extent. The first is the search for growing inner perfection, the search for increasingly general fundamental principles. In Einstein's works this direction acquires a geometrical character in the unified field theory. Einstein, like Weyl and several others, sought a more general geometry than the Riemannian in order to deduce from such a geometry not only a theory of gravitation but also a theory of other fields that would be explained as separate instances of unified equations. The other direction consists of the very rapid growth of the number of particles known to science and, correspondingly, of the number of fields, as well as the spectrum of the particle interactions. For several new particles and for the old particles in very intensive interactions, the transformation of particles of one type into particles of another type is characteristic, subject not only to the laws of quantum mechanics but also to the laws established by the theory of relativity. In this quantum relativistic sphere, many great experimental discoveries have been made. Here the most obvious is the search for and discovery of *appearances*: development went or at least seemed to go along the line of *existence*. Essence was clothed in new appearances which demonstrated existence.

Over the course of time, empirical material relating to various types of elementary particles, to their interactions and to the processes of formation and disintegration, acquired dimensions that obviously testified to the need for a unified theory of elementary particles. The entire history of the physics of elementary particles in the 1930—1970's demonstrates the connection between tendencies coming from opposite directions — the search for external confirmation and the search for inner perfection.

The fate of the unified field theory shows the same thing. It grew out of the search for inner perfection; here existence aspired to essence. Initially this tendency could be singled out. At the end of the 1930's Einstein said that the existence of the electron, which contradicted continuum electrodynamics, could become the starting-point for a general theory of elementary particles. Such a view corresponded to the actual genesis of the theory of relativity. I. E. Tamm writes about Einstein's remarks:

It is undoubtedly hyperbola but it is very characteristic of Einstein and it is instructive to contrast it with the prevalent view that the resolution of the fundamental problems of science must necessarily precede the accumulation of a great number of experimental data. In reality, the example of both the special and, in particular, the general theory of relativity demonstrates that the profound logical analysis of major experimental facts plays a decisive role in the construction of a fundamentally new theory. Of course the corollaries of the theory must later be verified many, many times over (Tamm 1956, p. 8).

During the 1940's a period of simultaneous and parallel growth began in the number of rediscovered immutable facts concerning elementary particles and their interaction and in the number of not altogether immutable hypotheses and schemata. These hypotheses and schemata by no means claim immutability and simplicity. They contain assumptions not so much about the structure of the world as about the subsequent development of science itself; in theoretical physics, the predictive motif resounds continuously. A noncontradictory and unified theory of elementary particles must appear which will supply inner perfection to the very precise, considerable and experimentally verified information gathered in the last two decades or so. The contours of this theory — still indistinct and in rough outline — are gradually being defined but no one claims to have a simple explanation for all the known facts. A high level of modesty is typical of the scientific literature of recent years, in contrast to the uncontrolled 'sound the thunder of victory!' which so often resounds in the popular literature on contemporary physics.

Let us attempt to transfer the modern situation, the modern clashes of theoretical physics, onto the plane of the most general concepts of existence, actuality and being. For this we shall recall the ideal toward which Einstein aspired in his search for greatest inner perfection in the theory of relativity.

It was a lofty ideal, the most lofty ideal of rationalism. It consisted of a model of the world in which there are no empirical constants. Constants exist in physics — the velocity of light, Planck's constant, the masses and charges of particles and a number of others which are beyond rational explanation and are represented as empirical. Einstein wrote of such a model of the world in 1949; we will later return to this very important epistemological construct (Einstein 1949, pp. 62—63).

At the beginning of the seventeenth century Kepler wanted to deduce the constants of astronomy — the median distances of the planets from the sun — by starting with a logical and geometrical schema and thus deprive the indicated constants of their empirical nature. This was beyond the power of the prologue of classical science.

But the idea of the logical deduction of the entire universe with all its empirical constants was the central idea of seventeenth-century rationalism in its highest embodiment, the system of Spinoza. If everything is necessary, if the very original conditions of all processes turn out to be deduced from others in the infinite rotation of *causa sui*, then empirical constants are subject to explanation. This was not a program; it was an ideal. It does not project the immediate goals of science; it demonstrates a tendency inherent in science.

Hegel expressed the idea of the logical deduction of physical constants. Such an idea is characteristic of his fundamental tendency — of the representation of actualizing essence and existence seeking ground. In our time this highest logical ideal of rationalism has become more than logical and more than an ideal. Heisenberg's idea of the calculated and not only observed spectrum of the masses of elementary particles is aimed at very fundamental constants of nature and is a program (for the time being only a program!) for the realization of the rationalist ideal.

But it changes during this process. Let us recall that in contemporary physics the tendency to explain the localized processes in the structure of the Metagalaxy is becoming more and more pronounced. At present in astrophysics elementary particle behavior is explained more or less successfully in terms of the behavior of galaxies and stars. But the Cosmos in its turn is introduced as a leading actor in a localized tragedy, enacted in spheres on the order of 10^{-13} cm and 10^{-24} sec. Previously, the rational analysis of being went from large objects to the small, into the depths of matter. The rational comprehension of a physical object amounted to answering the question of what does it consist. Now, with increasing frequency, in order to reveal the nature of a particle, one appeals to the Macrocosm and the Cosmos. Particles that are called elementary are elementary in the classical sense: subsequent searches for sub-particles no longer lead to knowledge of the nature of the particle. Elementary particles disintegrate into others not because they previously contained these others within themselves; such an explanation has been replaced by a new one. Not only the behavior but also the very being of each particle is determined by the influence exerted on it by other particles, perhaps all the particles of the Universe that interact with it. The explanation of the nature of the particle goes from the particles smaller in mass to the larger; a particle of small mass 'consists' of particles of large mass — perhaps the being of the Universe is concentrated in its being.

We shall soon return to these paradoxical conceptions. In all their physical complexity they are certain in one relation: no matter how the fate of each of these conceptions is determined, they all, as a whole, provide a basis for the certain deduction of a general tendency, the transition of modern physics from the classical 'of what does it consist?' to the more complex forms of what Hegel called 'essence'.

Such a tendency alters the logical nature of 'essence', 'existence', and 'actuality'. It takes away from the absolute limit of rationalist explanation, from the dead end of the causal chain. In the integral and stable cosmology of Aristotle such an end was the finite dimensions of the Universe and its eternally established center and the dislocation of natural locations. In Cartesian continuum physics such an end was the *a priori* laws of motion established by God, the impenetrability of matter and its very being. Only in Spinoza could being not provoke the question of subsequent grounding — it played the role of *causa sui*. In general, the finite world, like the infinite one, defied rationalist comprehension, because the Universe shown in this example could not be identified with others in a unified set, with a general law. Here the identifying function of reason ceased.

Atomism, i.e., the limitation of the divisibility of the world, also included the dead end of rationalist, causal analysis. Here the individualizing, not the identifying, activity of reason ceased. The atom (we mean the classical atom, whose name corresponds to its assumed indivisibility) is a small homogeneous cell in a heterogeneous world. It is the being of Parmenides in a Heraclitean world. Within the atom there is no heterogeneity, no physically distinguishable parts; space and time are no longer dynamic variables, characteristics of motion, or in general, characteristics of heterogeneous being. In this sense classical atomism was a theory of cells extended 'from without', but 'within themselves' devoid of heterogeneity. Space, devoid of heterogeneity, become devoid of content, pure being equivalent to *nothing*.

In the nineteenth century atomism became a doctrine of the hierarchy of the discrete parts of matter. Each level of the hierarchy is divisible, heterogeneous; it possesses physical, concrete being. In the twentieth century such a hierarchy has been pushed to the final level — elementary particles. But this is not the limit of rationalist explanation but rather a point of transition to a more complex rational comprehension of the microcosm, to an appeal, no longer to included smaller levels, but to inclusive, greater ones.

Such a transition is illustrated by the above-mentioned conceptions of the particle, whose being is determined by the being of all particles and by several even more complex conceptions. We shall emphasize Chew's 'bootstrap' theory: the particle owes its existence to other particles interacting with it. These are powerful interactions. The particles connected by such an interaction form a self-coordinated system which exists as a whole: it is its existence, and not the availability of the sub-particles, that forms the *essence* of each particle.

Here we already see the fundamental turning-point not only in physical representations but in the logic of physical thought as well. In classical science, primarily abstract, identifying analysis led to greater, inclusive systems, from the microcosm to the macrocosm. Concrete, individual, but not individualized, identified representations of the microcosm — molecules — form an abstract molecule that is absorbed in identifying sets which are the subject of macroscopic thermodynamics. Now the growth in the wealth of determinations goes from the particle to the Cosmos.

This existed also in classical science and, in the form of epistemological generalizations, within the framework of classical rationalism. But nonclassical science, in its constructs, which are new in principle, gives an unprecedented precision to this motion toward concrete being.

This tendency is illustrated still more precisely by representations of elementary particles as the direct synthesis of particles that are greater in mass. This paradoxical complexity of elementary particles is based on a deduction from the theory of relativity: as "the mass of a body is a measure of its energy content" [Einstein 1905, p. 639],[1] so intensive interactions take away part of the total energy of the particles that are united into a whole and decrease the mass of the system composed from them. This decrease, called a mass defect, allows the formation of a particle of small mass from particles of greater mass.

Gell-Mann suggested that particles entering into powerful reactions, i.e., all particles excluding photons, electrons, positrons and neutrons, each consist of three 'quarks'. The word 'quark' is taken from Joyce's novel, *Finnegan's Wake*, where it designated a sort of fantastic being. Quarks possess great mass and interact very energetically within the particle containing them, causing the absorption of a significant amount of energy; the mass of the particle then becomes less than the total mass of the three quarks forming it.

By taking into account the mass defect, it is possible to represent a particle of small mass that contains the Metagalaxy within it. This was Fridman's conception as expressed by M. A. Markov. It examines the purely relativist problem of the dependence of the mass of the Universe on the gravitational interaction of the bodies comprising the Universe. This gravitational interaction provokes a corresponding mass defect. It may be so great that the mass of all particles will be absorbed by their interactions and the entire mass of the Universe will be reduced to, or will approach zero. This is the closed or 'almost' closed Universe. What will the dimensions of the Universe be when it is 'almost' closed?

The geometry of an 'almost' closed system is very paradoxical. Metrics are deformed by the gravitational field. In such a non-Euclidean system, spherical surfaces described around some center grow at first and are then destroyed, are compressed into a point. Therefore, to the outside observer, the system may seem like a point, have microscopic dimensions, while in terms of its 'inner' dimensions it may be very large and include millions of galaxies. In Markov's opinion, the unstable Universe, with a comparatively large mass, will evolve and arrive at a stable condition when its mass becomes very small. Markov called this model a *Fridman* universe.

As was already mentioned in the Introduction, the very existence of the mass of an elementary particle, a mass which expresses the symmetry of the Universe and contradicts the dissymetricalized localized fields, may be connected with the Metagalaxy (see Kuznetsov 1970b, pp. 349—420; 1966, pp. 207—208).

When one goes from the particle to the Cosmos in order to explain such fundamental constants of physics as the masses of elementary particles, there is a significant change in more than just the traditional definition of the concrete and abstract as mutually exclusive paths of analysis. The definition of the abstract and concrete that was expressed by Hegel and existed for the most part in vague form throughout the entire evolution of classical rationalism, was also altered. For Hegel, abstract analysis, generally speaking, leads to the concrete, to the actual concrete. But both the actual and all the other categories with content in his *Science of Logic* express concrete content, viewing it as the other-being of the developing spirit. They bear the imprint of the 'beginning' category of being — of abstract pure being without content equivalent to pure *nothing*.

What does being represent from the point of view of the tendencies of modern physics mentioned in this section, beginning with the fundamental tendency of the theory of relativity to deny in a physical sense the position and motion of an absolutely isolated body?

It represents the Cosmos in its entirety, the entire accumulation, the entire self-coordinated system of powerfully interacting particles, it is other forms of All — of being as a whole and not of abstract matter, but of matter entirely concrete with some median density, expanded or not expanded, perhaps pulsating, perhaps possessing other, still more contradictory and complex properties.

If one takes the theory of relativity and all of modern physics with its tendencies that have not yet been transformed into simple representations, with its predictions and searches, then more than just the concept of actuality as the synthesis of existence and essence corresponds to it. The integral concept of being corresponds to it, not the pure, homogeneous being that excludes nonidentity, the first of the series of concepts in Hegel's *Science of Logic*, but heterogeneous being, which does not have *a priori* foundations and is the essence of experimentally observed processes.

The problem of the axiomatization of physics is connected with this.

In 1915—1916, David Hilbert published 'The Foundations of Physics' (Hilbert 1915—16), which has a dominant idea somewhat analogous to the dominant idea of his *Foundations of Geometry* (Hilbert 1899; 1971), where the subject is the axiomatization of geometry. 'The Foundations of Physics' is devoted to the theory of relativity and the perspectives of the complete transformation of physics into an axiomatized science. Hilbert spoke of the possibility of a relativistic theory of the microcosm. The physics of the second half of this century is approaching this and never before has there been such a pressing problem. Hilbert thought that the principles of the theory of relativity would ultimately allow the deduction of physical constants from axiomatized principles.

Such an ideal corresponds to the initial idea of a unified field theory. But the realization of this idea has altered it. Because of the theory of relativity, the axiomatization of geometry itself has acquired a 'beginningless' character. The original representation of an axiomatized physics recalls the Hegelian conception of *beginning*: axioms — the ultimate example of the identifying function of reason — serve as the beginning of the deduction of more concrete theorems. The axioms of geometry as *a priori* or based on 'agreement', resemble Hegelian 'pure being'; they are *physically* without content.

In modern understanding (based on the theory of relativity), geometrical axioms are themselves mediated, and mediated under physical conditions. The Euclidean postulate is correct in the absence of gravitational fields. In such an understanding, the axioms of geometry contain physical content, are concrete and no longer suggest abstract 'pure being', but concrete being. Generally, they do not destroy the world, but fill it. The physical axiomatization of geometrical axioms returns analysis to the Spinozan *causa sui*, to the snake biting its tail, and the 'first' principles of physics turn out to be the result of the concrete being of the heterogeneous evolving world.

Thus the theory of relativity, examined not 'on a slide' (*in vitro*), but in its vital dynamics (*in vivo*) is not only a theory of actuality, but also a theory of being, the concrete heterogeneous being embodied in the Cosmos; it does not ignore but rather explains the microprocesses and finds in them the explanation of its innumerable modes.

A certain new definition of being is connected with such a definition of the first (but by no means 'absolutely beginning') concept of physical explanation. It is not reduced to existence — to the synthesis of appearance and its individual, separate essence. The physical object acquires physical content, physical being, its distinction from nonbeing in its connection with the Metagalaxy. Therefore the definition of being includes the localized, the individual, the particular, and the world-line is the behavior of the object.

NOTE

[1] [This phrase is from Einstein's second paper on special relativity (Einstein 1905), which contains the relation, just stated, $E = mc^2$.–Eds.]

11. UNDERSTANDING AND REASON IN NINETEENTH- AND TWENTIETH-CENTURY SCIENCE

1. UNDERSTANDING, REASON AND THE GRADIENT OF SCIENCE

What is peculiar to nineteenth-century science, what distinguishes it from seventeenth- and eighteenth-century science would seem to be the triumph of reason over understanding. But seventeenth- and eighteenth-century science, unlike Aristotelian science, represented a similar victory. Each shift to new foundations of science is the shift from identity that opposes nonidentity to identity that is inseparable from nonidentity.

The distinction between reason and understanding was introduced by Kant as a kind of limit to thought based on understanding which comprehends finite objects. Reason comprehends the infinite. But for Kant it comprehends it only abstractly, identifying objects of thought based on understanding and grouping these identified objects in homogeneous and generally infinite sets.

Hegel writes of reason as understood by Kant: "And the result of all is that Reason supplies nothing beyond the formal unity required to simplify and systematize experiences; it is a *canon*, not an *organon*, of truth, and can furnish only a *criticism* of knowledge [cognition], not a *doctrine* of the infinite" (Hegel 1975, § 52, p. 86).

Hegel emphasizes the formal character of such a drawing out of cognition beyond the boundaries of the finite. Given a simple, abstract grouping which destroys concreteness by grouping it in a set of identified and depersonalized elements, nothing is changed: neither the set, nor the grouped elements that are related to the abstract set. Therefore Kant's philosophy exerted no influence on science. "Occasionally, it may be, in the first sections of a scientific work of that period, we find propositions borrowed from the Kantian philosophy; but the course of the treatise renders it apparent that these propositions were superfluous decoration, and that the first few pages might have been omitted without producing the least change in the empirical contents" (Hegel 1975, § 60, p. 92).

This charge against Kantian natural philosophy is naturally redirected to all natural philosophy (including, perhaps, even Hegel's own 'Philosophy of Nature'). Abstract, identifying grouping without the transformation of the objects grouped (when the symbol $\in$ is not the operator) does not change the material grouped; it *deprives* it of predicates rather than adds them. Such a path leads to ravaged, impoverished abstraction.

Now it will not be difficult for us to introduce an example of transforming inclusion, an example which is, moreover, physically and experimentally confirmed. The measurement of the coordinates of a particle brings about a change of velocity, i.e., the transformation of the whole world-line, of an all-inclusive set. We once again become convinced that classical philosophy prepared the way for nonclassical science and that it did this by generalizing the movement of science and not only its results.

Kant considers reason to be a manifestation of the primordial self-identity of the cognition 'I' and ascribes an identifying function to it. The activity of reason in the framework of Kantian subjective idealism leads to identification. For Hegel such a conception of reason leads him to understanding. Thought based on understanding takes as its object something finite. But it takes it only so as to identify it. The process itself of the apportionment of objects, the delineation of their borders, the differentiation of sides — in a word, the isolating function of understanding — leads only to grouping, to a translation from the concrete into the abstract.

It is not possible to differentiate between understanding and reason, ascribing to the former an identifying function and to the latter a search for nonidentity. Thought based on understanding in no way ignores the heterogeneity of being, the presence of differences and changes. Only the absolutization of the identifying function of understanding ignored them — the conception of Parmenides. Understanding grasps differences and nuances in nature, divides phenomena into classes, and the systematizing function of science can result from thought based on understanding.

Reason also possesses an identifying function. But it identifies phenomena at a higher stage. Understanding arrived at the identity of natural places, which are identical to themselves and do not change — this is the Aristotelian, static scheme of the universe. The result of thought based on understanding was the kinetic scheme of the Universe with determinate, identical correlations between intervals of time and paths covered.

But the function of reason, the result of rational thought that is not reducible to understanding, was the *shift* from the static scheme to the kinetic, as well as from the kinetic scheme of being to the dynamic. And each shift was to a higher and more meaningful identification. Such identification is the affirmation of difference, of nonidentity, and not only its denial. The establishment of uniform acceleration in itself is an identifying act. It rejects diversity in the behavior of a body, it establishes a certain invariant, which is order, law. But in relation to velocity, this is the affirmation of nonidentity.

By identifying being, reason discovers nonidentity. The way of reason is not to simplify the world, but to make it complex. Concreteness that is directly given does not possess a complex structure; it is without structure. It is the shift to a more complex structure as an object of thought that constitutes the way of reason. In affirming the changeability of what seemed immobile, in affirming the heterogeneity of what seemed unified, reason does not *isolate* objects but *individualizes* them.

Let pause for a moment at this distinction between the isolating function of understanding and the individualizing function of reason. For understanding, a separate subject is *only* the intersection of sets, the intersection of identifying lines, of predicate diversities. A given person is a man, an engineer, brown-haired, a Muscovite, a chess player, etc. Reason does not isolate the subject from such a predicate; without it the subject possesses only pure being, i.e., he is outside of real being, he disappears. But reason does not reduce inclusion in a predicate diversity to a subjective-nominalist function. It sees in such an inclusion a real operation. Therefore reason does not separate objects of thought from one another by ascribing to them absolute independence and transferring their connections to the sphere of purely subjective *rapprochement*. Reason reveals the effect of an object on the series that includes it and the effect of the series on the object. In this way the including series becomes not a simple summing up of identifying elements, but a complex structure whose complexity grows with each included link. This link in its turn really reflects the whole series and turns out to be a more complex structure. Reason reveals in nature the macroscopic being of the individual and the microscopic being of the set.

The distinction between understanding and reason is clearly visible in the conception of infinity or more accurately, in the conception of the finite and infinite as inseparable poles of being. In the original definition of reason and understanding Kant ascribed to the former comprehension of the infinite, and to the latter comprehension of the finite. But understanding is distinguished from reason not because it has no grasp of infinity, but because of the character of infinity that it constructs. Understanding constructs an infinite series of *isolated* and only subjectively linked elements that are independent of this series. The infinite here is separate from the finite, it excludes it, and the finite, in its turn, does not have within itself anything of the infinite. Hegel calls this split falsification.

The falsification of the finite and infinite by the understanding which holds fast to a qualitatively distinct relation between them and asserts that each in its own nature is separate, in fact absolutely separate from the other, comes from forgetting what the Notion of these moments is for the understanding itself (Hegel 1969, p. 145).

Hegel calls the infinite that results from such falsification 'spurious' infinity, and calls the infinity reflected in each of its finite elements 'genuine' infinity. Spurious infinity is the infinity of understanding, genuine infinity the infinity of reason (Hegel 1969, p. 137).

I do not want to illustrate Hegelian genuine infinity here with examples from mathematics or the natural sciences. The equivalents and results of reason and understanding cannot be differentiated in that way. Reason shows how the established identification shifts to a more complex one. But what we see in certain examples are just those established, fixed links of science; they illustrate not the activity of reason, which extends beyond the boundaries of understanding, but its activities within these boundaries. Reason amounts to change, transformation, the evolution of these links.

What I would like to do is to try to explain the meaning of the concept of genuine infinity by means of the following series of examples. Before us is a moving particle. It was first identified in terms of its velocity; then reason went beyond the boundaries of this identification based on understanding and discovered the nonidentity of velocity, that is, the fact that velocity changes. Change was also identified: constant acceleration was registered. Once again reason reveals its irreducibility to understanding, it registers the change of acceleration itself in different fields of force. Now the mass of the particle is identified, that is, the relation of force to acceleration; thus understanding again acquired status, and the motion of the particle is subject to a 'tranquil', ordered, identifying law. Let us go on to the next link in this series of clashes between reason and understanding. The particle moves so rapidly that the relativistic change of its mass is observed. Then understanding identifies its rest mass. But a further increase of the particle's energy brings about its transmutation; it is transformed into a particle of another type, with a different rest mass. The picture of motion becomes more and more complicated and concrete.

Each of these shifts reflects the entire infinite chain of world points, the entire world-line of the particle, in each of its elements, in each of its local states. Each shift shows that a new identification (after the identification of velocity, then acceleration, then mass, then rest mass) increases rather than decreases the richness of the concrete determinations of the particle's motion.

This may be considered a dynamic example; it shows a certain result of science that belongs in some kind of more general scheme. In our case, the example given requires that the scheme be generalized: it is now related not to the infinity that results from the addition of finite magnitudes but to the infinite that results from the division of a finite magnitude.

Obviously, reason that not only establishes laws (here it is within the boundaries of understanding) but also changes them, moves in a circle; it returns again and again to identification, to law, to the 'tranquil' reflection of the world, but each time it repeats this circle on a new level. Is there a beginning to this motion of reason? Hegel thought pure being to be such a beginning, equal to pure nothing. Science in its development came close to another conception, to the image of the snake biting its tail, to the idea of *causa sui*, to the causal chain that excludes beginning.

The movement of reason along such a spiral is a particular axis of the dynamics of cognition. Cognition shifts to transformations and invariants of increasing complexity; this is motion from one rank of complexity of the world to another, along an axis of ranks. Substance possesses infinite complexity. The cognition of substance is infinite.

How did the evolution of science and, on the other hand, the development in philosophy of the idea of the irreducibility of reason to understanding affect the relationship between science and philosophy?

In 1797, Schiller wrote a poem ('Naturforscher und Transscendental-Philosophen') in which he urged that natural scientists and philosophers keep apart:

Feindschaft sey zwischen euch, noch kommt das Bündniß zu frühe,
Wenn ihr im Suchen euch trennt, wird erst die Wahrheit erkannt.
(Schiller 1943, I, p. 331)

[What, already at war! Too soon you aimed at alliance;
As your paths divide will you acknowledge the truth.
(adapted from Schiller 1902, p. 323)]

Obviously this is a call not to war but to some sort of isolation, to the advancement of transcendental philosophy and natural science by different paths. Different, but not parallel. It is still early to think of union, but in the final analysis the paths intersect, and at their point of intersection lies the comprehension of truth. And then? Will the definitive truth be grasped? Will all future forward movement of natural science be reduced to the elaboration and illustration of the most recently discovered, definitive, fundamental representations of the world? Schiller's poem does not answer this question. The prognostic component of the poem is limited to an affirmation of the coming comprehension of the truth by differing paths. Transcendental philosophy, in developing and modifying its contents (perhaps even losing its transcendental character), moves towards truth, and does dictate *a priori* conclusions to natural science. Natural science moves towards truth, but does not borrow these conclusions from transcendental philosophy.

Schiller's appeal, as well as the prediction contained in his poem, met with a strange fate. In a broad sense both were realized. But not soon. Transcendental philosophy was not inclined to deprive science of friendly advice addressed to natural scientists. Often this advice was negative: It recommended that science not exceed its limits, which separated it from transcendental categories. It was well known that the basic message of transcendental philosophy — the *a priori* character of space and time — was off-limits to science: Thou shalt not go beyond the boundaries of Euclidean geometry. This ban was addressed not to geometry, which was free to build any scheme it wanted, but to science as a whole: Thou shalt not impart a real physical sense to non-Euclidean constructions. Obviously, this was not only a ban: logical transcendental philosophy in its teachings on space and time gave science a kind of positive conception — science comprehends *Euclid's world*. The overcoming of the *a priori* presumption of Euclid's world was in a certain sense natural science's answer to Schiller's appeal. But now we have gone too far ahead and have touched on the anti-Kantian subtext of the general theory of relativity. (This is a matter not only of the subtext but also of the text: Einstein wrote about the overcoming of *a priori* geometry as the precondition for the theory of relativity.) Let us return to Schiller's time and to the first generation of transcendental philosophers.

For this generation, friendship with natural scientists, which Schiller warned against, was not particularly pressing. For Kant himself the constructions of natural science related to the pre-transcendental, pre-critical period. His immediate disciples lacked the proselytizing temperament. But soon the situation changed. A broad natural scientific literature appeared in which the problems of mechanics, astronomy, physics, chemistry and biology were solved by means of purely logical constructions. Research on the part of natural philosophy and the logical construction of truths on the part of natural science became an essential component of that scratching of pens that, according to Jhering, forms the evening acoustical background characteristic of German university towns. ("What do they do, these young *privat-dozents*? They construct!" – Ihering 1909, p. 136). But by no means all of these young instructors were transcendental philosophers. Accordingly, the reaction of natural scientists and to a lesser degree the protests against natural philosophy lost their logical link with Schiller's appeal.

Incidentally, such protests did not agree with that appeal in the case of transcendental philosophy. Most often they were connected with the tradition of empiricism, with the old exhortation: "Physics, beware metaphysics!", with the inductivism of Newton, and even more so, of his imitators. Schiller was as far removed from empiricism as from the abstract-mathematical tradition; he was a friend of Goethe and, in his words, differed from the latter in that he did not proceed from phenomenon to idea but from idea to phenomenon. Empiricists who had protested against natural philosophy began from more simple and trivial notions.

An amusing, though frequent situation. At the time when protests against the natural-philosophical introduction into science of *a priori* schemes became particularly vocal, natural philosophy changed its initial premises, and then the protests often hit their target. In the middle of the nineteenth century they were directed with increasing frequency against Hegel's natural philosophy. Its supporters had no fewer claims to the *a priori* solution of natural science questions than did the Kantian-transcendentalists. Perhaps they had more. The laws of nature in Hegel's system were not separate from *a priori* laws and categories of thought; they occurred as their other-being, and science had to seek out these universal laws.

However, the situation changed when Hegel went from the illustration of laws of thought to their construction, particularly their fluid character, their evolution, their genesis. The farther Hegel was from already formulated laws, the closer he was to their dynamic generalization. The end of natural philosophy was the transformation of Hegel's dialectic, the denial of its *a priori* deduction, the representation of the dialectic hierarchy of forms of real motion.

Natural philosophy and empiricism reduced their task to the *coordination* and not to the hierarchization of scientific ideas, which shifted from one to another. It was reduced to the classification of scientific knowledge that was far from that prototype of the dynamic natural classification-hierarchy which had already been formed in biology. Both natural philosophy and the inductive-empirical systematizing tradition ignored the *individualizing* function of reason. Therefore the real turning point in the philosophy of natural science came with the idea of the hierarchy of forms of movement, which shifted from one to another, were connected with one another but could not be reduced to simpler and more general forms. Science turned out to be a ladder consisting of increasingly concrete steps, filled with an increasingly rich and complex content, with a growing multiplicity of definitions. The hierarchy acted against a simple distribution along *a priori* shelves; irreducibility acted against a reduction to a simpler lawfulness that was poorer in content; the ascent to the concrete acted against the unifying and decoloring transformation of the concrete into identical illustrations of general laws.

Such a tendency was characteristic, in a much less clear form, of seventeenth-century rationalism. Throughout its history (certain stages of which have been discussed in this book) the Fronde against *a priorism*, relying on empirical facts, was always active. Rationalism, uprooted from empirical sources, extends towards the abstract identity of concepts, finds an identical, unchanging, recurring substratum in changing impressions of being, but does not stop at this essential segment of cognition; it goes further, it absolutizes cognition and reaches fixed concepts that claim to be independent of empirical knowledge, it arrives at a metaphysical and absolute character, at concepts that preclude radical turning points in science. But there is another, opposite tendency, within the framework of eighteenth-century classical rationalism, in the systems of Descartes and Spinoza. Dialectical elements of classical rationalism express a hidden but sometimes visible sense of nonidentity, the changeability of being, the irreducibility of cognition to the fixed, abstract identity of logical conclusions. This tendency is carried over to classical German philosophy where it takes shape when the new, revolutionary, radical flight of dialectical thought makes philosophy a generalization of *developing* science.

This was the first period of developing *classical* science.

We must not think that paradoxical assertions that violate the purely logical deduction of essentially identical categories and concretize one and the same fundamental concepts emerge in science only with the advent of the theory of relativity and quantum mechanics. As we have said before, paradoxical assertions appeared at the very beginning of science, and their generalization fed the dialectical tendencies of classical philosophy.

2. ENTROPY

In the middle and in the second half of the nineteenth century, all the basic tendencies of classical science demonstrated, with considerably greater clarity than formerly, the reason of science and its irreducibility to understanding. As already noted, they are inseparable from science. Science, reflecting the fundamental complementarity of the statics and dynamics of the world, always uses the concepts of invariance and transformation or their pre-metrical equivalents.

Nevertheless, in the history of science it is possible to differentiate, in a very rough way, between the conceptions that formed the basis of the idea of nonidentity, the basis for change and for the transformation of science as a whole, and the conceptions that pushed thought toward the concepts of a 'tranquil' law, toward a conventionally apportioned territory of understanding. Conceptions of the type, 'everything is O.K.', when cut off from 'non-tranquil' conceptions, become the basis for the statistical generalization of science. The consideration of 'non-tranquil' tendencies tears asunder the framework of statistical generalization, leads to a dynamic generalization that goes outside of the framework of simply fixed representations and has an essentially prognostic character.

In classical science the first such 'nonclassical' conceptions were those of entropy and field. They were an objective prognosis of nonclassical science. The conception of entropy extended the theory of heat beyond the borders of continuum theory, pushing it toward the idea of discrete, quantized radiation. The conception of field developed fully within the theory of relativity, thus allowing us to see how the concept of field inevitably leads to conflict with the foundations of classical physics. In addition to paving the way for new, nonclassical concepts, both of these conceptions resulted in the representation of the fluidity of the fundamental principles of science. They still did not shake the classical foundations of science but they did shatter the confidence in their unshakeability. In this sense, science remained classical, but the classical confidence in any apodictic quality suffered a certain loss.

If we want to differentiate between the 'conception of understanding' and the 'conception of reason' in nineteenth-century science in a relatively approximate way, then the line of demarcation will pass through the law of the conservation of energy. In its prehistory, as a *mechanical* law of the conservation of life forces, it appears primarily as a law of conservation in a direct and narrow sense. A certain magnitude, a certain integral of motion, remains identical within the boundaries of a closed system. On the other hand, in its positive and qualitative interpretation, this law becomes a law of transformation, change, nonidentity. This is no longer a mechanical law, but a general, physical one. It contains a positive assertion: mechanical energy changes into heat and back again; each form of energy is capable of the shift to another form. If, in its negative form, the law of conservation points to a certain invariance, then, in its positive form, it has the character of transformation. From the middle of the nineteenth century, the law of the conservation of energy becomes the law of the *transformation* of forms of energy and of the *conservation* of its abstract measure.

As long as we are dealing with conservation, this abstract measure remains abstract. But as soon as we shift from abstract identity to the concrete, including nonidentity, as soon as the law of conservation becomes the law of transformation, it acquires a very concrete character. Now the concept of energy includes all concrete forms, which it leaves off and in which it is clothed. The dialectical broadening of the concept, which leads not to barren abstraction but to rich concreteness, appears before us in a clear physical form, and we are convinced of how close Hegel's logical balancing act is to concreteness.

The law of the conservation of energy in its positive form is inseparable from the law that indicates the direction of changes. Such is the law of nonidentity, nonconservation, nonreversibility and the increase of entropy.

Meyerson compares the second principle of thermodynamics — the law of entropy — with the first principle — the law of energy — and notes an almost total absence of even the most general formulations of nonreversibility before Sadi Carnot (Meyerson 1930, pp. 265—268). Indeed they did not exist in science. In philosophy, with the exposition of the concept of self-identity, invariance, substance, thought inevitably met up with the concept of nonreversibility. Brunschvicg reminds us of a similar concept in Kant who, in the *Critique of Pure Reason*, examined the concept of nonreversibility together with the concept of substance (Brunschvicg 1934, pp. 107—108). Here this concept is linked with the statement of mechanical processes, and nonreversibility characterizes only separate processes; further analysis might show a reversible course of the process. Such is the nature of the 'principle of causality', which is juxtaposed to the 'principle of constancy' in Kant's *Metaphysical Foundations of Natural Science*. The principle of the growth of entropy as a cosmic principle, the idea of the directed evolution of the world, are children of the nineteenth century. This principle could appear only on the basis of theoretically new, empirical information connected with the theory of heat engines. Attempts at the unlimited increase of the efficiency of the steam engine played the same role in the genesis of the second principle of thermodynamics as attempts to create a perpetual engine of the first type did for the first principle. In both cases, the *systematic* establishment of limits required a very general theory (Einstein 1956, p. 19). It therefore required some prior evolution of the most general definitions of being — the possibility of the identification of its elements and of its principal heterogeneity.

But the concept of entropy in its turn gave a new and very concrete meaning to these general definitions of being and at the same time to the definition of nonbeing.

Let us take a system in a state of maximal entropy, that is, one totally devoid of macroscopic temperature differentials. This is an ideally nonordered system. Here molecular motions are not identical; there is nothing that would force the large set of molecules to move in the same way and to create heat flux. In a system where there are no such masses of individuals identified by their motions, ideal chaos reigns, and macroscopic order, macroscopic *ratio*, which are impossible without identification, are totally lacking.

Without identification, the individual distinctions themselves, the individual nonidentity of the movements of molecules, lose their macroscopic sense. Macroscopically we can no longer say anything about a system with maximal entropy.

In principle, temperature cannot be measured unless heat is transformed into the work of raising a column of mercury or into any other form of energy. The set of microscopic objects that possess no macroscopic differences, no macroscopic order, have no macroscopic being.

Brownian motion — the shifting of very light bodies in a liquid due to molecular collisions — can be considered a physical demonstration of the existence of molecules. M. Smoluchowski wrote:

> It seems then that Brownian motion is indeed a phenomenon that clearly demonstrates the correctness of kinetic arguments and at the same time the incorrectness of thermodynamic notions, in that Brownian motion involves motion or rest in elements on a microscopic scale.[1]

"The incorrectness of thermodynamic notions" means their statistical character. Entropy becomes a rational concept only as a measure of the probability of the distribution of velocities of molecules. Brownian motion is a kind of analogue to *clinamen*, only now it is no longer speculative, but experimental. It shows that there exists a microscopic violation of macroscopic laws, a microscopic mutiny.

From here it follows that existence contains macroscopic and microscopic components. In the macroscopic ensemble there must be an ordered identification of micro-objects. But these identifications are statistical. The probabilities of the movements of any one molecule are identified. If there is an absolute, dynamic, non-statistical regularity, the shift from one form of motion to another disappears, the macroscopic existence of the ensemble itself disappears. With it also disappears the microscopic existence of particles. If the universe consisted of a uniform gas in equilibrium without macroscopic differentials, the very concept of the motion of molecules would become meaningless. The motion of particles makes sense if there can be reference systems in the world linked to macroscopic objects. Without one of its components, concrete being is transformed into *nothing*. In the next chapter we shall see into what kind of 'nothing'.

Such a shift from concrete being to nothing occurs when the individual motions of molecules, the atomic fabric that has been reduced to a purely macroscopic non-statistical scheme, disappear from the picture of the world. A picture without a macroscopic *ratio* is an empty picture. A picture with *only* a macroscopic ratio, with no microscopic individuals, one that is only statistically subject to macroscopic laws, one that preserves individual being, is also an empty picture. The two components of being that are inseparable from each other, that negate each other and lose meaning without each other are macroscopic *ratio* and the individuals that are only statistically subject to it.

Such a representation of being was permeated with the doctrine of entropy. Let us note only that the measure of the objective *ratio* of the macroscopic order was called *negentropy*. In contemporary science this is a measure of information, a concept close to order or to 'non-probability'. This last term scarcely needs explanation. The most probable distribution of the velocities of molecules is uniform-chaotic. If molecules move to one side, it means that some kind of macroscopic force is violating this uniform disorder. In itself, the identical movement of billions of molecules is much more improbable than the arrival at the same square of thousands of unordered people wandering aimlessly about a city. The measure of order is a measure of 'non-probability', it is a measure of the possibility of foreseeing the macroscopic result of microprocesses — and this is negentropy.

It can be shown that it is precisely the search for concrete being, the analysis of its components, that is characteristic of [the concept of] motion, from the classical theory of entropy to the nonclassical, quantum theory of radiation (Kuznetsov 1966b, pp. 233—245).

3. THE FIELD

Before Faraday, the field of force was a conventional term for the force that meets an experimental body at a given point, if the body appears in it. In other words, this was a characteristic of the possible interaction of bodies. Faraday considered the field an actual state of the medium, and the medium itself the sum total of elastic tubes of force. This new form of substance could be considered a kind of continuous body that is not, in principle, distinguished from other bodies. This was the conception of the ether, in Planck's phrase, "the child of classical physics, conceived in grief". This is a fair characterization. Contradictory properties had to be ascribed to the ether, and in the beginning of the nineteenth century it seemed that it did not fulfill the basic role of a mechanical object, that it could not serve as a reference body. After this its presence in the picture of the world ended.

Another line in the development of the field is its geometrization, the unique synthesis of Democritean 'being' and 'nonbeing', the endowment of space with geometric properties that depend on the sources of a field.

Finally, in the 1920's, quantum field theory achieved rather firm results. The field as a continuous medium possessing determinate properties at each point in space is considered a field of probabilities: at each point the probability of the presence of a particle is determined.

An historical circumstance is very interesting and paradoxical: before the appearance of the continuum theory of the electromagnetic field, there already existed the conception of the continuous distribution in space of some probability of physical processes that is determined for each point. Such a conception was one of the sources for the idea of the continuum field in classical electrodynamics.

An historical analysis of Maxwell's works on thermodynamics and electrodynamics establishes the effect of Maxwell's conception of heat on the elaboration of the classical notion of the electromagnetic field. The origins of Maxwell's macroscopic thermodynamics lie in the continuous distribution in space of average magnitudes, for example, the average velocity of the movement of molecules, i.e., temperature. If we shift to smaller and smaller volumes of space, for which the average velocity of molecules changes, then the indeterminacy of the real movement of molecules increases in relation to the macroscopic law.

The thought of a law that determines the probability of processes and gives indeterminate prescriptions for the processes themselves was a turning point in the history not only of physics but also of rationalism. From the beginning of modern times, thought was considered capable of tracing events from point to point and from moment to moment. Experiment, which must be, in principle, a local experiment, must confirm the local conclusions of a theory. Now two components of rationalism, which we have traced in this book, are divided. In principle, experiment provides an accurate answer for an infinitely small area. Logic and mathematics, speculation and deduction, arrive at an answer for a field and not for a point, and to the extent that the field is compressed into a point, the answers of the experiment and the results of logical-mathematical deduction diverge even further. Theory and experiment answer essentially different questions: experiment, the question of a physical event; theory, the question of the probability of this event.

This thought agrees with the traditional rationalist criterion of truth, since it was proposed that another, purely mechanical theory could, in principle, always determine the actual velocity of a particle at a given point and that this result would not differ from the result of the experiment. The notion of the continuous distribution of the values of a particular function of space and time was carried over from thermodynamics to electrodynamics. But in electrodynamics this function meant not the probability of events but the magnitude of the event itself. The statistical significance of the field was not transferred to electrodynamics, which developed at the time in conjunction with the idea of a real continuous medium.

But when it turned out that the electrodynamic field was a quantum field, the turn to a statistical conception was natural. By the nineteenth century, statistics had become a universal link between the atomistic and continual picture of being. Space, already in ancient times, was considered the sum total of the possible positions of a particle; now it became the sum total of more or less probable positions.

As long as the field theory was a continuum picture, it provided a continuous series of exact values, and experiment provided a continuous series of values, which, in principle, coincided with the calculations. Now the continuous series of values shows the change of probability.

Let us see how the continuum field theory influenced the relationship between physics and mathematics in terms of the new relation between Democritean 'being' and 'non-being'.

Without referring to the field theory and somewhat before the advent of electrodynamics, Maxwell had noticed essential displacements in these relations that were affected by the principle of least action. This principle, like the law of entropy and the field theory and unlike the Cartesian tradition, is one of the physical conceptions that interferes with the geometrization of the world. If we geometrize physics and reduce the properties of a body, as Descartes did, to its extension, then we will not be able to individualize the body. In the nineteenth century this difficulty assumed a new form. The laws of conservation geometrize the world in the sense that, following Noether's theorem (Noether 1918a, b), they correspond to the homogeneity of space (conservation of impulse), the homogeneity of time (conservation of energy) and the homogeneity of space-time (conservation of energy-impulse). The shift from conservation to homogeneity is the geometrization of the world. Opposed to this are laws that isolate matter from space, that save substance from the devastating [effects of] abstract geometrization.

The principle of least action belongs above all to classical physics. If this principle takes a Lagrangian form, it isolates the actual spatial trajectory of a particle from the number of other trajectories. In another form of the same principle, the actual world-line — the physical object — is isolated from the set of other world-lines that form space-time as a four-dimensional geometric object. We are dealing with the isolation of a physical object, with the differentiation of its predicates from the surrounding milieu, with the basic *aporia* of any physical theory that originates from identity, homogeneity, invariance. Geometrization threatens such a theory. Geometrization is the science of bodies that have ceased to be bodies, of Cartesian bodies that are indistinguishable from their places, having lost predicates other than spatial ones. Physics shifts to geometry when Democritean 'being' becomes indistinguishable from Democritean 'non-being'.

The law of entropy also separates matter — Democritean 'being' — from space and time. Opposed to the set of possible, thinkable processes that form Democritean 'non-being' is the real evolution of 'being', the evolution of a closed *physical* system that is distinguished by an increase in entropy. But the principle of least action and, to an even greater degree, the concept of field not only divide 'being' from geometric 'non-being', physics from geometry: They also bring closer together 'being'-matter and a new 'non-being' unknown to Democritus, a space that possesses physical properties. They bring physics and geometry closer together, but [by geometry we mean] the geometry of the *world*, the totality of geometrical properties that depend on matter and acting on matter.

In the course of the second half of the nineteenth century constant attempts were made to geometrize the principle of least action. For this it was necessary to ascribe certain new predicates to 'non-being', to space such as *curvature* and the interaction of particles; the elements of 'being' were replaced by the effect of 'being' on 'non-being'. This was already the conception of field.

Later on the general theory of relativity carried out this program. It geometrized (in the given non-Cartesian as opposed to the Cartesian sense) only gravitational fields. Of course, the word 'only' here must be put into quotation marks since gravity is the universal link between all concentrations of matter and fields with space-time.

The very principle of least action signifies that, in the absence of a field, a material point moves along a geodesic line of Euclidean space, i.e., along a straight line; and in a more general case — in the presence of a field — it moves along a geodesic line on a surface that possesses a certain curve.

In science of the second half of the nineteenth century, the concrete, contentful non-Cartesian geometrization of physics and the 'physicalization' of geometry proceeded by several paths. Statistical thermodynamics made its contribution to this process as did statistical mechanics. Here we can limit ourselves to mentioning Gibbs's statistical mechanics. He examined a great number of systems with identical properties except for configurations and velocities: for example, the set of systems consisting of one and the same particles, which differed, however, in terms of their position and velocities — in other words, they differed in terms of the 'phases' of systems at a given moment. Gibbs compared the possible phases for the systems examined to points of a kind of abstract, and, generally speaking, multidimensional space. Such a representation not only broadened and made more powerful the mathematical apparatus of thermodynamics and theoretical physics as a whole, but also gave a push to other abstract constructions in physics, to the mathematical development of multidimensional geometry and had a strong effect on the character of scientific thought of the following period.

The words just written — "abstract, and generally speaking, multidimensional space" — require a digression. The shift to more abstract spaces is, in this case, essentially a shift to more concrete spaces with more numerous determinations and physical equivalents. The term 'concrete' applied to three-dimensional space became extremely conditional when the theory of relativity made three-dimensional space an abstract section of four-dimensional space-time. The curvature of multidimensional space does not fit into the old representation of the abstract as something possessing a smaller number of determinations; it is closer to the concept of highest concreteness. Such castling [switching] of the terms 'abstract' and 'concrete' is the result of the new relationship between geometry and physics. The physical interpretation of geometric concepts in general is accompanied by a change in these concepts. In its turn the geometrization of physical concepts changes their specifically physical sense. Therefore, physico-geometrical transformations become non-commutative. If we geometrize a particular physical concept, and then physically interpret the result, we will get a new physical concept that is not identical to the original one. Such non-commutative transformations 'with return' become an increasingly essential method of physics.

4. PHYSICS AND MATHEMATICS

A similar variant of the link between mathematics and physics is characteristic of nonclassical science. Within the bounds of classical science the effect of non-commutative shifts becomes even more profound, i.e., it affects even more fundamental foundations of physics and geometry. Finally such situations arise when it is no longer possible to speak of independent, but interacting, trends of scientific development; geometry and physics could not be viewed as separate. As in the process of the dispersion of particles, we meet with such intensive interaction that it is necessary to reject the chronological analysis of processes and the differentiation between radiating and absorbing fields at the moment of dispersion.

The field theory created such a situation in nineteenth-century science. It was the prototype of twentieth-century physical geometry. Let us say a few words concerning the epistemological meaning of this concept of 'physical geometry', which has already been mentioned. Here several physical analogies will help us.

The first of them is virtual and real particles. The electron radiates photons, which it itself absorbs. These are virtual photons. In an analogous way, physics radiates mathematical ideas, which do not reach mathematics and are absorbed by physics itself. Physical theory in its development induces new mathematical concepts, but these concepts do not acquire autonomous value and autonomous being independent of original physical prototypes. Mathematics does not reach new stages in its development as a result of the radiation of similar mathematical concepts; the latter again are absorbed by physical theory itself. Such mathematical concepts, which do not acquire actual mathematical being, remain, as it were, virtual mathematical conceptions. One such example is the Dirac function, which serves as an example of a different relation between mathematics and physics. The initial 'virtual' concepts, which served only for the development of mathematical concepts, symbols and theorems radiating out of physical theory, can be transferred to other physical theories, lose their mutually simple link with a determinate physical prototype and in this way acquire autonomous mathematical being.

We can find the same scheme if we examine the development of mathematical theories. Mathematical theory sometimes develops by means of visible and, more often, invisible physical analogies, concepts and images, which are particularly important in initial, intuitive research and usually disappear during subsequent rigorous exposition and even more so during the axiomatization of mathematical theories. For a while such physical images do not merge with physical theory; they serve the development of the concrete mathematical conception that radiates them and are thus, as it were, are absorbed by it. Later on physical theory may appear and systematize these images; they will then acquire actual physical being.

In the analysis of the interaction between mathematics and physics, a truly linear representation is determined approximately. In examining the development of mathematics, we bear in mind a particular conception of the picture of the world, and in examining the development of physical ideas, we take into account a certain determinate set of mathematical means used by physics.

However, moments do occur in the development of science in which linear approximation turns out to be inadequate. Pursuing the same analogy, such moments can be called 'moments of dispersion'. Here it is no longer possible to isolate the development of physics and the development of mathematics by means of representations of the inner regularities and 'free course' of a given discipline that had received an external impulse. In the historical analysis of 'moments of dispersion', it is necessary to take into account the non-linear dependence of the development of mathematics and physics.

For nineteenth-century physics one such 'moment of dispersion' was the genesis of the theory of the electromagnetic field. The basic direction of this process was not reduced to the search for or even to the creation of *equivalents* — the mathematical equivalents of geometry and the geometrical equivalents of physics. Here there arose a new understanding of physics and geometry, where their link acquired ontological significance.

We should like to recall Maxwell's evaluation of the relationship between physics and mathematics given in 1870 in a lecture to the mathematical and physical section of the British Association; it dealt with the relation between physics and mathematics (Maxwell 1890, **II**, pp. 215–229).

Maxwell's lecture was a declaration of the physical significance of mathematical methods. The link between mathematical methods and their physical interpretation is not monolithic: there can be various physical theories relating to phenomena that differ in nature but use the same mathematical apparatus. But this does not mean that mathematical concepts have an *a priori* character and that physical conceptions are of empirical origin. Beyond the empirical extends the region of pure reason, but beyond that there is an even more considerable region where reason is united with the empirical. At the beginning of his lecture, Maxwell speaks of two lectures given at the preceding session of the British Association: Tyndall's lecture on the limits of physics, which explained sensually comprehensible physical models, and Sylvester's lecture, which was devoted to the nature of mathematical knowledge — the region of pure reason. Maxwell speaks of these lectures and of their connection:

> I have been carried by the penetrating insight and forcible expression of Dr. Tyndall into that sanctuary of minuteness and of power where molecules obey the laws of their existence, clash together in fierce collision, or grapple in yet more fierce embrace, building up in secret the forms of visible things. I have been guided by Prof. Sylvester towards those serene heights
>
> Where never creeps a cloud, or moves a wind,
> Nor ever falls the least white star of snow,
> Nor ever lowest roll of thunder moans.
> Nor sound of human sorrow mounts to mar
> Their sacred everlasting calm.
>
> (Maxwell 1890, **II**, p. 216)

But further on lies the region that links rational cognition with the empirical. This is the region of philosophy. Maxwell does not venture into it. But he sees that the daily work of the scholar leads to the same problem of the combination of Thought and Fact:

> But who will lead me into that still more hidden and dimmer region where Thought weds Fact, where the mental operation of the mathematician and the physical action of the molecules are seen in their true relation? Does not the way to it pass through the very den of the metaphysician, strewed with the remains of former explorers, and abhorred by every man of science? . . . In our daily work we are led up to questions the

same in kind with those of metaphysics; and we approach them, not trusting to the native penetrating power of our own minds, but trained by a long-continued adjustment of our modes of thought to the facts of external nature (Maxwell 1890, **II**, p. 216).

Where is that region of physics in which "daily work" leads to the same problems that are posed by the entire development of philosophy?

Maxwell built a geometrical apparatus — a system of *physical vectors* and their derivatives, which correspond to classical dynamics. But here the nonlinear and non-commutative character of similar physico-geometric reflections come into play. Maxwell's theory extends beyond the borders of mechanics and corresponds to physical processes that are *different* by nature. Physical processes are connected by the method of analogy. Let us state in passing that this method excludes the reduction of the electromagnetic field to mechanical models, the identification of their nature. Despite the wide-spread notion, Maxwell's electrodynamics are based on the irreducibility of the magnetic field to mechanical models of ether; therefore it naturally carried over into relativistic electrodynamics. But we are now interested by another question. Beyond the physical processes that are, by nature, different (if we remember the beginning of the 1870 Liverpool lecture, the world of physical processes that is explained there is the world into which Maxwell led Tyndall), and beyond geometry as the region of pure reason (this is the world into which Sylvester led him) lies the "region where Thought weds Fact". This is new physical reality where physical processes are expressed in the change of properties of space, and the change of forms of space is not the pure activity of reason but the expression of physical processes. This is the field of physical geometry. As was already said, classical field theory is a rehearsal for the genesis of physical geometry, realized in non-classical physics, in the theory of relativity.

Why only a rehearsal? In the history of the principle of least action, as in the history of the principle of entropy, and even more so in the history of the concept of field, we can clearly see the breakthroughs that lead out of the region of pure reason and out of the region of the empirical into the "more foggy region of the combination of Reason and Fact". If we use the same introductory remarks from the 1870 lecture, then we can speak [of moving] "from Tyndall's region" and "from Sylvester's region" into a region that we can reasonably label Maxwell's region. We know that, strictly speaking, neither pure reason nor the purely empirical ever existed nor ever could exist. But in science there have been organic periods when the constructions of reason, in a particular valid approximation, seem *a priori* in their foundations, and empirical comprehension seems immediate. In classical science such periods have been the fundamental background for the development, breakthroughs and radical turning points that demonstrate that the links between Reason and Fact were episodes.

In nonclassical science the continual change in the *original* constructions of reason and the dependence of experiment on these constructions seems to be a normal state, and organic periods, the exception. At the end of the nineteenth century William Thomson [Lord Kelvin] considered clouds in a clear sky a threat to fixed and finished science. Contemporary scholars are disturbed when they do not see similar clouds. Lorentz regretted that he had lived to see the collapse of classical physics. The contemporary scholar is afraid he will not live to see the advent of a unified theory of elementary particles that differs radically from the contemporary picture of the world.

The shift to 'Maxwell's region', the demonstration of 'the combination of Reason and Fact' depends not only on the content and level of scientific knowledge, but also on their gradient, on the curve of the rise of science, on the derivative in time, on its level. Therefore, every steep rise of classical science is a rehearsal of the style of nonclassical science.

5. QUANTA

If reason shows itself to differ from understanding whenever classical science reaches a sharp turning point, then nonclassical science is an ongoing demonstration of such a differentiation. In this sense, nonclassical science is science at its most rational, it is the reign of reason. However, it is a constitutional reign: the concepts [of reason] that express motion are inseparable from the 'tranquil' concepts of understanding that express rest. But, with this proviso, nonclassical science not only develops within the course of rationalism but also coincides with its flow, insofar as contemporary nonclassical generalizations are fundamental and encompass all being from elementary particles to the Metagalaxy.

In analyzing the evolution of rationalism as embodied in classical science, we can see its link with dialectics, with concepts that express motion. In mathematics, because of the illusory static quality of rationalism in the classical period, these concepts were termed irrational. But now it is very obvious that the rational is that which moves. Now it would be possible to rephrase Hegel's formula and say: all that moves is rational, all that is rational moves.

Let us consider Heisenberg's principle from this point of view, bearing in mind that the experiment in which the position of the particle is determined alters the particle's impulse, and the experiment that determines the particle's impulse alters its position. This means that once the motion 'is stopped' and the moving object is forced into a given 'here-now', we shift from the categories of reason to the categories of the understanding and discover immediately that we have no accurate experimental data concerning the movement of this object, the particle's impulse, and we can say something definite only on the basis of a wave question, that is, on the basis of the law of motion, once we calculate the value of the wave function, the amplitude of its oscillation at a given point and moment. In this way we can say something definite about the particle's motion, about its impulse. But this something definite is a determinate value of probability. The complementarity of corpuscular determinacy (the experiment registers the particle's *position*) and wave determinacy (Schroedinger's equation gives us the value of the *wave* function) is, then, the complementarity of the understanding; it is the complementarity of the 'tranquil' *determination* of the particle's position with the rational, 'non-tranquil' violation of the particle's original impulse, with the deformation of its motion, which is expressed in terms of the probable sense of the law of motion, the probable sense of the wave function.

The indeterminacy principle and all that is based on it is the physical and perfectly clear demonstration of the rationalist understanding of the world, in which experiment is inseparable from logico-mathematical deduction, the categories of rest are inseparable from the categories of motion, identity is inseparable from nonidentity.

The demonstrability of these general features of rationalism in non-classical physics, their visible character, the transparency of physical illustrations – all this is connected with the qualitatively determinable and experimentally registrable effects of the determination of the understanding on the determinations of reason, and vice versa, and of experimental paths of cognition on logico-mathematical paths of cognition, and vice versa.

In the scheme of quantum experiment that illustrates the indeterminacy principle, logico-mathematical cognition is represented by Schroedinger's equation, which determines the wave function. But the wave function is a description not of a thought process but of a real wave function. What is the wave function ψ: a particular mathematical abstraction, probability in the traditional sense of a subjective evaluation, or a real magnitude? The whole paradox of quantum mechanics is visible from this, since the wave function is a calculated probability and a real magnitude: waves of probability are diffracted while passing through an opening in a screen and in all ways generally demonstrate the reality of the wave process. Probability here is not a subjective evaluation but an objective situation. It is a real magnitude, taken not as something indirect but mediated; it includes 'memory' and 'prognosis'.

It is not the determinations themselves that are paradoxical. Only for the medieval Scholastics did mediated universals and immediately comprehensible objects exclude each other – only for them. In modern philosophy it became clear that both the immediate perception of the 'here-now' and the, in principle, non-detailed perception of homogeneous universals do not give physically meaningful results.

For Hegel logical mediacy became an objective process. Therefore the impression of the paradox of the synthesis of the logico-mathematical concept of probability with an experimentally recorded process is not connected with the rupture between such a synthesis and classical epistemological representations. This impression is created by the fact that very abstract epistemological antinomies turn out to be qualitatively determined, experimentally recorded and, moreover, practically applicable. In the *Brothers Karamazov*, the devil who appears to Ivan is paradoxical not because he is a devil, but because the cosmic spirit of negation appears in the most earth-bound, day-to-day, prosaic incarnation.

The unity of the rational and the empirical comprehension of the world, the inclusion of the empirical in *ratio* means that the formula $\alpha \in A$ (the individual α enters into set A) becomes not a purely logical construction, but an experiment, and this inclusion changes both α and A.

The inclusion of a local fact α – the presence of a particle at a particular point, in set A of momentary, potential spatial-temporal localizations that fill the world-line, its presence on the world-line – is the experimental determination of the particle's impulse.

Indeed, if this inclusion were reduced to the determination of coordinates, to the determination of the world point of a particle, then it would become tautological: 'the world point is located on the world-line passing through this point'. The nontautological inclusion of a in A means the determination of the direction of A, i.e., the velocity of the particle. This determination changes a, the given localization of the particle. If we wish to determine a more accurately, such a determination will change the potential world-line A of the particle. Similar changes can be measured qualitatively and recorded experimentally. From these changes flow numerous deductions and applications. The tower of pure thought to which Hegel withdrew turns out to be an experimental laboratory and even a workshop.

Now, in the 1970's, nonclassical physics itself, in the shape in which it was formulated almost half a century ago, has become the object of historical review. During these decades science has penetrated into the sub-nuclear region, and the riddles of processes that occur in regions on the order of 10^{-13} cm and 10^{-24} sec have become the basic riddles of nature. *Basic* — because efforts to solve them, in addition to everything else, determine the general style of contemporary physics. At the beginning of the 1930's, relativistic quantum conceptions and, on the other hand, the broadening of notions concerning the atomic nucleus, had already provided the basis for the prediction that the shift from atomic physics to smaller space-time regions would be linked with a transformation of the principles of science as profound as the shift to atomic physics itself.

In 1932 Fermi wrote:

> What determines the need to substitute quantum laws for classical laws is the change in the dimensions of the objects studied, as one passes from ordinary dimensions to atomic dimensions.
>
> In shifting from the study of atomic physics to that of nuclear physics, we are faced with a new change in the scale of phenomena . . . The experience of everyone working with atoms makes us assume that the laws that govern the behavior of the particles forming the atom cannot be applied to the behavior of the particles forming the atomic nucleus without a major transformation.[2]

The particles that form the atomic nucleus — protons and neutrons — are elementary particles, that is, physical objects for which the original determination ('elementary') is an unsolved problem; moreover, this problem is clearly connected with the most crucial problems of the Cosmos and the Microcosmos.

In addition, the physics of cosmic rays has come into being where elementary particles possess very high energies in whose presence particle transmutations occur, i.e., laws of behavior — the laws of the acceleration and dispersion of self-identical particles — are no longer in force here but rather the laws of being, the laws of the origin and disintegration of particles. These high energies (recently joined directly to the energy of cosmic rays) are not obtained in accelerators. On the whole, high-energy physics and elementary-particle theory, with all the extreme ambiguity and undecipherability of results received up to the present, represent a very simple and clear demonstration of the new character of scientific thought.

In classical physics the methods of reason were not interpreted in an epistemologically clear way; these interpretations led toward physical illustrations. Compared with classical physics, the theory of relativity is much more transparent; it allows us to see more closely and more immediately its epistemological credo. When Nernst said that the theory of relativity is not so much a physical as a philosophical theory, he was both right and wrong. Right, because the theory of relativity deduces simple, experimentally verified physical assertions from very general postulates concerning space and time, and in this sense it is closer to philosophy. Wrong, because together with the theory of relativity, *all* physics shifted to a new level, where concrete conclusions are not sporadic but are continually linked with general categories. This link brings the theory of relativity closer to philosophy, but does not separate it from physics if one bears in mind the basic style that is characteristic of twentieth-century physical thought.

In this sense, quantum mechanics attacks and makes flexible even more general concepts and even broader problems of the cognition and structure of the world. Without this new element that quantum theory introduced into epistemology, it would not have made its very imposing contribution to experimental and applied physics.

Finally, quantum-relativistic conceptions, high-energy physics and elementary particle theory lead directly and immediately to a new conception of being. As has already been mentioned, when the English called the thermometer (and many other things, even the balloon) a 'philosophical instrument', this was a tribute to traditional empiricism and a coming-down-to-earth of philosophy. If a particle accelerator were now called the same thing, this would be a recognition of the effect of physical discoveries on the apparently immovable foundations of philosophy, the penetration of experiment into the higher stages of cognition.

Sometimes the most fundamental trends in contemporary science are called big science or *megascience.* This refers not to the scale of construction and expenditure but to the scale of the problems. But in this case, unprecedented expenditures are connected with the nature of the problems. Elementary-particle theory now depends on accelerators and everything connected with them, and on astrophysical research.

Following the dialectical conception of the abstract and concrete, the word 'elementary' means that the wealth of determinations is extensive and intensive. Elementary particles are the most un-elementary, the most complex concept whose meaning is revealed in the area of 10^{-13} cm and 10^{-24} sec; this meaning is connected with the structure of sub-nuclear regions and with the structure of the Metagalaxy.

6. MEGASCIENCE

What aspect of basic research is of the most essential importance for the new evolution of rationalism? Perhaps the term *megascience* will be more accurate here; we would like to clarify how the characteristic peculiarities of nonclassical science as a whole have influenced contemporary rationalism, peculiarities that have been embodied most fully in research on elementary particles and on the structure of the Metagalaxy. This is not a question of the effect of one or another basic piece of research, but of what unites them and justifies the term 'megascience' or the ideal of megascience, which is the most general ideal of contemporary science.

The problem of the scientific ideal is connected with the problem of the freedom of science as its substantive determination.

As we said in Chapter 4 (p. 114 and p. 143, n. 2), Hegel calls the expressions 'independent thinking' and 'free thinking' pleonasms.[3] These are truly pleonasms, as is 'free science'. In providing a ground for such an evaluation, Hegel makes a very materialistic judgment concerning the liberation [of thinking] from the personal and from the power of the object of thought over thought itself. True, these judgments are compressed by the presumption of the original logical link that has been freed from the power of the object, the power of the concept of pure thought — of being equal to *nothing*. But this concept is found outside of the series of physically significant concepts of Hegel's logic. Significant concepts express the purging of thought of all that is personal and particular and the absorbing of thought in its object. In Hegel's words:

> For in point of contents, thought is only true in proportion as it sinks itself in the facts; and in point of form it is no private or particular state or act of the subject, but rather that attitude of consciousness where the abstract self, freed from all the special limitation to which its ordinary states or qualities are liable, restricts itself to that universal action in which it is identical with all individuals. In these circumstances philosophy may be acquitted of the charge of pride. And when Aristotle summons the mind to rise to the dignity of that attitude, the dignity he seeks is won by letting slip all our individual opinions and prejudices, and submitting to the sway of the fact (Hegel 1975, §23, p. 36).

In the next section (§24, p. 36), Hegel ascribes an objective character to thoughts and ranks among them forms, which, in ordinary logic, are considered only as forms of thinking. Hegel's logic describes forms of being and coincides with the doctrine concerning things. Further on (§24, *Zusatz*, p. 37), Hegel speaks of the concrete being of the kind – it is not only something common to all individuals, not only a name, it is a real community, a real set.

The power of the object constitutes the essence of science – it is the essence of thought addressed to the objective world, including thought experiments that seek an objective *ratio* in nature.

From this point of view, the freedom of science consists of freedom from the *a priori*.

The *a priori*, the original principles of science, grow out of conditional, relative, limited '*a priori*' representations, whose entire apriority is reduced to their relative stability. These are the representations of understanding. Their stability is a necessary condition for the rational comprehension of the world, but it can become so relative that the possibility for the absolutization of these notions of understanding disappears. A similar maximum power of the object is the maximum freedom of science. It is realized in nonclassical science, and in particular, in what is called *megascience.* Megascience encompasses the Metagalaxy and the microworld in the framework of a single conception, and its object is being as a whole, that concrete being which leaves no room for *a priori* sources of knowledge.

The freedom of science includes freedom from the *a priori* ideal. Such an ideal was the ideal of classical science as expressed in 1847 by Helmholtz in his lecture 'On the Conservation of Force' (Helmholtz 1847; 1971). This ideal consisted in the explanation of all processes in the world in terms of central forces. Such an ideal coincides essentially with Newton's program: to explain the forces of the distribution of bodies, and the position of bodies in terms of forces. However, the efforts to explain the processes of nature in terms of the activity of central forces is only half of Newton's program. And the explanation of the forces themselves? Until the advent of field theory, force was something that acted on a body and could be understood empirically – hence '*a priori*' empiricism: the empirical content of science opposed to *logical* mediacy.

But for science to yield to the power of the object meant least of all to follow empirical data and only them.

As we have already mentioned several times, the subject of science is not the empirical and is not logical constructions – they are impossible without each other. The subject of science is being with all of its components. The *a priori* – the obvious, whether notoriously logical or empirical – is something that is not subject to testing.

The freedom of science is freedom from the obvious, it is the freedom of paradox, of particular assertions to contradict the *a priori*, the logically obvious, and also the freedom of the 'flight from paradox' i.e., the construction of paradoxical theories in whose framework the experimental paradox loses its paradoxical quality.

In what can the ideal of science consist if no representations or schemes can claim the title of the final, the unshakeable, the obvious?

In 1949 Einstein expressed an ideal of the explanation of the world that is characteristic of nonclassical science, an ideal lacking an *a priori* character; it is not an ideal of perfection and hence the limit of scientific research. This ideal has already been mentioned above. We meet in nature a set of universal constants: mass and the charge of particles, the speed of light, astronomic magnitudes up to the radius of the Universe, the frequency of oscillations corresponding to different parts of the spectrum, etc. Einstein writes of them:

> Concerning such I would like to state a theorem which at present can not be based upon anything more than upon a faith in the simplicity, i.e., the intelligibility, of nature: there are no *arbitrary* constants of this kind; that is to say, nature is so constituted that it is possible logically to lay down such strongly determined laws that within these laws only rationally completely determined constants occur (not constants, therefore, whose numerical value could be changed without destroying the theory) (Einstein in Schilpp 1949, p. 63).

What does this ideal mean from the point of view of the history of science and philosophy; how does it determine and modify the classical ideals of cognition?

First of all, it means that the two halves of Newton's program are fused: (1) the determination of the position of bodies in terms of forces and (2) the determination of forces in terms of positions and, as was later made clear, in terms of the movement of bodies. Only instead of *bodies* it is better to say: in terms of the position of all concentrations of matter and energy, i.e. of ordinary bodies and fields. If universal constants enter into a single scheme of logical deductions, this means that fields and concentrations of matter are connected by a single conception and that neither form of reality enters into the picture of the world as immediate. This means that in the picture of the world there is no room left for either *a priorism* or empiricism.

Furthermore, Einstein's ideal means the fusion in one theory of cosmic constants (for example, the radius of the universe) and microscopic constants (for example, the 'radius' of the electron).

The universe presents the largest general set of modes, it is heterogeneous; moreover, all processes (and, accordingly, all constants) turn out to be true modes of a single substance, giving birth to itself, a substance which is simultaneously both *natura naturata* and *natura naturans*.

At the same time such a Universe is the physical equivalent of Hegelian concrete being. This is the synthesis of what was addressed to the future by Spinoza and Hegel.

Let us note that Hegel, in his doctrine on *measure*, i.e., on the unification of abstractly expressed qualities and quantities, spoke of the deduction of the "mathematics of nature" from qualitative determinations, from *proved* laws, i.e. from the relations of concepts, for example, the relations of space and time. "It is a great service", writes Hegel,

> to ascertain the empirical numbers of nature, e.g., the distances of the planets from one another; but it is an infinitely greater service when the empirical quanta are made to disappear and they are raised into a universal form of determinations of quantity so that they become moments of a *law* or of measure . . . (Hegel 1969, p. 343).

Let us note that the juxtaposition of these lines with Einstein's lines explains why Hegel's passages became incomparably more comprehensible than they were fifty years ago. Now natural-science conceptions not only fly together as illustrations of dialectical laws; they also, to a certain extent, change the sense of these laws, they deprive them of the aura and mark of apriority, they show their mediacy, they surround them with a cloud of possible interpretations.

The picture of the world without empirical constants is the fusion of inner perfection (the deduction of constants from the most general concepts) and external confirmation (the empirical testing of the sum total of deduced conclusions, consequently of the very first general concepts as well). The word 'first' does not mean here something close to Hegel's beginning, a direct, and therefore empty concept of abstract being. On the contrary, in the ideal picture of the world the most general logical foundations are deduced from nature itself.

Thus, the nonclassical ideal assumes a shift to a non-*a priori* logic that is rich in content and dependent on its subject. Formal logic can, with certain modifications, be applied to new physical ideas without changes in its fundamental laws. Such, for example, is the relationship of quantum mechanics to ordinary logic (see Bohr 1958, p. 313; Kuznetsov 1970b, p. 383). Every fundamental discovery will shake up contentful logic. Herein lie the criteria of the 'fundamental' and the membership of one or another results in megascience.

A similar characteristic of megascience that shakes up logical foundations renders rather inadequate the analogy in our Introduction, which compared the contemporary movement of science with the immediately perceptible image of a clock's second hand. The second hand makes the clock's course as a whole visible, but the clock is still a classical apparatus. Perhaps in order to characterize the contemporary dynamism of science, we should imagine relativistic clocks such that their moving hands would change in length: the clock face would become curved and even the fixed axes of the wheels and parts of the clock case would be curved.

This *explicit*, logical, paradoxical quality of megascience distinguishes it from the past. [The paradox is that it is] not a paradoxical quality: it always distinguishes science from metaphysics, and it is precisely the explicit paradoxical quality that is connected not with the sporadic, but rather with the practically continuous change in the logic of thinking about the world.

Let us sum up. Contemporary science answers the question posed by ancient Greek philosophy regarding being, substance, immutability and change in the following way.

It denies Eleatic substance, which is identical to itself in an absolute sense, homogeneous, immovable; it excludes non-being. It denies it in the Hegelian sense: it relativizes, limits, transforms.

It also denies the absolutization of nonidentity, *mors immortalis*, the replacement of an identical but changing and heterogeneous substance by a predicate without a subject, by motion without something moving, by change without something changing.

It includes in substance change in all its more complex forms, it moves along the axis of growth of this complexity, along the 'axis of ranks', to increasingly complex invariants of change, from coordinates to displacement, acceleration, impulse, energy, mass . . . Corresponding to this hierarchy is the hierarchy of the dimensionality of space, the order of derivatives, the rank of tensors which seems infinite. Corresponding to substance is the tensor of infinite rank.

Thus, the *ratio* of the world is not *mors immortalis* but the infinite complication of its structure — *vita immortalis*. Complication, the shift to increasingly complex transformations — to 'predicates' and to invariants — 'subjects' — herein lies the function of reason moving along the axis of the understanding.

Such is the contemporary representation of the relationship between the invariants of science based on understanding and the shift of reason to new invariants. By nature, by content, it cannot and does not want to be final. A similar representation discloses the relationship of being to becoming. And what is the fate of the third category, 'nothing'?

Contemporary science makes this category heterogeneous and gives it physical meaning. 'Nothing' is the result of the 'alienation of nature', of the absence of one of the components of being. This will be discussed in the following chapter.

NOTES

[1] "Il semble donc que le mouvement Brownien est bien un phénoméne qui démontre d'une manière évidente la justesse des raisonnements cinétiques et en même temps l'inexactitude des notions de la Thermodynamique, en tant qu'il s'agit du mouvement ou du repos dans des éléments de volume de grandeur microscopique" (Smoluchowski 1907, p. 1059).

[2] "Ce qui détermine la nécessité de substituer les lois quantiques aux lois classiques, c'est le changement de dimensions des objets étudiés, en passant des dimensions ordinaires aux dimensions atomiques.

"En passant de l'étude de la physique atomique à celle de la physique nucléaire, nous nous trouvons devant un nouveau changement d'échelle des phénomènes . . .

"L'expérience de tout ce qui est arrivé, dans le cas des atomes, nous fait supposer que les lois qui règlent la facon dont se comportent les corpuscules constituant l'atome ne sont plus applicables, sans modifications profondes, à l'étude de la facon dont se comportent des corpuscules constituant le noyau atomique (Fermi 1932, pp. 789–790).

[3] "'Think for yourself' is a phrase which people often use as if it had some special significance. The fact is, no man can think for another, any more than he can eat or drink for him and the expression is a pleonasm. To think is in fact *ipso facto* to be free . . ." (Hegel 1975, §23, p. 36).

12. NOTHING AND THE VACUUM

1. DETERMINATE NOTHING

Nonclassical science has given a precise character to the concept of heterogeneous being and has eliminated the concept of abstract, pure being from the world picture by depriving such being of physical meaning. Accordingly, it has filled the concept of *determinate nothing* with physical content and has deprived 'pure nothing' of such content. The concept of the vacuum is the modern, nonclassical equivalent of determinate nothing. From a nonclassical, historical perspective, we can now examine determinate nothing, the vacuum, as being that has lost one of its components: macroscopic *ratio*, or its microscopic filling.

Let us recapitulate, briefly, the conclusions of the preceding chapters.

Being is indivisible from motion, which means, first of all, the self-identity of moving objects and, secondly, the destruction of this self-identity. We are dealing not with the simple synthesis of identity and nonidentity, but with their complementarity: these concepts exclude each other and lose their meaning without each other. We ascribe to localized being a 'here and now' appearance in nonbeing, in the region of empty space and time, in the region of potential being, and thus we ascribe to the object a return to the 'here and now'. In other words, we ascribe physical content to the operation of the differential calculus: to increase and 'return', to the compression of the increase into a point and into a moment. Then, at the point, at the moment, at the 'here and now', not only is the very localization of the subject recorded, but also the subject as such; it possesses not only an address and a date, but also other predicates: the address and date are filled and the world-point becomes not simply a geometrical image, but a geometrical predicate of a physical object.

What is new here is what classical science contributed to the doctrine of being, namely the differential representation of motion, dynamism. Nonclassical science has made the results of the generalization of classical science more precise. Now we are familiar with concrete experiments that demonstrate that a particle cannot exist unless it moves, unless it possesses a world-line, a macroscopic existence. And a particle cannot exist unless it possesses some predicates other than presence on a world-line, some autonomy in relation to macroscopic representation, macroscopic laws, and the framework of world-lines. Thus, being that is devoid of these individual, localized predicates is being that has lost its autonomous here-and-now existence; undifferentiated and homogeneous being becomes the same as nothing. But if being as a whole is reduced to individual, localized, here-and-now existence, it also becomes nothing.

What do 'nothing', 'nonbeing', empty space and time represent in the light of such a conception?

Classical science — mainly field theory — and later, nonclassical science — the theory of relativity, quantum mechanics, the most recent concepts of elementary-particle theory — have excluded any analogies to Hegelian 'pure being' from the world picture. But Hegel did not ascribe physical content to this concept either, having identified pure being with nothing. But now the question arises: must concepts that, in principle, do not possess physical content be excluded from the world picture? Maxwell excluded from physics those mathematical concepts that do not have physical, i.e., experimentally observable existence, while Einstein excluded models of absolute motion, abolute simultaneity and ether, which also do not possess physical existence. Perhaps it is necessary to construct a system of logically deduced categories that exclude concepts lacking physical content? In that case, pure being becomes imaginary — logical 'ether' or logical 'absolute motion' — and must share the fate of its imaginary physical equivalents.

And pure *nothing*? Its equivalent in physics was empty space and empty time. These were physical equivalents because empty space and empty time, even in antiquity, meant the potential appearance of material bodies; we meet this conception in the ancient Greek atomists. But at present potential existence (there is no particle at present, but it was and will be at a given point at another moment) has been transformed into something quite concrete owing to the theory of relativity.

The theory of relativity differentiated (1) nothing as a concept lacking physical content and (2) nothing as the potential existence of a body. In order for a body to be at given point *a*, it must arrive from another point *b*, where it is located at present. Points *a* and *b* are united by a potential trajectory. Space as a concept containing physical content is an accumulation of such trajectories. But the theory of relativity excludes from the number of potential trajectories those segments uniting points *a* and *b*, in which the self-identical particle is found simultaneously. The particle, and in general any physical object, cannot move from point *b* to point *a* instantaneously or in a shorter amount of time than that which light requires to leave *b* and reach *a*. This means that it is not purely spatial concepts, but spatio-temporal concepts, not three-dimensional trajectories, but four-dimensional world-lines along which spatial position and time are altered that possess physical existence. Space and time together possess physical content, but not separately. And this is not all. In such space-time, the light cone cuts out the region where the particle's velocities do not exceed the velocity of light. Only this region possesses physical content; it is the region of the potential movements of self-identical physical objects.

But even this is not all. The theory of relativity demands that at each point and at each moment the moving particle be able to coincide with the potentially extended solid body – with a reference system. In time this presence must coincide with some readings from a clock. Thus space-time is not only a set of the potential localizations of particles, but also the accumulation of spatio-temporal reference systems.

Here quantum mechanics comes into play. It requires that at the point of the particle's potential presence, there be a device that records its position, for example, a screen with a narrow opening, and it asserts that the interaction with such a device alters the particle's impulse.

And what if there are neither these macroscopic reference systems nor macroscopic bodies of interaction in space-time? If there are no self-identical particles here? In other words, what if there is no macroscopic framework of world-lines?

Another question may also be raised: if only the framework is retained but its filling disappears, if the autonomous microcosm disappears, is being then preserved or does it, as pure being, become identical to pure nothing?

The point is that in the world of real processes, the components of being never disappear in an absolute sense. Let us imagine the disappearance of macroscopic order, for example, the complete symmetry of the probabilities of the microscopic shifts of a particle. It stands to reason that here there is no complete disappearance of statistical fluctuations; they are in equilibrium only in the transition to a macroscopic scale. This is a vacuum of microscopic differences. Let us take space where there are no photons. This is a vacuum of the electromagnetic field. Its energy cannot be equivalent to zero. When quantum physics reached the conclusion that the energy of the electromagnetic field could never be equal to zero, this was an essential shift in the doctrine of being and nonbeing. The concept of the vacuum appeared – of space where there are no macroscopic processes, where there are no nontrivially, self-identical, 'long-playing' physical objects. But this is by no means the pure 'nothing' of Hegelian logic.

Just as relative is the absence of another component of being – the microprocesses, the filling of the macroscopic framework. When it is stated that we have before us a solid body, a rigid system, where the position of the particle corresponds to the macroscopic structure and to macroscopic laws, all these assertions are relative. A solid body is not an absolutely solid body; modern physics excludes the real existence of such bodies. (Incidentally, they could realize a momentary dispersion of impulse and, consequently, rehabilitate absolute simultaneity.) There are no absolutely acting macroscopic laws, there is no absolute microscopic realization of macroscopic structure.

The theory of the vacuum brings to the characterization of rationalism a very essential circumstance.

Thought has the ability to operate by means of an *object that does exist* and compare what exists with what does not.

In one of his two 1897 doctoral theses, Brunschvicg wrote:

> If one could say: to think is to generalize, cannot one also say with just as much justification and more profundity: to think is to contradict? Thought is characterized by the ability to compare affirmation and negation. Being, the thing, is, in one sense, always positive; but thought is freed from being by giving meaning to nonbeing; that which is not is an object of thought just as much as that which is.[1]

When a modern physicist thinks of "that which is not", the object of thought becomes not logical abstraction, but a 'vacuum' object. Philosophy long ago arrived at the corresponding concept for which it was possible to find physical equivalents already in classical physics. The concept of *determinate nothing* means that we are dealing with a wholly determinate object, and we postulate its absence. But such a postulate makes sense if the absent object can, in principle, be encountered in a given milieu. The negative postulate, as a component of thought with content, is a negative answer to the question that can also have both a positive and an indeterminate answer.

Hegel says that "nothing which is opposed to the something is also the nothing of a particular something, a determinate nothing" (Hegel 1969, p. 83).

The modern representation of the vacuum corresponds to this concept of determinate nothing.

Nonclassical science was obliged to replace the classical void — the universal image of nonbeing — by the vacuum, i.e., by the absence of something determinate (the vacuum of an electromagnetic field, the vacuum of an electron-positron field, etc.). This is a concept full of content. The vacuum has determinate predicates, its energy does not disappear, the vacuum interacts with a particle and this interaction is experimentally recorded and measured. The vacuum in its modern sense alters the content of the concept of the 'potential'. In a mechanical representation, this concept has a subjective meaning: on the basis of certain assumptions flowing from objective statements relating to real things, we can now *represent* to ourselves the future behavior of the observed object, draw conclusions about the object in the future, the object that 'here and now' does not exist. The nonclassical representation of the vacuum is different: even now, in actuality, there exist predicates that make no sense without the future. We confront the complementarity of actual and potential predicates.

In the vacuum something is occurring now in actuality. There are actual processes, the birth and annihilation of virtual particles, processes that acquire physical meaning, i.e., can be experimentally recorded during the interaction with a 'real' particle possessing macroscopic being.

The modern concept of the vacuum was a continuation and modernization of the idea of the discreteness of the world. In classical science, until the middle of the nineteenth century, atomism played the role of Valhalla for the world of solid bodies where the solid bodies once again acquire being, but now already microscopic being and, remaining invisible, continue the same activities as they had before their destruction. Molecules and atoms were considered to be invisible solid bodies that behave in the same way as visible ones. Motion became molecular but was not essentially changed. Later, in the transition to chemical reactions, everything became much more complex, but chemists, and even more so physicists never abandoned the hope of reducing chemical reactions to the usual dislocations and redislocations of solid bodies.

In the middle of the nineteenth century, physicists began increasingly to see in atomism something that had been in principle ignored. Macroscopic thermodynamics is based on ignoring independent molecular movements. The actors became not the bodies, but the points of a certain field of average values. Then electromagnetic field theory, which was considered to be strictly continuous, appeared in physics. In such conceptions an ideal elimination occurs, an ideal transformation into nothing (nothing which has content, determinate nothing) of the more complex forms of motion and the more subdivided degrees of the atomistic hierarchy.

The situation changed in the twentieth century. The liberation of atomism from the power of macroscopic representation as well as the filling of the macroscopic frameworks by no longer ignored micro-processes occurred.

The first to fall was the illusory scheme of the absolutely solid body with fixed positions of motionless molecules. As has already been stated, such a scheme contradicts the theory of relativity. If a body, given an impulse, were to move as one whole, not altering the position of separate molecules, it would be possible to take a long rod and, by pushing one of its ends, obtain simultaneously a shift in the other end. Thus the purely timeless spatial process of the instantaneous transmission of a signal would be embodied in distance. It would be possible to fasten indicators at the ends of this solid rod with its molecules in fixed positions and they would move synchronously. We would obtain unified, absolute time.

But this is only a negative reaction against the immobility of molecules. The theory of the solid body arrived at a very complex scheme of the behavior of molecules at a temperature near absolute zero. Such an evolution occurred within the framework of quantum mechanics. The latter was an even more decisive break with the neglect of the individuality of micro-objects than the theory of relativity.

Let us imagine a set of micro-objects sufficiently large as to be considered a continual object. We move successively toward smaller and smaller spatio-temporal objects, toward smaller and smaller parts of this set. Ultimately we encounter the discreteness of the object and can no longer ignore the existence of individual micro-objects and their active individuality, their active actual existence, the fact that their microscopic behavior is not entirely determined by macroscopic laws.

This encounter with the actual existence of the microcosm also took place in the 1900's when Planck investigated black-body radiation, and Einstein, the phenomenon of photo-effects. Later, science encountered not only the *existence*, but also the *being* of micro-objects, i.e., their connection with the cosmos, their connection with cosmic *ratio*. The particle represents the cosmos; the Universe is reflected in it; its existence is inseparable from the existence of the cosmos and it, this existence, is perhaps the result of the self-coordinated system of interactions embracing the Universe.

2. THE ALIENATION OF NATURE

Now, after this generalizing and a partial concretization of the interconnected concepts of determinate nothing, potential being and the vacuum, we can turn our attention to the connection between these concepts and a concept that seems, at first glance, to be remote from them, the concept of *alienation*. Even if we limit ourselves to the philosophy of the nineteenth and twentieth centuries, the meaning of this term is represented in extremely different ways. What is common to all of them, what is characteristic of all versions, is the conception of the subjugation of the object of alienation to laws external to it or, perhaps, in an even more general form, the conception of the loss of the object's own behavioral traits, which are specific to this given object, traits that flow from its essence. But if the combination of individual traits with identified predicates is the basic criterion of concrete being, and if the loss of one of these components transforms actual being into determinate nothing, i.e., into potential being, then the category of alienation is similar to these concepts.

In such an approach, the concept of alienation is considerably generalized and rendered concrete. First, it becomes ontological, it becomes a category of being. Secondly, it becomes the subject of evolution: it ceases to be the age-old curse of man, as it was for Kierkegaard, or a purely fortuitous and nonessential predicate of man, as it was in Panglossian conceptions.

It is necessary to stress the connection between the concept of alienation so prevalent in nature and rationalism. Modern rationalism views the *ratio* of the world as an accumulation of real connections that embrace nature and human life. In nature, the condition of matter in a system with maximum entropy means the transformation of this matter into an accumulation of absolutely isolated individual movements that do not form part of macroscopic identified sets, but of absolute chaos. If, in nature in general, there were no macroscopic order, it would be impossible to say anything about these movements; in the absence of macroscopic reference bodies, their chaotic movements could not be determined as movement. Science views such a set of disordered processes as a vacuum, as potential being; it becomes actual being upon the appearance of the second component – the macroscopic order, macroscopic temperature transmissions.

Nature is also alienated as long as macroprocesses and the individuality of micro-objects are completely ignored. The anti-atomistic trend at the end of the nineteenth century is an illustration of such neglect. It became rootless when the discoveries at the beginning of the twentieth century demonstrated not only the inseparability of thermodynamics from kinetic representations (Brownian motion), but also the new confirmation of atomism, this time going beyond the framework of classical physics (the discreteness of electromagnetic radiation). In nonclassical physics, ignoring each component of being – both macroscopic *ratio* and microstructure – *obviously* leads to the alienation of nature. The inseparability of these components is revealed in theoretical thought; therefore empiricism in any of its forms is harmful to science, and more obviously to nonclassical science. The illusion of a linear connection between individual statements and a general representation of the world is typical of empiricism: separate statements are simply piled up in the world picture: neither they nor the picture are altered in this process. The positivist world picture suggests a picture composed of cubes with parts of the picture previously drawn on them. And the word 'previously' is important: the rejection of the search for essence transforms a phenomenon into an illustration of a previously composed representation of the world. Without apperception, without a general view, cognition of individual facts is impossible, and the rejection of a general view that penetrates the depths of experience in fact proves to be an unconscious application of an old apperception. Empiricism permits the replacement of old cubes with new ones, while disregarding the general coloration and composition of the picture and those things that change the tone and the pattern in the details of the picture. In essence, empiricism, like *Naturphilosophie*, operates through the understanding as opposed to reason, through the stasis of science rendered absolute as opposed to its dynamics. These roots of empiricism have become very weak in nonclassical science.

The *a priori* is also alien to nonclassical science, especially Hegel's *a priori*. The development of nonclassical science requires the differentiation of Hegel's *a priori* conceptions from his ultimately dialectical ideas connected with the generalization of the dynamics of classical science, ideas that even now retain their heuristic value. Nonclassical science provides such a differentiation with both the stimuli and the starting-points for an historical evaluation. It allows us to see in the entire history of philosophy and science an increasingly complete demonstration of nonalienated nature as a unity of macroscopic and microscopic being.

The first part of Hegel's *Encyclopedia* concludes with the words: "... this Idea which has Being is Nature" (Hegel 1975, §244, *Zusatz*, p. 296). But nature is alienation of the spirit. The spirit becomes nature, alienated from itself, rejecting itself, making a sacrifice of itself. [Hegel 1981, §§807—808, pp. 491—492; this English translation uses 'externalization' for 'alienation'; see also Hegel 1939—41, **II**, p. 311 — Eds.] For Hegel the spirit does not create nature; there are no Creationist ideas here, as Feuerbach thought (Feuerbach 1966, §30—31, pp. 48—51). He does not have even the representation of the 'emanation' of an idea or its 'degradation', along the lines of the Neoplatonists, as Renouvier thought (Renouvier 1896—97, **IV**, p. 18). Nature and spirit for Hegel develop in unity, to the degree that the abstract does not exist without the concrete and vice versa; the evolution of the spirit is the evolution of being. But nature is the motionless other-being of the spirit. In the spirit, in logical transition, lies the moving force of being, and when the spirit, rejecting itself, moves into nature, it becomes immobile. But the spirit, the ultimate reason of the universe, withstands alienation, other-being, by becoming transformed into nature. Nature does not become alienated because it is already alienated. When Hegel speaks of the alienation of the spirit, it is in fact nature that is alienated.

It is generally characteristic of Hegel that, in speaking of the spirit, he is in essence speaking of nature, while, when speaking of nature, he is really speaking of the spirit. It is easy to observe that the *Phenomenology of Spirit* and the construction of his *Science of Logic* are a coded story about nature: the scientific commentary arises here naturally; we are dealing with nature in motion that possesses *ratio*. However, when Hegel speaks of nature directly, as, for example, in *The Philosophy of Nature*, it is a story of the transformation of the spirit. Here Hegel limits himself to the behavior of the scientific categories in an *a priori* scheme of the evolution of the spirit and therefore he does not alter the content and meaning of the scientific categories, i.e., he does the same thing for which he reproached the Kantians. Nature for Hegel does not develop;

it moves in a circle. But its cessation of movement is, in fact, its alienation. When the immanent transition of the 'here and now' into the future takes place, when the immanent evolution of nature ceases as a unity of macroscopic and microscopic being, then the evolution of nature becomes a passive reflection of the spirit, which, *in nature*, is already alienated, is already in other-being. In rejecting the Spinozist *causa sui*, Hegel gives the spirit the prerogative of directed evolution, motion that is not circular. Nature proves to be alienated.

In the second half of the nineteenth century, the alienation of man was almost constantly central to philosophical and social thought, but the alienation of nature was not mentioned. The point is that the classical representation of nature only sporadically demonstrated the unity of its macroscopic and microscopic aspects. In social questions the connection between the individual and society, the search for the social group, trends and transformations assuring real being and the inability to ignore the individual were in the mainstream of scientific thought. But in natural science, ignoring molecules did not impede the resolution of concrete problems. This continued until the beginning of the twentieth century when it was finally realized that the particle could no longer be ignored and when, furthermore, the inseparability of the particle's being from that of the Universe finally acquired specifically physical ideas and became conclusions or a possible conclusion from experiment.

Deprived of the synthesis of the macroscopic and microscopic aspect, nature loses its freedom as a criterion of being. For Spinoza freedom consists in the fact that general necessity becomes immanent necessity, flowing from the very essence of the given process; it is not imposed from without. Thus it consists of the harmony of the macroscopic law and its microscopic embodiment in a localized act. For Hegel such an embodiment includes a contradiction: the macroscopic law and its microscopic realization contradict each other and, in addition, are inseparable from each other.

The theory of relativity contains a very general idea of the contradiction inherent in nature itself. Contradictions between (1) what is at rest, determinate, what has become, what is presently available and (2) the transitory, the moving, that which destroys localization — these contradictions became physical, but not logical contradictions. Rather, they became a contradiction of logic as a doctrine of being. Motion loses its meaning without the concept of a reference system, that is, without something immobile and, moreover, immediate, independent of any mediacy. At the same time, the theory of relativity emphasizes the conditionality of such a category of the understanding. In nature, in nature itself, without the interference of absolute spirit, we encounter conditional immobility and the motion eliminating it on all levels. However, at present both of these are not only logical statements, but real ones, or potentially real ones, lines, time, and everything equated with them.

Quantum mechanics also returns to nature the being alienated by the Hegelian absolute spirit. Like the theory of relativity, it eliminates the fiction of isolated determinate bodies. "Each thing", says Hyppolite, "is what it is and yet it is a determinate thing. But the empirical consciousness takes this unity as something immediate."[2]

This empirical consciousness, by seeing something immediate in this unity, is giving way to an illusion. A thing becomes determinate when it interacts with other things, when it enters a set and when this entrance is a physical process, a certain alteration of the thing and . . . the destruction of its determinacy. The determinacy of a particle is primarily a determination of coordinates. The theory of relativity relates these coordinates to macroscopic reference bodies while quantum mechanics demands that the determination of coordinates be the interaction with a macroscopic body, the result of an experiment. An important tendency, in an epistemological sense, is expressed in the rejection of determinacy without the coordinates of the position of the bodies. Science forces the concept to be clothed in an experimental form. Logical constructs must be real constructs, the logical deduction must be an experiment, and concepts entering the deduction must possess physical content. And this means that the Logos becomes experimentally represented and that without such a representation, it loses meaning. If we employ Hegelian concepts, then the transition of the 'spirit' — Logos — into nature is not a self-sacrifice but the sole guarantee of the physical content of the 'spirit'. And it is precisely in this content, in the ontologization of logic, that the central tendency of Hegel's dialectics lies. We are constantly being convinced that non-classical science discovered in Hegel's dialectics what is fundamental — what the category of pure being, the idea of nature as alienated spirit, and the alienation of nature in *The Philosophy of Nature* contradict.

The ontological interpretation of the term 'alienation' and its prevalence in natural science is by no means a simple analogy and not merely an external unification of different spheres in a single scheme. It is a determinate position in the question of the scientific roots of the alienation of man.

3. THE BEING OF NATURE AND THE BEING OF MAN

Neither the classical rationalism of the seventeenth and eighteenth centuries nor dialectical philosophy were a mere reflection of science. They had other roots as well, first and foremost the flow of epistemological and logical concepts, moral stimuli and intuitive selection criteria that grew out of the impression of the social being of man.

In ancient science this flow was not even required. There, social ideas found natural-scientific equivalents without any additional intellectual effort because physical and social concepts were both philosophical by definition. The integral character of ancient thought did not disappear in the subsequent differentiation of knowledge. Having split off from science, philosophy began to perform, for better or for worse, the functions of connecting and generalizing the branches of knowledge that had split off from each other, namely the study of nature and the study of man. Such functions were, or at least seemed to be, not so essential in the straight stretches of science, but necessary and obvious at the sharp turning points.

When the doctrine of nature changed course, when the criteria for selecting a path changed and the established ideals gave way to new, initially intuitive ideals of cognition, when the logic of science became, to a significant degree, psychology, the influence of social thought proved to be very profound and precise. Specifically, moral ideals and the sociological conceptions on which they were based could not alter the content of scientific results, but to a great extent they altered the direction of scientific interests as was discussed above. In the end the structure of the interests — the comparative intensity of variant tendencies — also appeared in the positive picture of the world: in each problem where the directions of forces play a role, the alternation of their absolute magnitudes generally leads to a shift in the component force.

Spontaneous migrations of ideas and analogies from the doctrine of man and human society to the doctrine of nature and vice versa (for example, the conception of 'the war of all against all' introduced into biology, or the concept of equilibrium shifted from mechanics into political economy) did not have the general and profound effect that they did in their generalization and systematic development in philosophy. In the beginning of the nineteenth century one could see a very intense generalization of the results that were only just beginning to shape in a precise way a new natural science and a new industrial system — the result of the Industrial Revolution of the eighteenth century. These results were expressed especially in the new scientific and social ideals.

Let us see how the correlations between moral and scientific ideals, between representations of man and representations of nature in Hegel's early works become concrete.

Historically, Hegel's dialectics were closely connected with that profound isomorphism which united the search for cosmic harmony with the search for moral and social harmony. Hegel proceeded from political economy, from criticism of Adam Smith and from social problems as a whole to the heights of abstract philosophy. He did not so much proceed from social problems as evade them, rising to where there was no alienation of man. Hegel understood, or perhaps only sensed, the connection between the alienation of the individual, his subjugation to something alien to him, the neglect of the individual, on the one hand, and the unshakeable, stern solidity of nature, as portrayed by classical science, on the other. Hegel got around this immobile nature; he saw beyond it the kingdom of the evolving spirit.

Here, beyond immobile nature, in the realm of pure spirit, Hegel sought refuge from alienation. But it was an illusory refuge.

Another, equally illusory, path of flight from alienation was the isolation of the individual: isolation from the cosmos and from all systems that include the individual, subordinate him to themselves, alienate his freedom without which he ceases to exist as an autonomous element of the universe. The apologia of the isolated individual and his existence became a protest against science, against the search for objective truth. The truth is subjective, says Kierkegaard, the most prominent of the representatives of nineteenth-century anti-rationalist protest and the forerunner of modern existentialism. The truth is inseparable from existence, and if it is rendered objective and universally significant, this is tantamount to terminating individual existence — suicide.

The way of objective reflection makes the subject accidental, and thereby transforms existence into something indifferent, something vanishing. Away from the subject the objective way of reflection leads to the objective truth, and while the subject and his subjectivity become indifferent, the truth also becomes indifferent, and this indifference is precisely its objective validity; for all interest, like all decisiveness, is rooted in subjectivity. The way of objective reflection leads to abstract thought, to mathematics, to historical knowledge of different kinds; and always it leads away from the subject, whose existence or non-existence, and from the objective point of view quite rightly, becomes infinitely indifferent. Quite rightly, since as Hamlet says, existence and non-existence have only subjective significance (Kierkegaard 1941, p. 173).[3]

The path of objective reflection, the path of abstract thought, in Kierkegaard's words, leads away from the subject and renders indifferent his 'to be' or 'not to be'. To some degree this is really so. Abstract thought unites subjects in a set of some identical elements, and each element, when identified with any other, becomes indifferent; its autonomous existence is negated. Émile Meyerson clearly observed and represented this annihilating function of the rationalist comprehension of the world with its identification of individual localized events and objects. Kierkegaard wants to extract man and his subjective truth from the levelling series. But will such an extracted and isolated element of the identified series continue to exist?

Let us turn directly to the nonclassical physical model. It would be absurd pedantry to reject the possibility of a rational interpretation of the philosophical controversies that modern science reveals as it so radically alters the form of these controversies and so energetically maintains their essence. Does the particle possess being when it does not enter the world chain, the particle without a world-line, the particle with purely localized, momentary existence?

Such a question has been raised more then once in the pages of this book; it is fundamental for an historical analysis of rationalism.

Obviously, such an abstraction will not have physical meaning. But the world-line means that the particle is viewed not in a momentary existence, isolated from the past and the future, but in space and time. We identify the particle at various points in space and at various moments of time and this is in fact that criminal identification that renders undifferentiated each localized 'to be' and 'not to be'. Kierkegaard protests against the comprehension of existence in space and time. The predicates become eternal and extend from localized existence in space and time (Gilson 1948, p. 230).

Of course, the reverse situation — the particle with only a world-line, only a spatio-temporal dislocation — also leads to an abstraction devoid of physical content. Real being demands the complementarity of these mutually exclusive determinations — the spatio-temporal and the localized.

What does the individual represent in not entering the levelling sets, the individual whom only the correlations of nonidentity connect with all the elements of the cosmos? Such an individual represents an abstraction devoid of any real human existence. The man whose brain has never held universally significant truths that do not meet, that are not transformed and do not oppose one another is, in truth, not alive. He is not alive at all. Absorption by the levelling set is suicide. Isolation from such sets is also suicide. Man's being, his life, possess a complementary nature. It demands the unification of the opposite poles.

The analogy between man (correspondingly, between human existence, social and moral harmony) and the particle (correspondingly, physical being, cosmic harmony) is not a mere analogy. It was never a mere analogy. Between the character of nature as represented by science in the given period, and the character of human relations there is a mediating link. This is human labor. Its content and conditions depend on the level at which man discovers harmony in nature, the connection between macroscopic and microscopic processes, the being of nature. The evolution of the concrete forms of labor, leading ultimately to eliminating the alienation of man, to his 'disalienation', depends on a further and more immediate search for cosmic harmony.

Long before Kierkegaard and his protest against objective truth in defense of existence, against rationalism and against the philosophy of Hegel, there was criticism of classical science and its embodiment in production — the factory system — from the vantage point of individual existence. In the time of Voltaire and Adam Smith, almost no one considered the laws of nature and their industrial application to be a force directed against man's individual freedom. And later Fourier included in his Utopian pictures the transmission of power over the world from the hands of God into the hands of Newton. However, disappointment came quite quickly. The classical ordering of Newton's Universe gave man more freedom and happiness than that of God's Universe — the *civitas dei* — but much less than man expected.

Scientific thought found itself at a crossroads. One path led to the rejection of the search for objective truth, to the rejection of the rationalistic spirit of science, at least where man was concerned. Man was led out of nature and contrasted to it, not as an active subject of the cognition and transformation of nature, but as a sacrificial victim seeking isolation. This conception of Kierkegaard had few followers, but many imitators.

The second path — in the opposite direction — was the isolation of reason from nature, which is transformed into the other-being of reason.

The third path — Marx's path — did not lead to the fiction of reason standing above nature as a self-developing, objective spirit. Here man does not isolate himself from nature; he subjects nature to himself and in the process of such subjugation is liberated from the power of the biological laws of being; his behavior and fate begin increasingly to be directed by specific laws that are not reducible to more general and simpler laws.

Let us examine in more detail the actual relationship of Hegel's dialectics to the problem of man's alienation. Kierkegaard did not understand this relationship. But this lack of understanding is characteristic not so much of Kierkegaard as of Hegel. Neither Hegel's most faithful students nor his fiercest opponents understood him. His very fate illustrates the fundamental trait of *understanding* — the combination of an identifying and classifying function with a statement about the unity and uniqueness of the object. Hegel's students (in the usual sense of the word) saw in his philosophy something absolutely self-identical; they were far from the understanding which corresponds to the spirit of Hegelian philosophy, from understanding as a process that modifies the understood conception. Hegel's opponents, for the most part, identified his philosophy not with the future but with the past. They did not maintain that the contents of *The Phenomenology of Spirit, Science of Logic, et al.*, are identical with every future rationalist philosophy, but said that it corresponds in essence to earlier systems of classical philosophy.

What remains of Hegel's philosophy in classical philosophy and what, in his philosophy, remains as its ultimate understanding, corresponding to the most profound meaning and the historical role of this philosophy in the evolution of human thought?

In an immediate and very similar image, the ideas at which Hegel arrived in the beginning of the nineteenth century were applied to the ideas of Fichte and Schelling and contrasted to them. The connection of these ideas with Kant's was more remote but equally unquestionable. If we are to speak of the influence of classical science on classical philosophy, then the fundamental point of distinction between Hegel and Kant, Fichte and Schelling is the *dynamic* generalization of what had been done in science during the seventeenth and eighteenth centuries. Fichte's philosophy, the philosophy of 'I', was a characteristic of the knowing spirit, but of a spirit that knew nature in its ordered, systematic, 'three-dimensional' representation, a spirit that had returned to nature as revealed by classical science. The philosophy of nature, Schelling's philosophy, was a philosophy of precisely such a classical nature, the nature that was known by science at the end of the eighteenth century. The 'nonclassical' elements of classical science, the very process of its transformation, its 'four-dimensional' representation, had not yet been placed within the framework of either the philosophy of 'I' or the philosophy of nature, i.e., in either Fichte's system or Schelling's system.

We already know that Hegel's philosophy was, to a considerable degree, a generalization of science, more so than the philosophy of Kant, Fichte, and Schelling. Now we can approach this question from a more concrete angle.

Hegel's panlogism is surprisingly similar to the artistic and sensualistic nature of Goethe. Perhaps Hegel is closer to Goethe than to his own immediate teachers and predecessors in philosophy. We encounter in Hegel a continuation of the line that passes, through the *Universal Natural History and Theory of the Heavens* and then through *The Critique of Pure Reason*, that apologia for the Newtonian scheme of the universe stripped of its initial impetus, generalized and rendered absolute, as well as a continuation of Goethe's revolt against Newton. Such a synthesis could not occur without a radical shift in both lines. The generalization of Newtonianism remained a generalization, but became a dynamic one, not only of conclusions, but also of the path from Aristotelian physics to Newton's system, to experiment, mathematics, to the differential representation of motion. In its turn, Goethe's anti-Newtonian revolt was transformed into a statement of mutability, evolution; it was a transformation of Newton's system, of contradiction as the basis of its evolution, the contradiction of the universe itself as the basis of cosmic evolution, the becoming of the Universe.

The generalization of the dynamics, contradictions and transformations in science did not stop at this. When the evolution of thought, with its contradictions, corresponds to the evolution of the world, when logic, in drawing into itself not only the identifying acts of understanding, but also the individualizing acts of reason, becomes the dynamic *ratio* of the world, then the possibility of the *rapprochement* and even the identification of logic with the essence of the world process is revealed. This is no longer the philosophy of 'I' and no longer the philosophy of nature. It is the philosophy of the knowing spirit, which has become the Spirit that develops the idea, the mobile *ratio* of the universe.

But such radical shifts and such broad transformations of concepts demand not only a logical explanation — the rapprochement and differentiation of ideas — but also a specifically historical explanation, an indication of the real impressions of being that have obliged thinkers to move on to other views.

Hegel finished writing *The Phenomenology of Spirit* the night before the Battle of Jena [14 October 1806]. Later, after Napoleon's abdication, he wrote to Niethammer that already at that time he had foreseen the reaction, "when the entire mass of the mediocre, unconciliated, with all their absolute leaden weight, relentlessly pushes forward until it pulls down the higher force and reduces it to the same level as themselves or lower".[4]

Hegel expected the reaction. It was the finale of a long period beginning with the great French Revolution. For Hegel the Napoleonic Wars, the dissemination of civil law as an expression of new economic and legal relationships, industrial life during the Continental Blockade and the ideological battle in all the countries of Europe, represented a historical continuation of the French Revolution. Hegel was one of those Germans "who thought about what the French were doing". In this case, reflection on the Revolution and on the new era gave an abstract form to concrete impressions and contradictions of being, led to "an escape from the hut of reality to the palace of free thought", of which Kierkegaard spoke while blaming Hegel for such an escape. But in this abstract form the impressions of historical events acquired significant depth; the events of which everyone was speaking proved to be connected with very broad historical, economic and cultural shifts. In Hegel's consciousness, the victories, defeats and collapse of the Empire were associated with economic doctrines and with real economic evolution. These associations and connections, sometimes precise, sometimes intuitive and unclear, are revealed in the extensive literature devoted to the ideological evolution of the young Hegel. An historico-philosophical analysis allows us to connect Hegel's representations of the world and the evolution of the world with his dissatisfaction with the results of the historical events that rose and fell before his eyes.

The representations of the world, as was already said, are not connected by their content with ethical ideals and contradictions. But the momentum forcing the thinker to seek new representations of the world is connected with ethics. For Hegel, as for other thinkers, the ethical impulse of cognition existed as the basic image of the changing representations of the world. Hegel searched for a way out of the conflict between the essential and the necessary, in connection with the distinction between the *existing* and the *actual*; the latter coincides with the *rational*. What is rational is, thus, what is actual; what is actual is, thus, what is rational. Hence the harmony, rationality and ethical value of being, and the ontologism, reality and objectivity of reason. But in the world surrounding him, Hegel did not find rational harmony. We will observe that he did not find such harmony even in the content, or the results of classical science; hence the failure of *The Philosophy of Nature*, the work of Hegel that is furthest from the dynamic spirit of nineteenth-century science.

What did Hegel do?

He made his philosophy the philosophy not of determinate being, but the philosophy of becoming. Harmony, the elimination of alienation, is the result not of a three-dimensional scheme of being, but of static understanding. It is the result of dynamic reason which knows not only how the world is constructed, but also about its lack of construction, its contradictions, the laws of transition, change, the becoming of the world.

In essence, Hegel did what the science of the new period did. Science moved from the prevailing harmony of the Aristotelian universe to the dynamic harmony of Galileo, Kepler and Newton, to the harmony of becoming. But for Hegel becoming is not reducible to a repetition of circular or elliptical movements.

The transition from the discord of being to the harmony of becoming means the rationalization of the object of human cognition; such an object becomes not the sensual, the observable, the real in a sensualist definition, but something more remote and ideal — the laws directing the evolution of the empirically real. There is the usual transformation of a curve into a tangent to it, the usual continuation of the scientific spiral of cognition and the appearance of the Spirit, the other-being of which is served by observable objects.

Such an evolution of concepts renders illusive the initial statement of the discord of being. This discord loses its connection with impressions of being and disappears in the kingdom of pure thought. Such a transition conceals the initial statement of the disharmony of the world so completely that Kierkegaard no longer sees this statement, and the next generation saw in Hegel an apologia for what exists and, with rare exceptions, only an apologia.

If we concentrate on the harmonious and rational evolution of subjective concepts, the immediately essential is no longer the object of influence. But for Hegel, the ideal rationalization of being, its transformation into the 'actual' is not only an ideal operation; it is a law of objective evolution, its logic has ontological meaning and, in essence, we have before us the 'algebra of revolution'; in a certain sense Hegel's conception leads not to reconciliation with, but to revolt against, actuality.

The initial statements are retained in spite of the transition of the problem of alienation from the 'hut' to the 'palace'. In reading Hegel, one is immediately struck by what may be called the thinker's social temperament. Perhaps this phrase 'social temperament' should be placed in quotation marks; for Hegel it has a very modest meaning. But it is precisely this modest political temperament, these sympathies for the French Revolution, these obvious Gallicisms (not only in a direct, linguistic sense), the unexpected remark about the morning paper replacing morning prayer for modern man — all this reveals the character of the ethical impulse that directed Hegel's interests and thoughts along new lines.[5]

The initial ethical sources of Hegel's philosophy are clearly evident in a letter that he wrote to Schelling in 1795. In it he speaks of the political events of those years; the letter is imbued with a revolutionary enthusiasm as well as a sense of the liberating, even revolutionary effect of classical philosophy. Hegel writes of the heights to which man has risen, of his aspiration toward freedom. "I think that there is no better sign of the times than the fact that mankind holds itself in high regard. It proves that the halo surrounding the heads of the oppressors and lords of the earth is disappearing. Philosophers demonstrate this dignity; the people will learn to feel it and will not demand, but will themselves assume their trampled rights. Religion and politics have played under *one* roof; the former taught what despotism desired: contempt toward mankind, toward his incapacity for any good whatsoever and for an independent existence".[6]

Of course, this is the usual reflection on what is happening in another country. But such a very traditional evaluation of German thought about the French Revolution is superficial in its sarcastic undertones. *To think* means primarily to extend the object of thought in time and expand it in space. Thought always contains a spatial generalization and a prediction. In this case, to think of the events of the end of the eighteenth century in France meant to think of analogous events in other countries and in the next (newly begun) nineteenth century. The expansion and extension goes further. Hegel was thinking of the most profound and general basis of the events that had occurred and, by extending their direction further, he became convinced that his youthful hopes and the hopes of all mankind had not been realized. Here, in the statement of the neglect and alienation of man, which were eliminated neither by political revolution nor by the system of economic liberalism, is the initial stimulus for the search for the real harmony of being. This search led Hegel to an illusory resolution of the problem.

4. PHILOSOPHY AND POLITICAL ECONOMY

Let us see how late eighteenth- and early nineteenth-century socio-economic processes and some sociological and economic conceptions were expressed and generalized in Hegel's works.

Jean Wahl, in his book, *Le malheur de la conscience*, says that the entire early period of Hegel's work illustrates the empirical, psychological and ethical aporias of his philosophy. Wahl writes:

> Hegel's philosophy cannot be reduced to logical formulas. Or rather, these formulas hide something that is not purely logical in origin. Before becoming a method, dialectics is an experiment by means of which Hegel passes from one idea to another.[7]

One of the most essential components of dialectics as an "experiment by means of which Hegel passes from one idea to another" was economic analysis, classical political economy and its criticism. According to Jean Hyppolite, Hegel was in part following Adam Smith in his representation of the dialectics of the concrete, the concreteness of truth (Hyppolite 1955, p. 89).

Hegel's concrete logic, the idea of the set as a *concrete* set rich in determinations, vital, contradictory, developing, is perhaps really logically connected with the classical free-trade conception of society. For Adam Smith society is by no means a set of unified, identical elements that have lost their individuality. Macroscopic laws act through the *will* of individuals, the differences between these wills, the individual deviations of these wills from macroscopic law, the neutralization of these deviations; all this produces the mechanism of social harmony, equilibrium, the correspondence between the needs of society and the allocation of social labor.

Society, liberated from state guardianship as portrayed by Smith, is a set of individuals who are not absorbed by the whole; the whole is not a collection of identified elements, i.e., something devoid of many determinations, but is, on the contrary, a set with a wealth of determinations that is greater than the sum of its parts. But this concrete picture, rich in determinations, was to a considerable degree illusory: man did not realize his potentials under the conditions of economic liberalism. For the overwhelming majority of people, these conditions were the source of the devastation and alienation of labor.

Hegel went beyond the idyllic picture of the universal harmony established despite the differences and autonomy of individuals, but only in a negative sense. He saw that economic harmony is problematic, that the elemental macroscopic laws are realized through the suppression of individuality. In reality, the equilibrium of social production originated from the variation between individuals and individual wills, but it was realized through their suppression, through their alienation. Marx saw in economic development the forces that would lead to a different harmony, the ultimate harmony of the individual being of people and their social being in a concrete set that retains an infinite wealth of determinations.

Hegel did not see these forces. The flight of his thought, as it attempted to embrace the infinite complexity of being, took place not in political economy, but in philosophy. The evolution of Hegel's interests, his shift from economic problems to specifically philosophical ones, was a negative result of reflection on economics. This negative result — in economics there is no internal mechanism that realizes social harmony — was expressed also in the content of philosophical constructs. The idea of concrete sets was expressed in an abstract form.

Hegel's economic ideas were first expressed in 1799 in the commentaries on James Steuart's *Principles of Political Economy.* This manuscript has not been preserved, but Rosenkranz, who had held it in his hands, said that, "In it were concentrated all of Hegel's ideas concerning the nature of civil society, need and labor, the division of labor, the division and the property of social classes . . ." Further, Rosenkranz notes Hegel's numerous observations about the historical process and the connection of political ideas with economics. These observations and, to a considerable degree, the entire contents of the commentaries, were directed against mercantilism as advocated by James Steuart. "Hegel attacks what is dead in the system of mercantilism and attempts to find in competition and in the mechanism of labor as well as trade the salvation of man's soul."[8]

Apparently, the lost manuscript of 1799 was a fundamental critical rebuke of mercantilism: statism, the influence of the state on civil society, deprives man of 'inner life'. Subsequently, Hegel's economic conceptions were attacked from the opposite direction in economic thought, by anti-statist ideas and by economic liberalism. In the works of 1803—1807 (*System of Morality, Philosophy of Mind, The Phenomenology of Spirit*) and in the later *Philosophy of Right*, Adam Smith's ideas formed the starting-point for Hegel's conceptions. But only the starting-point. Hegel's economic views are opposed to Smith's ideas; they are opposed not only to the free-trade program but also to its deepest theoretical foundations.

Hegel does not agree with Adam Smith and with liberalism as a whole on one fundamental question. Civil society cannot, by itself, without political form, without a state, guarantee equilibrium and existence. Hegel's idea was to find a general harmony that realizes the macroscopic laws of social life, which would not suppress individual fates and individual aspirations in a mechanical manner.

Hegel's thought is thought that is extremely, perhaps maximally, abstract in its level of generalization and extremely concrete in its drive to find concrete sets that preserve the individuality of their elements. As we know, this thought is connected with the spirit of classical science, with its dynamic, reconstructing drive, with the gradient of the transition from medieval thought to new thought about the world. It is connected with the idea of ordered sets in which individuals do not lose their individuality, are not identified, do not drown in the universal abstract repetitive image. In the drive to find the concrete, nonidentical filling of universal concepts, to find the universal that does not devour its elements, Hegel often selects illusory solutions. The nature-scientific conceptions in *The Philosophy of Nature* are relevant here. Hegel does not see that the dialectical tendency of his philosophy embodies revolution, change, transformation, and not the crystallization of the new science in its frozen and almost final form. Hegel's relation to political and social revolution and to the almost final consolidation of its results seems somewhat analagous.

The French Revolution had much more radical tendencies than the positive results to which the bourgeoisie wanted to limit it. The embodiment of its halting, obstructive limiting forces was the Thermidor and the Eighteenth Brumaire. It would seem that for Hegel Napoleon was the 'understanding' of the Revolution, rather than its 'reason'. But here the historical limitation of the empirical and moral impulses of theoretical thought comes into play.

In the Preface to *The Philosophy of Right*, Hegel wrote that philosophy generalizes the ideas of its age. ". . . every individual is a child of his time; so philosophy too is its own time apprehended in thoughts" (Hegel 1967, p. 11).

Philosophy really is the daughter of its time. But the concept of *its own time* is a complex one. 'Its own time' is a combination of processes that include the continuation of the past and the eventual future, the differential indicators of the processes that are occurring. It is a combination of processes that flow very slowly and create an illusion of the eternity of what 'its own time' has found and the eternity of processes that must be developed, of questions that will be answered later. Which processes are generalized by philosophy, how does it view 'our own time'? Does it pay attention to what has already solidified or to what is presented as a question, to what has not yet acquired a solid form?

In the chapter concerning the relationship of Hegel's philosophy to classical science, this question was presented in a historico-physical plan. The basic dialectical ideas of Hegel's philosophy reflected the eventual future of science, the labored constructs of *The Philosophy of Nature*, the illusory immobility of science. Now this same question relates to historico-sociological, economic and moral evaluations of 'its own time'. The general spirit of Hegelian dialectics reflected the tendencies of the French Revolution as they poured over the dam holding back the demands of the Third Estate. Evaluations closest to the historical events extolled the barrier that held these tendencies in check.

Hegel's world-view hardly suggests a simple sum of conservative and reactionary ideas. The structure of his world-view is nonlinear: each element alters the whole, the whole alters each included element. Having revealed the illusory liberation of man in the realization of the ideal of the physiocrats and Adam Smith, Hegel rushed to the side of statism, at first extolled Napoleon and then the Prussian monarchy, and became the official Prussian court philosopher. This rush resounds along the entire chain of logical deductions, right down to the initial determination of being. In their turn, the political and moral evaluations included in the chain of Hegelian philosophy change their meaning, acquire a multiple meaning, and the apologia for statism becomes an accusatory act. When the existing is represented as the rational, new criteria are added to it and some elements of the existing turn out to be irrational and, consequently, nonexistent. It remains only to put into effect the verdict of Reason, to discontinue the illusory existence of these elements.

Such a conclusion was the result of a transformation in Hegel's very own logic. It ceases to be an *a priori* movement of reason (until Hegel, logic was, for the most part, considered as *a priori* immobility); it becomes, *from the very beginning* (i.e., after bidding farewell to 'pure being') mediate and dependent on empiricism, on experience, on active experience. The task of philosophy becomes the transformation of the world. The concept of transforming activity — experiment and *labor* — enters philosophy.

Hence it follows that the doctrine of the social forms of influence on nature is a part of the theory of the cognition of nature. At present, this is a very obvious feature of production: modern production constantly creates not only ordinary values, but dynamic values as well — the conditions for the future acceleration of the tempo of civilization. These conditions are scientific representations and produce a dynamic effect of a higher order than constructions and technological practices. It must be said that while epistemology, too, in its turn has not yet become a part of economic science, it is obviously joining the ranks of dynamic factors there.

For Marx, the problem of alienation became a dynamic problem, and alienation itself, a process. Twentieth-century science views entropy, i.e., the vacuum of macroscopic being, its transition into nothing, as a process, and thus portrays the mechanism of the growth of negentropy as the origin of macroscopic differentials; it portrays macroscopic structure as a result of a chain reaction produced by a separate, individual physical object. By analogy, Marx discovered the path of the 'disalienation' of labor and the objective laws of such a process.

There is no need to analyze Marx's economic theory here and to find in it the categories of alienation and 'disalienation' that have been reviewed and to compare these concepts with the philosophical and physical equivalents (determinate nothing and the vacuum, respectively) similar to them. We shall limit ourselves to one fundamental category of *Capital*, namely, commodity fetishism.

Commodity fetishism transforms the ratio of labor expenditures, i.e., the category characterizing the social division of labor, into a predicate of commodity, into its value. Hence the illusion of a certain independence of value from labor; hence the illusory hopes of finding in the physical, chemical, and other properties of a thing the basis of its value.

Such illusions, leading away from the truth and actual cognition, were an unjust continuation of a certain section: labor is crystallized into value; a characterization of labor — homogeneous, abstract, assigned labor — becomes the characteristic of a thing. The real subjugation of man to elemental social laws is expressed in this idea. The statement of alienation becomes an accusation and a verdict against these laws.

In *Capital* the term 'alienation' disappears because concepts of alienation and its overcoming are embodied in economic categories. Thus they become historical. The search for an economic theory of the alienation of labor was a continuation, a part, of an aspect of Marx's philosophical evolution. One of the most important stages of this evolution, which inevitably passed into the genesis of a new economic theory, was the economic and philosophical manuscripts of 1844. The organic and fundamental transition from philosophical (and in connection with them natural-scientific) interests to economic interests is evident here. In essence this is not even a transition *from* philosophical and natural-scientific interests, but rather a transformation of these interests, an organic and natural transformation flowing from Marx's philosophical positions, from the growth of his thought beyond the boundaries of Feuerbach's doctrines.

In these years, Marx's ideas began to rest increasingly on the *history* of science as well, on the analysis of its transformation, on what was called above the dynamic generalization of classical science. Examined in such a way science becomes a human matter. Man appears as man comprehending nature and this comprehension expresses "the needs of man as man". According to Marx, as natural science progresses, it increasingly includes this aspect and ultimately the history of nature and the history of man are merged. "Natural science will in time incorporate into itself the science of man, just as the science of man will incorporate into itself natural science: there will be *one* science" (Marx 1964, p. 143).

Such science no longer opposes man as a foreign force that alienates him. On the contrary, it reveals the essence of man. But the relation of man to science acquires an irrational character. Influence on nature, labor, ceases to be an expression of the essence of man, an expression of his freedom and the basis of his real being. The self-alienation of labor consists of the fact that ". . . labor is *external* to the worker, i.e., it does not belong to his essential being; . . . in his work . . . he does not affirm himself but denies himself . . ." (Marx 1964, p. 110).

Such alienation is inherent in private property: "*Private property* thus results by analysis from the concept of *alienated labor*, i.e., of *alienated man*, of estranged labor, of estranged life, of *estranged* man" (Marx 1964, p. 117).

Hegel and Kierkegaard posed the problem of alienation in different ways but this difference is significantly less than that which separates both of them from Marx's position. They both rejected the single real path to the elimination of the alienation of man, to actual freedom. For Hegel, real, earthly and concrete human practices cannot replace the logic of the motion of ideas. For Kierkegaard, freedom and the autonomy of the individual also remain illusory. The kingdom of freedom, says Kierkegaard, is the future. But the future is determined by the past; it disappears as a zone of freedom if the world is controlled by strict scientific causality. Therefore Kierkegaard rejects science, causality, rationalism, the spatio-temporal and cosmic response of the individual. This is a negative conception of freedom; it is based on absenteeism and isolation.

But the isolation of the individual from macroscopic activity, from interaction with the macrocosm, just like its neglect, leads to the alienation of the individual, to his transformation into a passive and ignored element of the macroscopic scheme, indifferent in the sense of 'to be' or 'not to be'. In essence he is no longer an element of being; it is as if he were out of touch with the macroscopic scheme (for example, with the framework of world-lines), with its ultramicroscopic filling. It is precisely this being-out-of-touch that causes alienation. What was guessed at by classical philosophy, that received precise form in Marx's philosophical and economic manuscripts, what has not acquired precise physical shape in nonclassical science, is the ontological meaning of alienation. A local process cut off from the rational scheme of the universe becomes illusory, abstract; it is deprived of physical being. The rational scheme cut off from the local filling is also alienated and becomes 'foreign', no longer physical, but geometrical, devoid of physical being. The human individual, cut off from labor and thus cut off from its connection with nature and with society, loses his strictly human existence. Labor that does not express man's own impulses becomes empty labor, alienated labor that has lost its *actuality* in a Hegelian sense, that is degraded from being to existence.

Labor as the basis of man's own being, the alienation of labor as the basis of the alienation of man, the liberation from the alienation of labor as the basis of the liberation of what is specifically human in man — this entire conception is found on the way to the constructs of *Capital*, the theory of abstract labor, the entire economic and sociological doctrine of Marx. However, this road goes from the analysis of not only the Hegelian concept of alienation, but also 'The Difference between the Democritean and Epicurean Philosophy of Nature' (see p. 105, n. 8) where the general ontological function of individual destructions of macroscopic laws is developed. The nonalienation of individual autonomy and localized processes as a guarantee of being unites in Marx the conception of nonalienated, active nature not dissolved and not frozen into immobility in the world spirit — a conception close to Spinoza's *natura naturans* — with the conception of nonalienated man possessing a genuinely human being.

5. PHANTOM AND REALITY FOR EINSTEIN AND DOSTOEVSKY

Let us now approach the problem of determinate nothing in terms of a somewhat unusual aspect. This problem is examined here from a nonclassical historical perspective: the concept of determinate nothing is likened to the modern concept of the vacuum. On the other hand, the concept of determinate nothing has both a natural-scientific and a social ring when compared to the concept of alienation. The demonstrated aspects of the single problem of determinate nothing, the vacuum, alienation, may be illuminated by comparing the poetics of Dostoevsky with the ideas of Einstein. In addition, such a comparison allows us to shed light on one of the basic problems of the history of rationalism. Classical rationalism merged with nonclassical science in the world-view of Einstein where the ontological tendency of rationalism received full and obvious expression, that is, its transformation into a doctrine of being, a concrete, mobile being that includes concrete determinate nothing.

Einstein said that Dostoevsky gave him more than any thinker, more even than Gauss. This evaluation is not easy to explain; it is connected with very profound questions that are not reducible to the interactions of scientific and moral ideals. For Einstein, the fundamental tendency of Dostoevsky was ontological; for the scholar the work of the writer was by no means a moral sermon.[9] It was a fight against destruction, against the annihilation of being, against its transformation into a vacuum, into eventual being. What we are dealing with here is not an evaluation of Dostoevsky's work as a whole, but with the determination of that aspect of his work that is relevant to Einstein and connects Dostoevsky's images with Einstein's ideas. But for any general evaluation of Dostoevsky's work, this ontological aspect is also essential. Dostoevsky is often viewed as an advocate of irrationalism, his rejection of reason being connected with the moral inferiority of harmony based on reason, on the objective *ratio* of the world, if only a paradoxical 'non-Euclidean' *ratio*.

There is a certain basis for such a view. Nevertheless, a more attentive analysis from the vantage point of modern science, the representation of Dostoevsky's art as a question to which science must respond, leads to a different opinion. Dostoevsky is included in the line of rationalist thought, and his major interests are considered within the framework of the ontological problem. Such an opinion was directly expressed by Einstein in a remark to be discussed below.

Dostoevsky's poetics are rationalist poetics. The images of his characters, their speech, situations, background — all these express primarily a mental conflict. The very melody of Dostoevsky's works is a turn of thought, sick, paradoxical, but nevertheless thought. The driving motive of his characters' actions is their striving to *discover*, to measure the height of a heroic deed or the depth of a fall, to discover whether universal harmony exists and of what it consists, to discover the meaning of being, to resolve problems of morality, philosophy, religion — in other words, to resolve, understand and comprehend their meaning. Raskolnikov tells Sonia his motives for murdering the old woman: "It was something else I wanted to find out, it was something else that goaded me on: I had to find out then, and as quickly as possible, whether I was a louse like the rest or a man. Whether I can step over or not. Whether I dare stoop or not? Whether I am some trembling vermin or whether I have the *right* —" (Dostoevsky 1968, pp. 432–433).

Dostoevsky's purpose is anti-rationalist. He wants to demonstrate the predestination of reason that rejects traditional faith and claims a sovereign role. Ivan Karamazov, Stavrogin, Raskolnikov are adepts of reason opposing faith, and the purpose of Dostoevsky's novels consists in demonstrating the ruin of emancipated reason. But Dostoevsky's poetics is entirely different and, as is so often the case, leads to other conclusions which the author did not make and which were muffled in his purpose. In all of Dostoevsky's novels it is impossible to encounter a single confrontation that does not flow from thought: no matter how complex, paradoxical and entangled, it is in fact thought. Moods, feelings and passions are born of thought and they recede into thought; they are inseparable from it. Dostoevsky is a writer for whom Logos and Eros are not in opposition to each other but coincide or, in any case, are mutually pervasive.

Hence Dostoevsky's musicality. Any twist, however unexpected, any retort, however paradoxical, any character's action, however impossible, illogical, or far beyond the bounds of the normal — all this flows from thought, from contradictions, from twisted paths, from paradoxes of thought. As a result, the melody of the narration is unusually harsh, but it is a rationalist melody.

After all, even thought embodied in modern science is paradoxical in that it contradicts 'normal' logic; it is based on 'mad' logic, but it is strictly logical thought. Rationalist thought. There is not and cannot be a direct logical movement into the 'madness' of nonclassical science from the 'madness' of the melody in Dostoevsky's novels. But there is isomorphism here, and it is possible for something to be influential not in a logical but in a psychological way. These influences are unclear, unrecorded, but real.

It was said above that the passions of Dostoevsky's characters flow from thought and pour into thought. Indeed! After all, the major and most universal passion is the passion for knowledge. Alyosha Karamazov says of Ivan: "He's one of those who don't want millions, but a solution of their problem" (Dostoevsky 1966, I, p. 92). And such are the others too, both Raskolnikov and Stavrogin, whose entire consciousness, entire life, is filled with the drive to find a "solution of their problem".

They answer them with an experiment, which has received the designation *cruel*. Dostoevsky wrote of Edgar Allen Poe:

> He almost always takes the most exceptional reality, places his hero in the most exceptional external or psychological position and with tremendous force of insight, with deadly accuracy, he tells of the condition of his man's soul (Dostoevsky 1930, p. 523).

But Dostoevsky himself places his characters in even more exceptional conditions. In fantastically difficult situations, in instants separating them from suicide, madness, ruin, in moments of the most agonizing ordeals, in experimental conditions as remote from the everyday as relativist experiments are from the classical, Dostoevsky's characters aspire to understand, acknowledge or reject cosmic harmony.

If there is no harmony in the Universe, the Universe is a phantom. Such an idea pervades Dostoevsky's rationalist poetics. This is rationalism as we have tried to represent it in this book — knowledge based on experiment that seeks in nature the objective *ratio* without which the cosmos would remain chaos.

In *Crime and Punishment* Svidrigaylov says to Raskolnikov:

> We're always thinking of eternity as an idea that cannot be understood, something immense. But why must it be? What if, instead of all this, you suddenly find just a little room there, something like a village bath-house, grimy, and spiders in every corner, and that's all eternity is. Sometimes, you know, I can't help feeling that that's probably what it is (Dostoevsky 1968, p. 305).

But what Svidrigaylov fancies is illusory being; it is a phantom. Not the pure *nothing* of Hegel. Not the very thought of mere being, i.e., thought in which 'to be' ceases to be a copula and becomes a word lacking a subject and content. Eternity, "something like a village bath-house", is an image of *ruptured* being, devoid of connection, scattering into local elements. Thus the phantom expresses virtual being here, a vacuum of being, Hegelian determinate nothing. But ruptured being is not the only phantom. Being, where the macroscopic *ratio* excludes the individual and is independent of the entire *clinamen*, is also a phantom.

In his conversation with Ivan Karamazov, the devil complains of a universe without 'events'. It is "an endless religious service: it would be holy, but a little dull" (Dostoevsky 1966, **II**, p. 755).

There is an individual aspect in the same ontological problem. Is the stream of consciousness real if it does not pass through the crucible of doubt? This question is posed by the same Karamazovian devil. Apparently, for Einstein the problem of the reality of the stream of consciousness was the essential problem in Dostoevsky's art. In Einstein's conversation with Solovine, made public in 1930, the latter spoke of Dostoevsky as a thinker interested primarily in the problem of suffering. Einstein replied:

> I don't agree with you. The idea is different. Dostoevsky showed us life, to be sure; but his aim was to turn our attention to the riddle of spiritual being and to make it clear even without commentary.

The issue is how to think of the problem of the spiritual being of man in terms of his dependence on the whole. In classical science *absolute* existence, independent of the existence of the rest of the world, was ascribed to a separate particle. The same postulate of absolute being was encountered outside of natural science as well, if we just recall the fiction of the 'unity', which was not unique to Stirner. For Einstein not only are coordinates and velocity, i.e., the position and behavior of the particle, concepts without physical content, devoid of meaning without other bodies, but the very being of the particle cannot have meaning without the entire system of interacting bodies. In quantum mechanics as well as in elementary-particle theory, the dependence of the particle's individual being on the cosmos that interacts with it became, as we have seen, an even more basic tendency in scientific thought. As far as spiritual being is concerned, Einstein leaned toward Spinoza's idea: man's true spiritual being demands the comprehension of truth; his consciousness is imbued with the drive toward such comprehension.

But individual being cannot be reduced to a passive reflection of the whole. It includes something individual, and without the individual it becomes illusory being. This is the ontological aspect of the rejection of cosmic harmony when such harmony annuls the being of the individual and ignores his fate. Such a rejection was the critical point in the internal life of the intellectual confrontations in *The Brothers Karamazov*.

In a conversation with Alyosha, Ivan Karamazov says that God has created the world "in accordance with the Euclidean geometry". But, he continues, non-Euclidean geometry exists: some people suggest that being corresponds to it and that somewhere two parallel lines will meet. This non-Euclidean cosmic harmony does not suit Ivan and he says:

> Let me put it another way: I'm convinced like a child that the wounds will heal and their traces will fade away, that all the offensive and comical spectacle of human contradictions will vanish like a pitiful mirage, like a horrible and odious invention of the feeble and infinitely puny Euclidean mind of man, and that in the world's finale, at the moment of eternal harmony, something so precious will happen and come to pass that it will suffice for all hearts, that it will allay all bitter resentments, that it will atone for all men's crimes, all the blood they have shed. It will suffice not only for the forgiveness but also for the justification of everything that has ever happened to men. Well, let it, let it all be and come to pass, but I don't accept it and I won't accept it. Let even the parallel lines meet and let me see them meet, myself — I shall see and I shall say that they've met, but I still won't accept it (Dostoevsky 1966, I, pp. 274—275).

Why does Ivan Karamazov reject cosmic harmony and 'most respectfully return the ticket' to the kingdom of this harmony? Primarily because it does not atone for the sufferings of the individual. Does this mean that cosmic harmony is to be put on trial before the justice of morality and compared with moral ideals? This would be too tranquil, traditional and classical a resolution of the question. For Dostoevsky the moral ideals themselves are not invariant, absolute, the final line in the analysis of world harmony, something previously given.

In 1946, in a conversation with Ehrenburg, Einstein said:

> I was greatly impressed by *The Brothers Karamazov*. It's one of those books that shatter the mechanical conception of man's inner world, of the limits of good and evil (Ehrenburg 1966, p. 77).

If we compare this brief remark of Einstein with his response in the conversation with Solovine, then most likely, he means the following:

"The mechanical conception of man's inner world" operates once and for all with given definitions of good and evil, with immutable logical and moral canons of consciousness. It is analogous to mechanical processes, i.e., motions of self-identical, nondisappearing and non-arising bodies governed by immutable laws. 'Spiritual being' includes a change in the questions posed, a reconsideration of the "limits of good and evil". The basic inspiration of *The Brothers Karamazov* lies in a reconsideration of traditional boundaries, in the 'cruel experiment' necessary for their reconsideration.

Einstein, as we have seen, did not agree with Solovine's remark about Dostoevsky as a thinker for whom the basic problem was human suffering. Probably the representation of Dostoevsky as an apostle of traditional moral ideals was foreign to the profoundly paradoxical and anti-traditionalist thought of Einstein. Traditional moral ideals were the 'tranquil' matrix (in the Hegelian sense), the noncontradictory and stable mold, into which many wanted to force the extremely agitated, dynamic and contradictory conceptions of the writer.

Dostoevsky's moral philosophy was based not on understanding but on reason. Every local action alters the entire system of moral philosophy. And not only moral philosophy. It provokes a re-evaluation of *all* being as a whole, not only from a moral standpoint (does this world correspond to the moral demands of man living in it?), but also from the ontological standpoint: is this world real or is it an apparition, as, for example, St. Petersburg seems to Dostoevsky? When Dostoevsky returns the entrance ticket to the kingdom of harmony, he thus removes from harmony the entrance ticket to the kingdom of genuine reality. For Dostoevsky not only is everything actual rational, but everything actual also corresponds to a moral ideal or it ceases to be actual. The moral ideal — the non-ignored individual — is also the ontological ideal. The city where the 'negligible' suffers as an ignored detail of statistical harmony is a phantom-city.

As we have seen in analyzing the evolution of classical rationalism from a nonclassical perspective, the filling of the rational scheme with local events is an inalienable property of rationalism when it becomes a doctrine of being. But the filling destroys the rationalist scheme. This nonlinear character of the inclusion of localized actions in a set also transforms the scheme into reality.

From such a point of view, normative and absolutizing morality cannot satisfy Dostoevsky. This is also related to the trivially regimented morality of the Middle Ages, to the morality of the Reformation that flows from the ideal of the Gospels and to the Kantian categorical imperative. In Dostoevsky's moral philosophy, as embodied in his poetics, the main motive is the autonomy of moral action in terms of the general norm. Thus we are dealing not only with compassion toward the 'negligibles'; even, perhaps, not very much with this compassion. Dostoevsky's characters are attracted by moral experiment. Raskolnikov and Ivan Karamazov and Stavrogin wanted to convince themselves experimentally of the possibility of deviation from the social norm, from macroscopic stable harmony. And this is not only a moral but also an ontological experiment. The Universe in which the tormented child is crying is devoid of ontological justification because it is not filled with local fates; these fates are ignored. The fate of the tortured child is simultaneously the fate of the Universe. And this is not only a problem of the moral acceptability of the Universe; it is also a problem of being. But the Universe where 'parallel lines meet and all men's crimes will be atoned for' risks being 'an eternal hosannah', i.e., the same phantom, boredom, 'a room like a village bath-house'.

The peculiarities of Dostoevsky's poetics flow from the strained, morbidly strained, interest in cosmic problems in their local, individual aspect. The combination of the Spinozan 'telescope' and the Leibnizian 'microscope' is typical of Dostoevsky, his poetics and his moral philosophy. Dostoevsky's gaze penetrated into the microscopic details of landscape, image, scene, into the most trivial turns of dialogue; it included the Universe.

Dostoevsky's moral philosophy is active and dynamic, but not evaluative. Dostoevsky is as far removed from accusatory pathos as from exhortation. When Ivan Karamazov speaks of bestial cruelties, it is not a tranquil narrative; he hates executioners, and his story calls forth Alyosha's response, "shoot him!" (Dostoevsky 1966, **I**, p. 284). But both brothers, Alyosha and even more so Ivan (for whom, as for the author, the whole conversation is a 'cruel experiment'), are not swallowed up by evaluative statements: 'what villains!' and 'what unfortunates!', but by others: they have an excruciating desire to understand what is this world in which all that has been told can occur.

In any case, it is not a homogeneous world, and, what is especially important, not a world with stable laws. In such a world, as already stated, cognition of the truth becomes inseparable from the transformation of the world. Morality loses its evaluative force; it acquires a cognitive and reconstructive force. The paradoxical and unresolved local clash is just as characteristic of such morality as are the experimental results in radically evolving science that are unresolved within the framework of the traditional picture of the world (classical science in its sharp shifts and nonclassical science practically constantly). For Dostoevsky this is more than an analogy. For him the unresolved clash before which Raskolnikov, Ivan Karamazov, Stavrogin and others stopped, is an experiment, and not only the experiment of Raskolnikov or Stavrogin. It is the experiment of the author himself.

It is an experiment that is cosmic in its object, cruel in its method. In science such experiments have appeared only in this century: problems of the order of the existence of space that is separate from time or the discreteness and continuity of space and time are resolved under the conditions of superhigh velocities, energy and stress. Both in literature and in science such experiments are connected with the operative character of the formula $\alpha \in A$; the local individual event is not simply contained in the Whole, but transforms it.

Such experiments are connected with a certain feature of the psychology of creative work, which clearly appears in art, but which also exists in science. It is the ability to view the world as if for the first time, without the burden of traditional associations. But it is not the inadequacy of vision that transforms the past into a compact, unfilled, canvass primed for a new picture. It is a lofty and sharp insight in relation to the world as well, to its colors, and to cognition itself — to the blueprints of the universe.

In Einstein's creative work, what role was played by sharp insight into the world, its colors, its heterogeneity, its microscopic aspect? Is Einstein's sharp insight connected with the rejection of non-Euclidean harmony?

For the theory of relativity as a system of simply resolved problems, for the theory of relativity *in vitro*, such a question does not make sense. The theory of relativity sees the *ratio* of the world, the basis of cosmic harmony in the framework of world-lines that curve because of the pull of gravitational fields, in the four-dimensional non-Euclidean geometry of the world. But for the theory of relativity *in vivo* and, accordingly, for Einstein's life and for the immortal life of his ideas, the framework of world-lines was inadequate. It does not possess the physical being that distinguishes the cosmos from a geometrical scheme. Particles without a macroscopic world-line are ruptured being, a vacuum. However, the world-line without certain equivalents of Eupicurean *clinamen*, without ultrarelativist filling, is not true being, but rather 'determinate nothing', indicating processes that do not figure in the geometrical scheme, but which may already be determined. The macroscopic correlations of the theory of relativity indicate these still absent ultramicroscopic details of the scientific picture of the world. The modern stage in the development of science is characterized by an intense search for a simple representation of this ultramicroscopic world.

Such a search is also the starting point for an historical analysis of classical rationalism, which reveals in its evolution a sequential *rapprochement* to physical being.

NOTES

1 "... si l'on a pu dire: penser c'est généraliser, ne peut-on pas dire avec autant de raison, et plus de profondeur: penser c'est opposer? Ce qui caractérise la pensée, c'est la faculté de mettre en parallèle l'affirmation et la négation. L'être, la chose est, dans un sens toujours positif; mais la pensée se libère de l'être en donnant un sens au non-être; ce qui n'est pas est objet de pensée au même titre que ce qui est" (Brunschvicg 1897; 1964, p. 12).

2 "Chaque chose est ce qu'elle est, et elle est pourtant une chose déterminée. Mais la conscience empirique prend cette unité comme immédiate ..." (Hyppolite 1953, p. 147).

3 This passage is cited by P. P. Gaidenko in her book on Kierkegaard: *Tragediia estetizma* (Moscow, 1970), pp. 9—10.

4 "Die ganze Masse des Mittelmäßigen mit seiner absoluten bleiernen Schwerkraft drückt ohne Rast und Versöhnung so lang bleiern fort, bis es das Höhere herunter, auf gleichem Niveau oder unter sich hat" (Hegel to Niethammer, 29 April 1814 in Hegel 1969—81, **II**, p. 28).

5 "Das Zeitungslesen des Morgens ist eine Art von realistischem Morgensegen" [Reading the newspaper in the morning is a kind of realistic matins], from 'Aphorismen aus der Jenenser Zeit' in Hegel 1936, p. 360.

6 "Ich glaube, es ist kein besseres Zeichen der Zeit als dieses, daß die Menschheit an sich selbst so achtungswert dargestellt wird; es ist ein Beweis, daß der Nimbus um die Häupter der Unterdrücker und Götter der Erde verschwindet. Die Philosophen beweisen diese Würde, die Völker werden sie fühlen lernen, und ihre in den Staub erniedrigte[n] Rechte nicht fodern, sondern selbst wieder annehmen, — sich aneignen. Religion und Politik haben unter *einer* Decke gespielt, jene hat gelehrt, was der Despotismus wollte, Verachtung des Menschengeschlechts. Unfähigkeit desselben zu irgend einem Guten, durch sich selbst etwas zu sein" (Hegel to Schelling, 16 April 1795, in Hegel 1969—81, **I**, p. 24).

7 "La philosophie de Hegel ne peut pas être réduite à quelques formules logiques. Ou plutôt ces formules recouvrent quelque chose qui n'est pas d'origine purement logique. La dialectique, avant d'être une méthode, est une expérience par laquelle Hegel passe d'une idée à une autre (Wahl 1929, 'Préface', p. 9).

8 "Alle Gedanken Hegel's über das Wesen der bürgerlichen Gesellschaft, über Bedürfniß und Arbeit, über Theilung der Arbeit und Vermögen der Stände ... concentrirten sich endlich in einem glossirenden Commentar ... Mit edlem Pathos, mit einer Fülle interessanter Beispiele bekämpfte Hegel das Todte desselben, indem er inmitten der Concurrenz und im Mechanismus der Arbeit wie des Verkehrs das Gemüth des Menschen zu retten strebte" (Rosenkranz 1844, p. 86).

9 I have tried to illuminate the connection between the moral and ontological aspects in the chapter 'Einstein and Dostoevsky' in my book, *Einstein. Life, Death and Immortality* (Kuznetsov 1972a). The section that follows contains a brief summary and some development of the conclusions reached in that chapter of my book. [See also Kuznetsov 1972b.]

AFTERWORD*

Tradition allows an Afterword certain violations of genre, something personal, a certain, albeit very limited, lyricism. I should like to relate some impressions and thoughts evoked by the cultural monuments of Paris, which are also reflected in the contents of this book.

During the autumn of 1970, when the most of this book had been written, I spent time in the Bibliothèque Nationale reading the works of the great thinkers of the seventeenth century, the most recent authors, and modern philosophers and physicists. In the mornings, I would wander along the streets of the Latin Quarter, along the quays of the Left Bank and Île de la Cité, around the Louvre, and most often around Notre Dame, absorbing impressions of Paris and jotting down almost everything that came into my head. What I read helped me to see Paris, and what I saw helped me to understand what I had read. Perhaps these notes, put into some sort of order for this Afterword, will help the reader to see several aspects of the *rapprochement* between Reason and Sensibility, identity and nonidentity, the macroscopic organization and microscopic filling of the world, which pervade *Reason and Being.*

The impressions and thoughts are grouped into eleven sections, the first of which I was bold enough to give the name that Heine gave his book that collects some of his newspaper articles written in the form of letters from Paris.[1]

1. *LUTETIA*

From the Pont St. Michel, I look at the Seine and its banks. The sun has not yet burnt off the morning mist, which blurs the distinct and so familiar form of the quays and bridges. This is so typical of a Parisian scene: the rational composition and the soft, elusive coloring. It transforms the topographical scheme into something real, not contained in the scheme. The harmony of line and color create the impression of

* [This Afterword was written originally as the Introduction to the (unpublished) French translation of *Reason and Being*. — Eds.]

what might be called the intensiveness of being. It is precisely being, and not only existence. According to Hegel, the essence revealed by reason transforms existence into reality and allows one to see in the universe not chaos, but cosmos, to ascribe an ordered character to real being. The view of Paris suggests the fundamental problems of being. But this is thought with its eyes opened; in the landscape of the city are embodied not only the universals of medieval realism, but also the traditions of nominalism with its intense attention to objects and colors. However, why would any one want to close his eyes here? Within one's field of vision is the most enchanting embodiment of the contradictory harmony of Paris. Along the Pont St. Michel young girls are walking with the *tenue* of Brigitte Bardot and the eyes of Descartes in the portrait by Hals.[2] These eyes look at Paris, but Paris also looks at them; gay, and at the same time, sad; all-comprehending irony becomes an element of the Parisian landscape.

Even at first glance, it is four-dimensional. The Parisian landscape cannot remain three-dimensional, purely spatial; time enters into it immediately. Temporal associations involve not only information about the time when the cathedrals and palaces were built; these associations are inseparable from the visual impressions. Since Helmholtz's time we have known that visual impressions are impossible without a certain apperception. Here in Paris this includes the idea of the past and the idea of the future. Hence visual impressions cannot remain only impressions; they become deductions, and space, which your gaze penetrates in Paris, is a four-dimensional space-time continuum. Of course, this does not prevent the Parisians from considering the four-dimensional world of the theory of relativity a difficult and abstract category.

The four-dimensionality of impressions, the inclusion of the past in the present, has also forced me to call this sketch by the first, Roman name of the city and thus to repeat the name that Heine gave to his Paris letters. There is no pretension here; I simply wanted to emphasize what has always produced the greatest impression on me during my strolls around Paris.

The four-dimensional character and the intellectual accompaniment of the impressions of this city lead me to meditate on the fundamental, pervasive problems of scientific and philosophical thought. Even in antiquity, philosophy had confronted the initial paradox: the past no longer exists, the future does not yet exist, the present is of zero duration and is therefore a nonexistent boundary between the past and the future. The paradox of being is resolved by the inclusion of the past and the future in the present. Such an inclusion is by no means a momentary enlightenment in philosophical thought; it is the prolonged and fundamental calculus and the differential laws of physics serve to illustrate very clearly the inclusion of the past and the future in the present, in the 'now' of zero duration. A moving particle possesses at each moment not only position in space (it passes through a given point, its existence at this point is momentary), but also velocity (the maximum ratio of the path traversed to the time elapsed and the first derivative from the position in respect to time) and acceleration (the derivative). The generalization of such mathematical and physical concepts shows that the present possesses being to the degree to which it includes the past and the future. This inclusion may be greater or lesser and the degree of inclusion is the intensiveness of being.

I shall limit myself to this fleeting hint at the constructions set forth in comparatively systematic form in this book. I should like to return from the realm of pure thought to Paris, to see the streets already flooded with the new sun, to see the patches of light on the surface of the Seine, to hear the discordant murmur of human voices. In essence, this concrete polyphony did not disappear from my consciousness even during the construction of philosophical categories, during the construction of the specificity of these categories in their inalienable sensual background. With the return to Paris, the logical schemes in their turn are preserved with the constant, rational, ideological resonance of the specificity of the living impressions of Paris . . .

I crossed the Pont St. Michel to the Île de la Cité, turned right at the place du Parvis Notre-Dame, where, at present, construction is going on, and went along the narrow passage by the little public garden, with the statue of Charlemagne, to the cathedral. It always produces a very unified effect that cannot be divided into separate components, yet the effect is also complex. The historical recollections are so natural that they seem a direct impression. Here what I just said is especially distinct: one literally *sees* the past, the entire history of the cathedral from the heathen temples to the present, feels the continuity and the four-dimensionality of being. The cathedral does not seem an embodiment of immobility, or self-identical and, thus, lacking in history, purely spatial, beyond time, and so, devoid of true being.

On the contrary, it seems a symbol of dynamism. It seems connected not only with the past, but with the future as well. And here is why.

Vladimir Vernadsky and Pierre Teilhard de Chardin introduced the concept of the noosphere of the Earth.[3] Together with the lithosphere, the hydrosphere, and the atmosphere, there exists a sphere created by the labor and reason of man, where the natural structure is altered, and the natural processes are arranged according to the aims of man. But there is still another sphere of the Earth's surface. It might be called the cultural sphere, but this name is already taken: there exists the concept of the cultural layer of the lithosphere. What we are dealing with is the totality of the cultural values kept on Earth which connect the past with the present and the present with the future. It is crystallized information, answering the questions: *why* do certain processes occur (this is science, scientific values forming the causal scheme of the universe); *how* must the natural processes be arranged in order to realize the goals of man; and *to what end* is it necessary to arrange them; of what do the moral and aesthetic ideals of man consist. At issue are the answers to a very old question, 'what is the true?' and to equally old questions, 'what is the good?' 'what is the beautiful?', in other words the totality of scientific, technological, moral, and aesthetic values, accumulated by mankind and determining the further progress of culture. They are accumulated particularly in the form of books, works of art, buildings — the monuments of culture. What is this sphere of the Earth's surface to be called? Perhaps the name should emphasize its basic function, the connection of the generations, the realization of the continuity of cultural progress, critical, refining, changing and still preserving and continuing the *legacy* of the past. Therefore, we shall call the totality of the cultural monuments the kleronosphere from the Greek *κληρονομία* .(inheritance). Earth and water received Latin names, but the sphere created by man was given a Greek designation: the 'noosphere', from *νοῦς* [*νόος*] — reason. We shall observe parenthetically that the Earth's kleronosphere needs to be protected from pollution and destruction no less than its atmosphere and hydrosphere. It is the basis of intellectual and moral potential: the ability of mankind to accelerate further cultural progress.

Of what does this progress consist? Such a question is intimately connected with the definition of culture. Among the many different definitions, the following is appropriate to the present.

The intensiveness of being is measured by the merging of the flow of phenomena with their essence, with laws; it is measured by the ordering of this flow, by the transition from chaos to cosmos. In thermodynamics, the intensiveness of being is the ordering of thermal motion, the existence of macroscopic temperature drops; it is *negentropy*, opposing *entropy* — the measure of disorderly, chaotic molecular motion, devoid of macroscopic ordering. And in addition the intensiveness of being is measured by the autonomy, by the individual being of the molecules and their movements. Here it is a question of the representation of the real ordering of being, the rational scheme of the universe, the system of scientific knowledge of nature. Here the true is the ideal, and the infinite approach to inexhaustible, objective truth is the stuff of scientific progress. Science appears here as the monologue of nature. But when we examine science as a *component of nature*, its development is determined by the goals of man, the effect of research, social and moral ideals, aesthetic criteria. Science merges with these ideals, with the monologue of man. Culture is the dialogue of man and nature, a dialogue which alters both man and nature. The level of culture, its intensiveness, is measured by the intensiveness of the connection between its components, by the degree of the transformations of science, production, social relations, morality, and art into such components. Culture does in fact consist of such a connection; culture is the synthesis of the Good, the True and the Beautiful. In such a synthesis, they are realized; science is embodied in economic and social values, the necessary becomes the substantial, and the good becomes the criterion of the real. The effect of the kleronosphere, the influence of the great past on the present and future, the influence that makes them great, consists, in fact, in the intensiveness of such a synthesis and such a realization.

In Paris, both the past as crystallized in the material monuments of culture and the subjective cultural legacy — the peculiarities of the customs, traditions, thoughts, and psychology that make the past the present — are distinguished by a profound connection, by an intensive interaction between reason and emotion, between the substantial and the necessary, between the true and the good, the true and the beautiful.

It is precisely here, in Paris, that Henri Poincaré introduced aesthetic criteria into mathematical thought, and the next generation of mathematicians and physicists saw in mathematical elegance a representation of the real structure of the world. And for one hundred years before this, the dry and precise lexicon of Lagrange had a powerful influence on the aesthetic criteria of art. As far as the synthesis of the substantial and the necessary, the true and the good, science and social and moral ideals is concerned, Poincaré spoke of their logical independence. The indicative mood is the sphere of the substantial. The imperative mood is the sphere of the necessary. We cannot logically deduce aesthetic norms, norms of the good, from statements about the substantial, just as we cannot logically deduce statements, propositions in the indicative mood, from aesthetic norms, from phrases in the imperative mood. But if such a *logical transformation* is impossible, the *historical transformation* of the indicative mood into the imperative, and vice versa, occurs constantly. Paris, with its cultural monuments, its cultural traditions, is a remarkably obvious and brilliant demonstration of such transformations. When Newton's system — a system of the indicative mood, the monologue of nature, a system of statements — penetrated into the Paris salons, and then from the salons into wider circles, it became a program for social transformation, it moved into the imperative mood. The transformation from one component of culture to another, the intensiveness of the general cultural synthesis is even more brilliantly revealed in the history of nonclassical science. Here, in Paris, Frédéric Joliot-Curie and his collaborators, convinced that the decomposition of the atomic nucleus of uranium under certain conditions produces a chain reaction, at once thought of an ethical problem: should research be continued that threatens mankind with atomic explosions, and what must be done so that the thrust of scientific creation produces just such a thrust in the battle for social and moral harmony. The monologue of nature — the statement of the objective physical process — moved into the monologue of man, into a declaration of his goals, and then into a dialogue between man and nature, into a program of an arrangement of objective processes that would realize the goals of man.

The logic of scientific analysis brings it to the points of transition, where logic moves into emotion, deduction into aesthetic and moral intuition. After such a transition, logical analysis continues, but it originates no longer only from observations, but from the goals of man as well, from his social, moral, and aesthetic ideals.

In meditating on the correlations of rationalist deduction, on the one hand, and on moral and aesthetic intuition, on the other, I approached the railing dividing the place du Parvis Notre-Dame from the square Jean-XXIII, crossed the square and sat down on a bench not far from the memorial of the *Martyrs de la déportation*.[4] The bench stands under a huge willow whose branches graze the surface of the Seine. Now the risen sun illuminates the flying buttresses of the cathedral, and the gargoyles [*chimères*] stand out against the backdrop of the pale blue sky.

2. THE GARGOYLES OF NOTRE DAME

When the sun illuminates the southeastern side of the cathedral, the statics of the structure, the system of buttresses supporting the building from the outside and allowing it to be carried upwards, become especially distinct. But the cathedral is not transformed into a mechanical scheme. Notre Dame remains a living embodiment of the beautiful and the good no matter what irrational forms hide the aesthetic ideals of the cathedral's builders. When one walks into the cathedral, one continues to feel the harmony of the architectural details, the stained glass windows, the sculptures, and the system of arches carried upward, and the cathedral and the consciousness of its visitors. Herein lies the soul of the Gothic. For the medieval consciousness, *height* was not only a topographical concept, it was also a synonym for the *highest*, morally and aesthetically, for the highest *ratio*, the highest good, the highest beauty (language here prompts the connection of ideas – we meet the linguistic similarities that Hegel called 'the joy of reason'). M. M. Bakhtin, in his remarkable book, *Rabelais and His World* (1968; 1984b) demonstrated the correspondence between topographical and moral meaning in the concepts of 'height' and 'depth' in medieval and Renaissance culture (cf., for example, p. 21).

The Gothic, like every style, is a specific form of the inclusion of elements in the set uniting them, in a system characteristic of an artist, a country, a tendency, an age. The style of the artist, the architect, and the writer are the special features of creative work that are preserved in the transition from one work to another, from one subject to another. The style of an age is an invariant as is the transition from the work of one artist, architect, writer, to the work of another, within the framework of the dominant tendency. There exists a style common to an age, embracing science, poetry, music, architecture, painting . . . It is the invariant of inter-genre transformations. It is not dependent upon genre, upon the elements that enter the set; it characterizes the intensiveness and forms of inclusion. It is possible to employ the usual mathematical symbols and signify the included element by α, the including system by A, and the inclusion itself, using the symbol $\in$, by the expression $\alpha \in A$. Then the general characterization $\in$ will be called the style, invariant in the presence of change α and A. The Gothic is characterized by the maximum inclusion of α and A: the maximum subordination of concrete sensory impressions to the general concept, to the universal (medieval realism); the maximum subordination of details in painting to the whole, to the general composition of the picture; the maximum subordination of architectural details to the general mechanical as well as aesthetic and moral ideal.

The sculpture of Notre Dame is also governed by the Gothic style. Here one is not allowed to reach the topographical height, but the figures and faces, if only those found in the *Galerie des Rois*, are imbued with a surpassing synthesis of thought, feeling, and beauty.

Now the sun strikes other sculptures as well — the gargoyles of Notre Dame. Their faces and bodies express whatever you like, but are not subject to the Gothic universal. But to what precisely? They undoubtedly think; one of the gargoyles is even called that: 'Le Penseur' ['the thinker']. But what thought is imprinted on the fanciful faces of the gargoyles? What are they thinking about? What is the cultural and historical function of these ironic grimaces, sometimes sad, sometimes indifferent, and always self-absorbed?

And this is just the word: *self-absorbed*. It is precisely here that the content of the gargoyles' meditations lies. Here individual consciousness is not swallowed up by higher principles that level and identify individuals, transform them into simple embodiments of a universal concept. The gargoyles each declare their own independence, their nonidentity to other individuals, to other elements of the set; they do battle with the levelling power of the Whole.

Once in a small village in the Champagne, I saw a church from the twelfth century — the same age as Notre Dame. An anonymous sculptor had surrounded the outer walls with a border consisting of the faces of the inhabitants of the village. Here individuality appears not in the fantastic-symbolic images of the paradoxical gargoyles but in earthy (and in this sense opposed to the Gothic spirit) sculptured portraits. Later, the carnival culture (this concept was introduced by M. M. Bakhtin [in his (1968; 1984b)]) of the Renaissance inherited both forms in the struggle of individuality against the identifying and levelling tendency of the Gothic; the carnival features were the descendants of the gargoyles, and the deliberate crudeness of the carnival spectacles raised and continued the oppositional, earthy tradition of the Middle Ages.

Neither the gargoyles of Notre Dame nor the physiognomies of the peasants decorating the village church claimed to replace the traditional universal with a new universal, with new general principles and ideals. This occurred later, during the Renaissance, and still later and to a greater degree during the Baroque, when decorative elements in architecture became an inalienable part of the ensemble, and in science, motion — the destruction of the peripatetic, static harmony of natural places — proved to be the basis for a new, dynamic harmony. The gargoyles and the entire oppositional, carnival culture of the Middle Ages only suggested the limits of the Gothic subordination of elements to the including system, just as nominalism, until the fourteenth century, only suggested the limits of the omnipotence of universals. Such a suggestion had considerable significance for the concept of being. Long before Émile Meyerson, it had already been suspected that the being of a system demands both the identity and nonidentity of its elements as well as their subordination to a system, to levelling, to identification, and, at the same time, to autonomy, individualization, 'rebellion' against identity.

The nonidentity of the elements gives the system real being. If a given moment were not distinguished from past and future moments, if they were identical, time would be gathered into one 'now', one nonextended moment, and would disappear. The same thing would happen with space, if there were no nonidentity in spatial points. Nonclassical science demonstrates the 'alienation' of physical macroscopic concepts when their microscopic filling — the fact that separate particles possess their own peculiarities of behavior that do not flow from macroscopic laws — is ignored. Let us take the 'world-line' of a particle, i.e., the macroscopic totality of the space-time localizations of the particle, the points where it is found at each moment in time, and the moments themselves. Let us imagine that the particle is entirely subordinate to such a set, that its behavior is identical, that only the spatial positions and the moments of time change. In other words, the being of the particle is reduced to entrance into a world-line. Then the world-line becomes a geometrical and not a physical concept, a concept of four-dimensional geometry and nothing more. Quantum mechanics maintains that the particle possesses uncertainty of position and uncertainty of velocity, that it interacts with macroscopic bodies and only owing to such interaction does it acquire a more or less determined position (because of the uncertainty of velocity) or a determined velocity (because of the uncertainty of position). The world-line proves to be blurred, and this expresses precisely its physical character, its irreducibility to geometry, its physical being.

What is this — an external analogy, connecting cultural and historical clashes with those of physics? Let us wait with the answer to this question.

3. *VANITAS VANITATIS*, THE BRETON MEGALITHS AND THE PROBLEM OF INFINITY

The gargoyles of Notre Dame suggest the nonidentity of the elements of being, their sovereignty as well as the fact that the complete subordination of elements to the including system, the complete subordination to the Whole, the complete medieval realism in the spirit of Plato and all its later modifications, deprive general concepts of physical content. And what about complete insubordination, the complete autonomy of localized elements?

Let us begin with physics. We shall continue with the previous example. The world-line remains a purely geometrical concept if there are no local, ultramicroscopic events. But such events do not have physical meaning without the macroscopic skeleton of world-lines, without the Whole. Local events can end in transmutations, in transformations of particles of one type into particles of another type. However, such a transmutation means that the mass, the charge, the continuity of the life of the particle, are altered. In other words, the world-line of the particle, its curvature in a given field, its length, changes. But the concept of the world-line means subordination of the particle to some law, some identity of behavior at various world-points. Without the macroscopic ordering of being, the individuality of the micro-object is a concept without content. In nonclassical physics, a 'unique' particle does not exist.

By analogy, models of a Robinsonade [a novel with a subject similar to that of Robinson Crusoe], Stirner's 'unique', any concept of man's being outside of society is devoid of content. The connection with the whole is the condition for the being of the individual.

Does the local, the individual, the finite, exist? The idea of infinity raises this question in a very sharp form. In the face of infinity, the finite becomes null. The sensation of the nonbeing of the finite in infinite space and time provoked Pascal's pessimistic tirade: "I see only infinity on every side, hemming me in like an atom or like the shadow of a fleeting instant" (Pascal 1966, §427 [Brunschvicg §194], p. 158; see also p. 177, n. 10 of this volume). The words of Ecclesiastes were heard much earlier and resounded even more pessimistically, relating to *everything*, to the ensemble, even an infinite one, of finite objects and actions. Such an ensemble is the vanity of vanities, *vanitas vanitatis*.

I recalled this conception not long ago in a village in Brittany near a field of megaliths. Dozens of huge stones form a path running from east to west. According to Breton legend, these stones were thrown by dissatisfied fairies who had brought them to build Mont Saint-Michel. The megaliths have been lying there for around eight thousand years. Is this long?

On the temporal scale of stones, cliffs, and mountains, it is not long. I then remembered a prose poem of Turgenev. It is called 'A Conversation'. The conversation takes place between two Swiss Alps, the Jungfrau and the Finsteraarhorn. The Jungfrau is asking what is happening below, in the valleys. "A few thousand years go by: one minute. And the Finsteraarhorn roars back in answer . . . " He informs her that below little insects — people — are swarming. Thousands more years go by: one minute. The Jungfrau repeats the question and discovers that the insects have already disappeared (Turgenev 1970, pp. 242–244).

The megaliths are the 'sacred stones of Europe'; they were created by man, but their history is infinitesimally small on the temporal scale of the Jungfrau and Finsteraarhorn.

This trend of thought, like the views of Pascal, I now note in my Paris diary, because the Breton meditations are continuing now, on the southwestern bank of the Île de la Cité, where before me tower the most famous 'sacred stones of Europe' — Notre Dame de Paris. Is not the thousand-year history of Notre Dame (and all that happened around it, on the Île de la Cité, in Paris, on the whole Earth), a vanity of vanities, if one turns to the geological scale of time? And if one turns to infinite temporal scales?

The attempt to answer this question was connected with a somewhat unexpected association. In 1854, Bernhard Riemann gave a lecture in Göttingen, 'Ueber die Hypothesen, welche der Geometrie zu Grunde liegen' [On the Hypotheses that Lie at the Foundations of Geometry]. In it, he connected the concept of infinity with that of curvature. Their correlation is easy to understand in the case of two-dimensional space, for example, on a flat and spherical surface. Flatness possesses zero curvature and can be infinite: the shortest line connecting two points in this case is a straight line that may be extended as far as is wished on both sides; flat space is infinite. On the surface of a sphere, the shortest line between two points will be a curve, for example, a meridian, uniting, in the shortest manner, the north and south poles. Such a line can no longer be extended as far as one wishes; it will return to its point of origin. Curved space may be finite, precisely like the surface of a sphere. Riemann generalized the correlation of infinity and curvature to three-dimensional space, where it is a bit more difficult to represent curvature; to four-dimensional space-time, where it is more difficult still; and to any abstract space of a different dimension. At present, in physics and cosmology, the representation put forth by Einstein of curved, finite space and noncurved, infinite time is wide-spread. The curvature of space is determined by the density of matter and energy in the Universe and, thus, the problem of the infinity of the universe is resolved by means of local measurements and experiments.

For man, for his thought, for a way out of his pessimistic self-awareness as "an atom or the shadow of a fleeting instant", what is essential is not strict noncurvature and, accordingly, the infinity of the universe. Man becomes practically nonexistent, "an atom or the shadow", even when he measures himself against finite, but gigantic periods of billions of years and against spaces of billions of light years. But for the ideas and feelings of man, it is essential that infinity — strict or not strict — become a local determination characterizing the 'here-and-now'.

Nonclassical science in several of its assumptions examines the being of an elementary particle as the result of its interaction with the entire Universe. At present in nonclassical science and in everything connected with it, Hegel's idea of *true infinity*, which does not oppose the finite, but is embodied in it, which determines it and — this is already a modern idea — which itself depends on its finite elements, has acquired an obvious, physically tangible character. The concept of true infinity is now acquiring a very distinct character; it is easy to illustrate it with any physical model. But physical models thus become models for correlations remote from physics, down to correlations directly determining the pessimistic or optimistic reaction of man to the infinite space-time surrounding him.

For modern man, infinity does not provoke dizziness. Hegel laughed at the lines in Kant where the dizzy grandeur of infinity was discussed. Hegel said that in this case, the dizziness was provoked by boredom: the concept of "spurious infinity", i.e., the simple increase in scale is already very empty. This resembles the image of eternity that appeared in Dostoevsky: Svidrigaylov says to Raskolnikov that eternity seems to him like "a village bath-house, grimy, [with] spiders in every corner, . . . that's all eternity is" (Dostoevsky 1968, p. 305). This is an image of spurious infinity. Hegelian true infinity is dynamic; it determines finite processes and events; it takes the form of a scientific law, a universal subordinating to itself as great a number of processes and events as desired. Man relates to this infinity *sans façon.* He not only uses infinities in the differential and integral calculus, but he also arranges universal laws, determines the sphere of their action, and here there is no more place for dizziness, apotheosis, or for the feeling of one's own insignificance ('like an atom or a shadow'). At present in non-classical science and its applications, man is transforming the most fundamental laws of being, and infinity is proving to be a plastic material in his hands. Not only do the eight thousand years of the existence of the megaliths or the one-thousandth anniversary of Notre Dame cease to be perceived as something infinitesimally small in the face of eternity, or in the face of the geological scale of time. The interval of time on the order of 10^{-24} sec — a trillionth part of a trillionth part of a second — may be the duration of a process which in a radical manner alters the flow of geological processes. One should think of how to free the participants of Turgenev's 'Conversation' — the Finsteraarhorn and the Jungfrau — of the consequences provoked by such processes.

4. FOR WHAT DID THEY GIVE UP THEIR LIVES?

On the walls of Paris one often encounters plaques with memorial inscriptions: someone fell here for the freedom of France. And then the date the hero perished, most often the days of the Resistance fighting in the summer of 1944. Sometimes, it is briefer: he fell for France. These formulae are identical: freedom is a component of being, the destruction of freedom would be the destruction of France.

Spinoza included freedom in the determination of being. Freedom is behavior flowing from the essence of the subject; as Spinoza says, the fact that the sum of the angles of a triangle is equal to two right angles is not the result of coercion, no one demands this of the triangle; geometrical definitions flow from the essence of geometrical objects. Thus, scientific statements determine what Spinoza calls freedom. For man, freedom is the free manifestation of man's genuinely human essence not distorted by external impulses alien to man. When man is free, his thought and activity are determined by the immortal human essence, by what goes beyond the limits of individual existence; freedom communicates infinite being to man. Therefore, says Spinoza, "a free man thinks of death least of all things, and his wisdom is a meditation of life not of death" (Spinoza (1982), *Ethics*, Pt. IV, Prop. 67, p. 193; see also p. 107, n. 20 of this volume).

Freedom is thus communion with infinity, with that which is greater than man, but which does not oppose him as something alien and inimical, but is embodied in his thoughts and activities, and gives him genuine reality. Such communion occurs through the ideals of man, the ideals of the good, the true and the beautiful, which are embodiments of the infinite.

National self-consciousness, the idea of the reality of France, the sense of such reality, the sense of the connection with the infinite life of the people, is not a chain of logical deductions. Poincaré was right in part: logic cannot deduce imperatives from the indicative, the necessary from the substantial. But such a transition, as has already been mentioned, exists; it realizes the unity of the substantial and the necessary, the synthesis of the good, the true and the beautiful; and the synthesis of the past, the present, and the future. The French who perished at the walls where the memorial plaques now hang; all those to whose memory the inscription on the *Fountaine St.-Michel* is dedicated;[5] all the heroes of the Resistance and all who fought for France, were thinking of her past and of her future, and in their consciousness, the past and the future entered the present. And perhaps they did not even think, but felt it; their emotions reflected that great past and a premonition of the great future, and the present, the genuinely real, not "an atom or . . . the shadow of a fleeting instant", but the communion with infinity, the motion from the past to the future. Here, for the sake of infinite motion, for freedom as a condition of being, for the infinite realization of the good, the true and the beautiful, the heroes of the Resistance and the liberation of France gave their lives.

Therefore, the memorial plaques on the walls of Paris enter into a single ensemble with the memorials of the past, with the "sacred stones of Europe", as witness to the continuity of progress.

National self-consciousness precludes nationalism, because it is not a barrier, but a bridge to all human consciousness; it exteriorizes the individual 'here-and-now' both in space — it opens the heart on the side of the whole Earth — and in time — on the side of all of history, including prehistory, including the megaliths.

5. *LA GIOCONDA*

I return to the place St. Michel, continue to walk along the quay. I cross over to the Right Bank by way of the Pont-Neuf, reach the Louvre, mount the stairs to the Victory of Samothrace and soon find myself in front of the picture that has always seemed to me to be the masterpiece of rationalist painting. Or, if you will, of painted rationalism, of the idea merging with the aesthetic perception of the world, of truth merging with beauty. This is *La Gioconda* [Mona Lisa]. Leonardo da Vinci's creative work seems to me to be the way to sixteenth-century Italian *Naturphilosophie*, to seventeenth-century classical rationalism, and the classical science of the seventeenth and eighteenth centuries (cf. pp. 115—122 above). This relates both to artistic practice and to the aesthetic program, especially to the contents of the *Treatise on Painting*, where Leonardo writes: "Painting is proved to be philosophy because it treats of the motion of bodies . . ." (Leonardo da Vinci 1956, **I**, §7, p. 5; see also p. 144, n. 11 of this volume). In fact, the new philosophy rejected the static harmony of the immovable natural places in Aristotelian physics and cosmology; it examined the world as motion and it was precisely in motion that it saw the condition for the real being of the world. V. P. Zubov, in his book on Leonardo, in expressing his ideas, writes that for him the object of art is the present connected with the past and the future: "Life is possible only where there is a 'before' and an 'after' and a connection between the two. Therefore, time is not only a 'destroyer of things', but a prerequisite for their true existence" (Zubov 1968, p. 263). What exists, lives, possesses real being is what changes, what is nonidentical to the past and the future and at the same time is identical to the past and the future. Leonardo's paintings create this impression of exteriorization in time. And exteriorization in space. On several of Leonardo's canvasses and frescoes, exteriorization in time is distinctly apparent: we divine the motion that will conclude the pose and the action that will end the intense expression on the face. But we also see the connection between the expression and pose and the composition — the exteriorization in space; we need only recall *The Last Supper* with the symmetry of the figures and the architectural details, and with the asymmetrical variety of the expressions on the faces. This symmetry moves into the background, into the pose and expression on the face of Christ, which are completely realized only in the general composition. In these cases, we divine also the connection between the content, the composition, the color scheme of the painting and the ideals of the age. Exteriorization functions not only in time and space; it also connects the painting with the multidimensional and complex world of the emotions, the necessary, of moral ideals.

Exteriorization looks different when we have Leonardo's paintings before us, expressing at first glance only the beauty of an individual image. But only at first glance. Exteriorization exists here as well; the paintings ultimately express not only the beauty of the individual image . . . but more about this later. One does not think of them as one stands before the *Gioconda* in the Salle des États. Here one sees only its beauty. The strange landscape, against whose background Mona Lisa is represented, has already played its role; it does not detract from the encounter with beauty — with beauty with a small 'b', with the beauty of the *Gioconda*, and with Beauty with a capital 'B', with the eternal beauty of the world, with the embodiment of its infinity. But this is one and the same. Eternal beauty has been embodied in the individual, individual beauty has become eternal, the infinite has been embodied in the finite and there is no more talk of the finite individual being as 'an atom or a shadow', as an infinitesimally small existence.

Of course, this is isomorphic with the the new science, which perceives truth as a scheme of differential laws — the correlation of infinitely small increments of time and increments of space, velocity, and acceleration. Of course, this is isomorphic with the new morality, which rejected the static scheme of traditional, static norms; it is isomorphic with the Good, which has become dynamic, changes its norms, and no longer opposes the actions of individual people in the form of eternal tablets ignoring their will. But isomorphism in both cases does not achieve obvious expression; here there are no continual and obvious transformations; here everything is realized not by logic, but by intuition, by illumination, by inspiration. And if Mona Lisa is smiling at something definite, then, most likely, while smiling, she is stating the irreducibility of Beauty to logical deductions.

In this irreducibility, in the *implicit* connection between the ideals of the true and the ideals of the good, lies the essential peculiarity of the beautiful, perhaps its definition. Let us recall Leibniz's definition of music: "The soul calculates, itself unaware of this". For Leibniz, as for Galileo, Descartes, and Newton, the truth was expressed in calculations; the definition quoted may be generalized to include all the forms of art that seek the beautiful, and to all the forms of science that seek the true. On the other hand, perhaps, art is the equally unconscious quest for the necessary. Perhaps the beautiful is connected with the true and the good as an unconscious transition from the substantial to the necessary, and as an unconscious, implicit form of the criteria of the true and the good. Unconscious and implicit, but real. When one looks at the *Gioconda*, both the indicative and the imperative disappear. There remains only the lofty ascent of spiritual forces, that inspiration which is an invariant for the transformation of the true into the good and of the good into the true, an invariant in the translation from the indicative mood to the imperative and vice versa.

6. *BEAU* AND *BELLES*

When dealing with the *beau*, it is necessary to be, to some extent, a disciple of Tartuffe, in order not to mention *belles*. A disciple of Tartuffe in the country of Molière! . . .

Therefore, we will leave the Louvre, return to the place St. Michel, and climb the boulevard St. Michel past the Sorbonne to the Luxembourg Gardens. We have returned to Descartes's eyes in the 'Hals' portrait mentioned above (cf. p. 396). Yet these are not Descartes's eyes but inimitable eyes in general. Each pair of eyes is inimitable and each gaze of these eyes (and, what is more frequent, of course, each gaze into these eyes) is an inimitable event. This is a very nominalist spectacle, the universals here are formed with difficulty, and they change with each event.

Heinrich Heine confronted a somewhat analogous difficulty in the last century. In *Florentine Nights*, he wrote that it is difficult to answer the question as to whether the Parisiennes are beautiful. "For all their features are in constant motion; each Parisienne has a thousand faces, every one more laughing, more *spirituelle*, more charming than the other, and he would be well bewildered who under it all could detect the fairest, or the real face at all. Or are their eyes large? What do I know? We do not long examine the calibre of a cannon when its ball decapitates us. . . . Is her mouth great or small? Who can tell where the lips leave off and laughing begins? To form a correct judgment, the one judging and the object judged must be in a condition of repose. But who can rest by a Parisienne, and what Parisienne ever rests, herself?" (Heine 1892, p. 69).[6]

I recalled these lines more than once on the boulevards of Paris and now I would like to enter in my diary several considerations that will probably anger both scientists and poets. I want to draw a parallel between the difficulties about which Heine was speaking and the fundamental difficulties of philosophical and scientific thought. Some will find in such a parallel a profanation of science, others will see here a lowering, a betrayal of genre, a profanation of what is the subject of the lofty, the lofty lyricism itself. But in this city, Rabelaisian laughter still resounds, directed against the *immortels* [the forty members of the *Académie Française*] and against the absolute differentiation between the 'high' and the 'low'. The risk can be taken.

The search for real beauty in the flow of changing facial expressions suggests the long search for the object at rest, immobile, 'present', in philosophy and science. Such a system at rest and therefore 'perfect' was the static scheme of natural places in Aristotelianism. But this is only a part of the general tendency of ancient and medieval culture. The canonical was considered perfect: the true — an approximation of the canonized scholastic scheme; the moral — that which corresponded to the canonized, fixed tradition (for example, 'fair profit', 'fair percent' — concepts of medieval morality and, at the same time, economic concepts). Canonical beauty was considered Beauty, beauty was what identified faces, movements, melodies, images, made them subordinate to the canon, the ancient or another, but to the canon.

Galileo devoted more than a few pages of his *Dialogue* to proof of the new view: the perfect is not a static, but a dynamic scheme. Science did not find at once the invariants of motion (velocity, and then acceleration, and then the coefficient of the ratio of force and acceleration — mass); it did not find at once new, dynamic canons. These did not penetrate at once into the aesthetic perception of the world, into aesthetic criteria, into the content of the concept, *Beau* and into the characterization *belles*.

But not only does the inclusion of motion, change, noncanonicity, nonidentity, individuality in the aesthetic canons draw Heine's difficulties closer to the clashes of science. In the lines cited from *Florentine Nights*, the last sentences also speak of the change in both the observer and the object of observation. Here, aesthetic-scientific analogies are so natural that it is not necessary to name them.

Now we may add something to what Heine wrote in the 1830's. The encounter with *belles* may change the canon, *Beau*. Contrary to the traditional meaning of the word 'canon', canons become polivalent, changeable, variant. This is already an instance of the nonclassical correlation of the individual and the general. In the relation $a \in A$, the individual element a changes the general concept, the set A that includes it. An isolated process, for example, the spontaneous origin of a neutron in a block of plutonium, may produce a chain reaction. This is a very general scheme of nonclassical science, its development, generalization, and application. The individual is not only an embodiment of the infinite Whole. It is an Archimedean fulcrum to move the Whole, to transform infinity. Infinity not only determines each of its finite elements, it also changes its character in relation to the individual element. This correlation is expressed most brilliantly in contemporary cosmology: as was already mentioned, the problem of the infinity of the universe is resolved in any event, the scheme of infinite space in the modern picture of the universe changes in any event, depending on the local curvature of space and the local experiments and observations determining it.

Is it necessary to ask forgiveness of those to whom this passage is dedicated for likening poetic souls and the poetic manner to the prosaic data of science? Perhaps they are not prosaic at all, perhaps in the poetry of science lies its most dynamic aspect? Perhaps it is precisely poetry that unites the quest for the true with the quest for the good and the beautiful.

7. LOGOS AND EROS

The rather widespread representation of the conflict between Logos and Eros and the displacement of the former by the latter seems to me extremely archaic, just like the newest forms of irrationalism. They are archaic not in the sense that they were once correct and have now lost their meaning; they were incorrect earlier as well, but now this has become obvious. If the terms 'Logos' and 'Eros' have any definite meaning, if real objects and processes and noncontradictory concepts are connected with them, then these concepts are at present coming closer together and are becoming complementary in several instances.

'Logos' signifies the word, reason, logic. But it is primarily the Word, as an expression of reason, thought, logical deduction — a very Cartesian concept, the basis of classical rationalism, polished and comparatively simple. Which does not prevent and even helps one to see a certain historical evolution of what is connected with this term. On the other hand, 'Eros' is sometimes understood in a sense that contradicts its initial meaning and is internally contradictory. The Greek Ἔρως [Eros] and the Latin *Amor* are the god of love. The natural generalization of Eros is the totality of emotions with a determinate individual concern. The incorrect generalization is eroticism. It gives rise to the reverse translation and to a definition of Eros as a symbol of eroticism. But although there is some basis for the first designation, for 'eroticism', there is no basis for the opposite translation, 'eroticism-eros'. Love is distinguished from eroticism by individual concern; it is always love for *la* or *le*, and not for *une* or *un*. Such is the 'Eros' of Giordano Bruno and the '*amor intellectualis*' of Spinoza. Here the one as well as the concrete Universe serve as the object of love.

Modern science has given this concept a wholly rational character. In essence, the relativist statement of the curvature of world-space that depends on local observations serves as the basis for the new interpretation. In nonclassical science, the 'universe as a whole' is systematically used to support the resolution of local physical and astrophysical problems. The applications of nonclassical science do not embrace the Cosmos, but they do embrace the Earth; their results demand planetary and age-old calculation, and modern Eros or *amor intellectualis* becomes a necessary condition for the preservation of the ecological balance on our planet. It is realized in the *rapprochement* of moral and aesthetic criteria with the specifically scientific, with Logos, with the growth of what might be called the intensiveness of culture.

8. THE INVARIANTS OF CULTURE

The images of Paris correspond to the general rationalist, sensual, and if we return to the tendencies of medieval thought, then to the realistic-nominalist conceptions of being. Separate impressions do not dissolve in the Whole, they are inimitable. Hence, that 'evening sadness' with which such universal masterpieces of poetry as the *Odyssey* and the

Divine Comedy are imbued: consciousness bids farewell to the inimitable day — tomorrow will be different, with a separate local impression, with an inimitable image. Moreover, the separate impressions of Paris merge into a single whole and the sorrowful melody of farewell to the individual and to the transitory is interwoven with the optimistic note of the entry of passing impressions into a single, although heterogeneous, image and their preservation in it. In this single image, the present merges with the past and future, the specificity of each genre of culture merges with the influence of one genre on another, the independence of cultural values merges with the organic assimilation of the foreign. It is a synthesis in time, a synthesis of genres, a synthesis in space, a synthesis realizing unity and preserving autonomy, based on the identity and nonidentity of the united elements.

There are mathematical concepts that express very precisely the ambivalence of being, the inclusion in the determination of being of the identity of its elements, their nonidentity, the ordering of the universe, and reveal its macroscopic structure, the existence of systems, the negentropy of the universe and, on the other hand, the autonomy of the elements entering the system. This is the concept of *transformation* and the *invariant of transformation*. They are complementary and one without the other loses its meaning.

One hundred years ago, Felix Klein in his famous 'Erlangen Programme' (see p. 66, n. 5, of this volume) posited the concept of the invariant as the basis of the definition of various spaces and geometries. In everyday elementary geometry, transformation consists of the transfer of a geometrical figure to another place, or the rotation of this figure, or both. The size and form of the figure remain unchanged during such transformation. In this case, we have before us metric space governed by Euclidean geometry. Its invariant is the distance between the points of the geometric figure. In projective geometry, geometrical figures change their dimensions, but remain similar to themselves, preserve the same ratios between distances. Only the form of the geometrical figure serves as the invariant of such a transformation. In 1918, Emmy Noether connected the invariants of several expressions with the preservation of physical sizes. The invariance of a determinate expression during transformation in time, in the transfer from one time to another, signifies the preservation of energy. The invariance of this expression in relation to transfers in space signifies the preservation of momentum (see Noether 1918a and 1918b).

Concepts of transformation and the invariant allow us to lay down with more clarity several historico-cultural conceptions, including the impressions evoked by Paris.

In the history of culture, we confront temporal transformations, transitions from one time to another, from one age to another. The concept of such transition, change, transformation would lose meaning if there were not some invariant, certain definitions of culture that remain identical in the process of its change. Of what does such an invariant consist? What transforms the group of consecutive monuments of culture, dating from various ages into its history, into the modification of the self-identical subject of change?

The history of culture also includes transformations dating from the same age. They are transitions from one sphere of culture to another, from one genre of art to another, from science to literature, from art and science to morality, etc. These transitions also transform science, literature, morality into components of culture. But what is preserved during this? Of what does the invariant of transformation consist, which might be called (with a certain generalization of the concept of genre) intergenre?

And the third question: of what does the invariant of the transition from one geographical and linguistic cultural sphere to another consist? Such transformations make culture into universal culture. They may be called *translations* into a more distinct form of the international transformation of cultural values. What is the invariant of translation?

Generally speaking, temporal, intergenre, and spatial transformations cannot be completely isolated. The history of culture revealed much earlier than physics the four-dimensional character of the studied continuum. In the history of culture, there was never a purely spatial, momentary action, exclusive of time, over distance, *action at a distance*. The illusion of momentary action at a distance allowed classical physics to separate space and time. The theory of relativity demonstrated that each real physical process occurs with finite velocity, that it cannot be momentary, outside of time. It replaced space and time with the four-dimensional space-time continuum. Accordingly, it does not examine isolated spatial and temporal transformations and isolated laws of the preservation of momentum and the preservation of energy. In the theory of relativity, there exists a general law of the preservation of momentum-energy. But in a certain classical approximation, for small energies, the preservation of momentum and the preservation of energy figure separately.

In the history of culture, no one ever doubted that spatial transformation always occurs in time. Not only in the sense of the necessity of a known period of time for the migration of values, methods, ideas, criteria, but in a less trivial sense. Even in the purest spatial transformation — in literary translation — the work that has appeared in the new cultural medium strengthens or modifies certain historical traditions, hinders others and, thus, provokes a temporal transformation, a change in time of the functions of cultural values, their evaluations and effect.

It is time to answer the questions raised about cultural invariants.

Let us first take isolated transformations — temporal, genre, and spatial. The determination of their invariants is quite trivial, which should not surprise us: after all, unlike the situation in classical physics, in the history of culture, the isolation of transformations does not have a great heuristic effect on a great number of phenomena. Here, if we again use the relation $\alpha \in A$, both the ordering details α, and the ordered set A, may be the invariants, and the character of the arrangement, $\in$, i.e., the general stylistic peculiarities (what is especially characteristic of the transformations from one genre to another in the framework of the same country and the same age).

These analogies with the classical transformations of time and space do not correspond very closely to the real history of culture, but analogies to relativist transformations of the four-dimensional space-time continuum do. Analogies of such relativist transformations are transitions of cultural values, ideals, and methods of creation from one country to another, which signify at the same time a transition in time, a transition from one age of the cultural development of the world to another age and embrace all the basic genres and spheres of culture. Here both the including sets and their local elements and forms of inclusion change in a nonlinear manner (i.e., with the interaction of the changes so that the general change can no longer be the sum of the individual changes). What in fact remains self-identical, what is the invariant of such general transformations?

It is the link of the local with the infinite, it is the ideals of the good, the true and the beautiful — their link, their influence on one another, their realization. The content of the true changes, but it changes in the process of infinite progress, the infinite approach to objective truth.

Aesthetic norms and aesthetic canons change, but this change is not *mors immortalis*, but *vita immortalis*, not the destruction of cultural values, but their preservation by means of modification. This preservation also expresses the existence of the invariant of culture — its integral invariant.

9. ON INSPIRATION AND THE POETIC MIST IN THE LANDSCAPE OF PARIS

The fundamental transformations of culture include transitions from the indicative mood to the imperative, from statements to emotions, which are impossible to represent as chains of logical deductions. Here, logic gives way to intuitive illumination, which is called inspiration and is a necessary component of the poetic perception of the world. Is it possible to give any precise definition of inspiration? It would seem impossible, since inspiration in fact signifies something by its very nature indefinite. Einstein said that a precise depiction of mist is impossible. It is impossible if one uses macroscopic as well as metric images, geometrical figures with determinate positions of points and determinate distances between them. But in the microscopic aspect (schemes of the ordering of the molecules of water) metric representations become applicable to mist, and in the macroscopic aspect, it is possible to give a very precise description of mist by means of a system of numbers that measure humidity, temperature, visibility and other variables.

A description of mist is also possible using macroscopic and metric images, but this description will be negative: mist is something not transparent, or translucent, without clear outlines, lacking visible structure, in other words, homogeneous . . . All this may be expressed both by words (the Cartesian criterion of clarity) and in painting (the criterion of distinctness for Leonardo da Vinci).

In physics, negative definitions are widely used, several similar to the anecdotal 'without which jam do you want your tea?' In philosophy there exists the concept introduced by Hegel of "determinate nothing", i.e., the absence of a determinate 'something'. One such a "determinate nothing" was the Cartesian vacuum (the glass is empty when there is no water in it, a fish pond is empty when it has no fish) and the vacuum of modern physics (for example: 'the vacuum of an electromagnetic field'). Cannot this concept yield something of heuristic value for the conception of genre transformations and invariants and for the definition of 'inspiration'?

I was occupied with this question when looking at Paris during my morning strolls along the quays of the Left Bank and around the Île de la Cité. I have already spoken here of the morning mist. When one looks upstream and downstream from one of the bridges of the Seine, the mist preserves the clear-cut, but also softens, the geometrical outline of the banks, the quays, the trees bending over the Seine, the buildings of Notre Dame, the Palais de Justice, and the Louvre. But further, this outline is blurred and ultimately transformed into a play of color, where real impressions merge with fantasies, where the buildings seem to be clouds, and the clouds, mountains, where transitions from the image of one nature to the image of another nature flow neither from immediate observation nor from logic, where statements move into what one wants to see, and vision merges with creation.

Perhaps similar impressions called into being the term 'poetic mist', belonging, apparently, to the number of expressions that Hegel called the unexpected joys of the spirit. These words are sometimes applied in a direct sense: the poetic mist in the landscape of Paris is an almost immediate visual statement, we see precisely a *poetic* mist, and it is precisely that we *see* it. They are also applied in a figurative sense: 'in the poetic mist of memories', 'in the poetic mist of imagination', etc. It ought to be noted that the very transition from the literal meaning to the figurative is shrouded in a certain poetic mist, where abstract semantics is indistinguishable from thought in images. But this already relates to the problem of poetic translation, about which discussion will follow.

Now, if the reader would allow me to return briefly to the sphere in which the mist (I will attempt to show that it is a *poetic* mist) partially hides, partially expresses physical concepts. I can do nothing: the word 'impressions' demands after it as an object not only the name of the object (impressions of Paris), but also the name of the subject of the impressions (the impressions of the physicist, the philosopher, etc.). Impressions are impossible without rational apperception everywhere, and more so in Paris, where the landscape always remains the landscape of the capital of classical rationalism, where the gaze inevitably penetrates the phenomenological envelope of existence and the sphere of being.

In this diary, I have already spoken about elementary particles, which, from time to time, are annihilated, becoming transformed into particles of a different type. These are fundamental transformations, where there is a change not only in position, velocity, acceleration, and the behavior of the particle in general, but also in the particle's rest mass and charge that determine the being of a given type of particle. The ultramicroscopic sphere where such transmutations take place is sometimes called a 'black box'; it is the place where at present it is difficult to explain the course of events, not only because of experimental difficulties (they will probably be eliminated to a significant degree by increasingly powerful particle accelerators), but also because of the peculiarities, in principle, of the ultramicroscopic spheres. Here, concepts of position, continual motion, velocity, acceleration prove to be inapplicable – concepts that characterize the macroscopic world and, with well-known limitations, conditions, and modifications, the atomic world. But under all conditions, the description of ultramicroscopic events must include macroscopic concepts: the transmutation of a particle signifies the change of its virtual world-line, its behavior in macroscopic spheres.

In the meantime, in elementary-particle physics, a unique situation has arisen. The descriptions of observations here are, as a rule, inseparable from hypothetical, sometimes conventional, almost always ambiguous conceptions. Here separate, prescriptive explanations 'on credit' come to the fore, until a general theory appears which will give them what Einstein called 'inner perfection' and until new experiments give them another of Einstein's criteria, 'external confirmation'. The expectant and predictive style of scientific thought is characteristic of elementary-particle physics. Here, before the eyes of the experimenter, a picture is revealed that suggests the far edges of the Parisian, or even any other, landscape, where the eye sooner divines than observes the outlines submerged in foggy darkness.

This is not an external analogy. Both in one place and the other, there is a field of ambiguous intuition that must become a clear picture when one approaches (let us say on a *bateau-mouche* [Parisian pleasure boat]) those places that before were shrouded in mist and when the physicist, armed with new experimental and mathematical methods, approaches the ultramicroscopic spheres.

But even then the mist does not disperse. For one who sees the Bois de Vincennes and the *château* close up, the misty distances of the course of the Marne are revealed. The scholar for whom the hypotheses and predictions of elementary-particle physics are replaced by a simple picture, will see other, as yet unclear, spheres. The poetic mist was always and will always be a necessary component of scientific progress.

It is precisely a *poetic* mist. Poetry has always been viewed as something that goes beyond the framework of logic, beyond the framework of a group of deductions uniting in a simple manner the assumptions that follow one after another. Thought gallops here through the sequence of simple deductions. It moves toward the Whole from the individual, toward the infinite from the finite, toward the extended in space and time from the local 'here-and-now'.

And not only thought. Here thought is inseparable from emotion. If it is not moved forward by the simple logic of consecutive deductions, then the moving force will be moral and aesthetic criteria; the good and the beautiful invade the course of creative thought. In reality, this is not a replacement, but a merging; thought and emotion are indistinguishable. Such an invasion of emotion into the course of thought, the invasion of the Whole into the local, is in fact called inspiration. Mozart defined it beautifully: it is the moment when in one instant one hears the entire, still uncomposed, symphony.

10. THE INVARIANTS OF POETRY

The components of culture are united by implicit transformations that may be called poetic translations from one language of culture into another, if one is to generalize, as is done at present, the concept of language. Poetry in such a translation appears not as a genre of art, but in a broader sense – in the sense of its emotional accompaniment that is inseparable from any information.

Information here means the system of certain [in the sense of reliable] statements, representing the structure of being, the existence in it of structural sets. But the determination of the structure of being is unclear. It demonstrates not only the identification of elementary objects and processes, their subordination to general laws, their entrance into unifying sets, but also the autonomy and nonidentity of these elements. The representation of being in consciousness is not reducible to information about its ordering, about its negentropy. Man's consciousness also represents the individuality of the elements, their nonidentity, the limits of their subordination to general laws. All this is represented not only, and not even so much, in logical deductions, as in concrete images. The images are created and grouped with the participation of intuition in the framework of a poetic, emotional vision of the world. The inseparability of poetic perception from simple information flows thus from the inseparability of identity and nonidentity as a component of being.

Why do 'poetry', 'the poetic', 'poetics' figure in modern language in such a general sense ('the poetry of labor', 'the poetry of science', 'poetic mist' — right down to 'the poetics of prose')? Why has the name of one of the genres of literature received such a broad application? What are the roots of such an apparently natural and lawful generalization?

The whole idea is that poetry in the special, narrow sense is the most 'noninformational', romantic, emotionally charged genre of literature, close to music, and at the same time undoubtedly an 'information' genre. Poetry is ambiguous, it includes the emotional perception of the world and, at the same time, it uses words as a means of achieving the ordering of the world, its *ratio*, its logic, the identity of elements entering the macroscopic world systems that create structure. The term 'poetry', generalized and applied to the most disparate tendencies in knowledge and the transformation of the world, signifies that in these tendencies the unification of information with the emotions acquires an obvious character.

But let us return to poetry in the special sense. Poetic creation may be viewed as the transformation of one set of statements and emotions concentrated in the consciousness of the poet, emotions fused and inseparable, into semantically, phonetically, and metrically ordered systems of words that provoke a fusion, analogous to the initial one, of statements and emotions in the consciousness of others. It is also an invariant of poetic creation as transformation. Is it an invariant of the transformation of language and all that is connected with language? Is it an invariant of translation?

Translation is the transformation of semantics, phonetics, the mist of associations surrounding the word, metrics, intonations . . . In translation there may be separate invariants and *rapprochements* – semantic, phonetic, metric, etc. Depending upon the existence and absence of certain separate invariants, it is possible to define a certain type of translation. But a *poetic translation* is obtained if the system of separate invariants, no matter what it is, leads to the general poetic invariant, to the preservation of the information that is inseparable from its emotional effect and of the emotions that are inseparable from information. Information and emotions depend on the fact that in translation they can be changed from semantics, phonetics, associations, rhythm, metrics, intonation. But such a combination of changes and the preservation of the variables connected with language is possible when the initial informational and emotional complex is preserved. This is, in fact, a poetic translation. Such a definition of the invariant of poetic translation flows from the definition of poetry as emotion expressed in content (containing information) language, from the acknowledgment of the importance in language not only of the information contained in the assumptions, but also of phonetics, metrics, etc. Ultimately all this flows from the acknowledgment of the ambivalence of being.

The invariant of poetic translation coincides with the invariant of initial poetic creation. In this sense, they are inseparable from each other. The author of an original work is no more free than the translator, he is bound by the initial inner '*avant texte*', by the statements and emotions that are imbued with the immediate impressions of being, by memories, logical constructs, events . . . This *avant texte* disappears in the work and its informational and logical content is preserved in an ambiguous manner in the created text. The translator is not less free than the author of the original work. He can select any new, semantic, phonetic changes or invariants; only if he preserves the initial informational and emotional complex, will his creation then be a translation (the informational groundwork is preserved) and a *poetic* translation (the emotional filling is preserved).

When one thinks of this poetic link between thought and feeling, the city of classical rationalism and classical sensualism ceases to be a system of merely visual impressions. A cultural legacy is reflected in language, literature, poetry — these most important elements of the kleronosphere.

11. THE POETRY OF SCIENTIFIC PREDICTION

Ambiguous transitions, intuitive illuminations, 'moments when the entire, still uncomposed, symphony is heard', form the poetic aspect of science, because they promise the transition to certain information. This is the poetry of epistemological optimism as opposed to the pessimistic oath of Du Bois-Reymond: "*Ignorabimus*!", in which there is not a shred of poetry. The poetic mist disperses, we will know the mechanism of transmutations, we will explain the causes of the observed spectrum of the masses of elementary particles, we will explain with certainty the nature of quasars, and we will answer in a simple manner a number of other questions. And yet we will never attain Laplacean highest reason, which knows the coordinates and velocities of all the particles of the Universe — in this reason there is no more poetry than in '*Ignorabimus*'. As was already said, the mist that has dispersed in the Bois de Vincennes will cover the valley of the Marne. The process of 'depoetization' cannot be completed. This is nothing. In 'science as a whole' (and this concept at present is becoming as real as the 'Universe as a whole'), the role of poetry, to the extent that knowledge moves forward, does not decrease, but increases.

The appeal to increasingly general problems in the resolution of individual, local questions is characteristic of nonclassical science. The modern, close interweaving of astrophysics and cosmology, on the one hand, and elementary-particle physics, on the other, are a sign of the times. This tendency indicated will become increasingly powerful — such a prediction very closely approaches certainty. Problems that can be resolved in a simple manner are connected with problems where the 'poetic mist' of complexity was by no means dispersed. This expresses Einstein's criterion of 'inner perfection': scientific theory must be deduced from the most general principles in the most natural manner, without special, artificial admissions of the *ad hoc*. But it is precisely in the sphere where, at present, the most general principles are arising and being modified — both in elementary-particle physics and in astrophysics and cosmology — that the greatest complexity remains.

Hence, from the sphere of maximum romanticism, the sciences move elsewhere, raising the intellectual potential of science as a whole. And thus, they make scientific and technological progress increasingly dynamic. Scientific and technological progress may occur with a certain constant velocity (the positive first derivative with respect to time from the level of science and technology and ultimately from the level of productivity of labor), with constant acceleration (the positive second derivative), and even with an increase in acceleration (the positive third derivative). Technological and inventive works guarantee undiminished velocity for scientific and technological progress; specifically, scientific research leads to acceleration; megascience — the most basic research — serves as a guarantee of the increase in acceleration. But the higher the stage of science, the greater the function of intuition, the more ambiguous the logical methods, the more dense the poetic mist of science. The role of poetry in science is proportional to the degree of dynamism; the dynamic effect of science may serve as the yardstick of this function.

The poetry of science is created by the appearance of aesthetic and moral criteria in scientific intuition. Modern science unites the criteria of the true and the good, when goals are set before it, when it becomes planned by expedient activity. Aesthetic criteria ultimately are reducible to the generality of the initial principles, to that same 'inner perfection' of Einstein. But for science as a whole, this criterion is realized when all the branches of science are developed according to a plan originating from the fundamental predictions and practical goals that signify the victory of moral and social harmony on Earth.

The reader sees that the impressions from which I began these notes have led me into the sphere of meditations on culture and its invariants, on inspiration and poetry in science and on its future, very remote from immediate impressions. What is to be done — the impressions about which I am speaking have a very long logical and psychological course and far-reaching consequences. It only remains for me to ask forgiveness for the chaos of these notes. The person who publishes such notes becomes the object of observation that may be analogous to biological observation *in vitro* ('on a lens'; in an ordered preparation) or observation *in vivo* (of the living tissue, the living organism). As distinguished from an object of anatomical or physiological observation, the author has the opportunity of choosing the variant *in vitro* or the variant *in vivo*. In this case, he has chosen the second variant.

NOTES

1 [Heinrich Heine was a German poet of Jewish origin born in 1797; he lived from 1831 until his death, in 1856, in Paris where he wrote articles for the *Augsburger Allgemeine Zeitung* or *Augsburger Zeitung*. He collected and edited the articles written between 1840 and 1843 into a volume called *Lutezia* (1854). For full bibliographical information concerning this and other editions, including an English translation, see the Bibliography of Works Cited at the end of this volume — Eds.]

2 [Frans Hals (1584—1666) painted a portrait of René Descartes about 1649 before the latter left Holland for Sweden; *this* painting hangs in the Copenhagen Statens Museum for Kunst [Royal Museum of Fine Arts] (on loan from the Ny Carlsberg Glyptothek). The famous Descartes portrait that hangs in the Louvre is considered to be a copy, by an anonymous artist, not Hals, of this painting (cf. Slive 1970—1974, **I**, pp. 164—168; **III**, pp. 89—91) — Eds.]

[3] [Vladimir Vernadksy (1863—1945) was a mineralogist and one of the founders of geochemistry; he studied the geochemical aspects of the earth's biosphere. Pierre Teilhard de Chardin (1881—1955) writes in *The Phenomenon of Man* (1959):

> Much more coherent and just as extensive as any preceding layer, it is really a new layer, the 'thinking layer', which, since its germination at the end of the Tertiary period, has spread over and above the world of plants and animals. In other words, outside and above the biosphere there is the noosphere (Teilhard de Cardin 1959, p. 182) — Eds.]

[4] [The memorial of the *Martyrs de la déportation* (opened 1962) was designed by the architect, Henri Pingusson. Located in the square de l'Île-de-France at the eastern extremity of the Île de la Cité, it stands in memory of the 200,000 Jews deported from France during World War II — Eds.]

[5] [The *Fountaine St-Michel*, on the south side of the Place St-Michel, was erected in 1860; at its base is a memorial to the inhabitants of the Fifth and Sixth Arrondissements who lost their lives during Resistance operations in the summer of 1944 — Eds.]

[6] "Denn alle ihre Gesichtszüge sind in beständiger Bewegung, jede Pariserinn hat tausend Gesichter, eins lachender, geistreicher, holdseliger als das andere, und setzt denjenigen in Verlegenheit, der darunter das schönste Gesicht auswählen oder gar das wahre Gesicht errathen will. Sind ihre Augen groß? Was weiß ich! Wir untersuchen nicht lange das Caliber der Kanone, wenn ihre Kugel uns den Kopf entführt. Und wen sie nicht treffen, diese Augen, den blenden sie wenigstens durch ihr Feuer, und er ist froh genug sich in sicherer Schußweite zu halten. Ist der Raum zwischen Nase und Mund bey ihnen breit oder schmal? Manchmal ist er breit, wenn sie die Nase rümpfen; manchmal ist er schmal, wenn ihre Oberlippe sich übermüthig bäumt. Ist ihr Mund groß oder klein? Wer kann wissen wo der Mund aufhört und das Lächeln beginnt? Damit ein richtiges Urtheil gefällt werde, muß der Beurtheilende und der Gegenstand der Beurtheilung sich im Zustande der Ruhe befinden. Aber wer kann ruhig bey einer Pariserinn seyn und welche Pariserinn ist jemals ruhig?" (Heine 1979, p. 41).

BIBLIOGRAPHY OF WORKS CITED

Aristotle. 1972. *Metaphysica*. Translated by W. D. Ross. 2nd ed. Oxford: Clarendon Press. (*Works*, vol. 8.)

Asmus, Valentin F. 1956. *Dekart*. Moscow.

Bailey, Cyril. 1928a. *The Greek Atomists and Epicurus, a Study*. Oxford: Clarendon Press.

Bailey, Cyril. 1928b. 'Karl Marx on Greek Atomism'. *Classical Quarterly* **22**, 205–206.

Baillet, Adrien. 1691. *La vie de monsieur Des-Cartes*. Paris.

Bakhtin, Mikhail M. 1965. *Tvorchestvo Fransua Rable i narodnaia kul'tura srednevekov'ia i Renessansa*. Moscow: Khudozhestvennaia lit-ra. See also Bakhtin 1968 and 1984b.

Bakhtin, Mikhail M. 1968. *Rabelais and His World*. Translated by Hélène Iswolsky. Cambridge, Mass.: M.I.T. Press. Translation of Bakhtin 1965. See also Bakhtin 1984b.

Bakhtin, Mikhail M. 1984a. *Problems of Dostoevsky's Poetics*. Edited and translated by Caryl Emerson. Minneapolis: University of Minnesota Press. This is a new translation of Bakhtin's *Problemy poetiki Dostoevskogo* (Moscow, 1963), which is a much expanded second edition of *Problemy tvorchestva Dostoevskogo* (Leningrad, 1929). An earlier translation, by R. William Rotsel, was published in 1973 by Ardis in Ann Arbor, Michigan; it has the same title.

Bakhtin, Mikhail M. 1984b. *Rabelais and His World*. Translated by Hélène Iswolsky. Indianapolis: Indiana University Press. A re-issue of Bakhtin 1968.

Baudoin, Charles. 1924. *Psychoanalysis and aesthetics*. Translated by Eden and Cedar Paul. New York: Dodd, Mead. (Translation of *Le symbole chez Verhaeren: essai de psychanalyse de l'art*. Geneva: Editions Mongenet, 1924.)

Belkin, Abram A. 1956. *F. M. Dostoevskii v russkoi kritike; sbornik statei*. Moscow: Goslitizdat.

Bohr, Niels. 1958. 'Quantum Physics and Philosophy. Causality and Complementarity'. In *Philosophy in the Mid-Century — A Survey / La philosophie au milieu du vingtième siècle — Chroniques*. Edited by Raymond Klibansky, vol. 1, pp. 308–314. Florence: La Nuova Italia.

Borovskii, Ia. M. 1945. 'Poetika dokazatel stva Lukretii' [The poetics of proof in Lucretius]. In Lucretius 1945–47, vol. II, pp. 198–219.

Bruno, Giordano. 1879–1891. *Opera latine conscripta* ... 3 vols. in 6. Naples: Morano.

Bruno, Giordano. [1950] 1978. *Concerning the Cause, Principle, and One*. Translated by Sidney T. Greenburg. In *The Infinite in Giordano Bruno*, by Sidney T. Greenburg, pp. 77–173. New York: Octagon Books.

Bruno, Giordano. 1958. *Dialoghi italiani*. Edited by Giovanni Aquilecchia. 3rd ed. Florence: Sansoni.

Bruno, Giordano. 1964a. *The Candle Bearer* (1582). English version by J. R. Hale. In *The Genius of the Italian Theater*, edited by Eric R. Bentley, pp. 196–314. New York: New American Library. Translation of Bruno 1969.

Bruno, Giordano. 1964b. *The Expulsion of the Triumphant Beast.* Translated and edited by Arthur D. Imerti. New Brunswick, N. J.: Rutgers University Press.

Bruno, Giordano. 1964c. *The Heroic Frenzies.* Translated by Paul E. Memmo, Jr. Chapel Hill: University of North Carolina Press.

Bruno, Giordano. 1969. *Candelaio.* Edited by Giorgio B. Squarotti. Turin: Einaudi. See also Bruno 1964a.

Brunschvicg, Léon. 1897. *La modalité du jugement.* Paris: Alcan. See also Brunschvicg 1964.

Brunschvicg, Léon. 1934. *Les âges de l'intelligence.* Paris: Alcan.

Brunschvicg, Léon. 1949. *La philosophie de l'esprit.* Paris: Presses universitaires de France.

Brunschvicg, Léon. 1964. *La modalité du jugement.* 3. éd., augmentée de *La vertu métaphysique du syllogisme selon Aristote* (1897), Latin thesis translated by Yvon Belaval. Paris: Presses universitaires de France. *La vertu métaphysique* . . . is a translation of Brunschvicg's *Qua ratione Aristoteles metaphysicam vim syllogismo inesse demonstraverit.*

Cassirer, Ernst. 1963. *The Individual and the Cosmos in Renaissance Philosophy.* Translated by Mario Domandi. Oxford: Blackwell.

Chuquet, Arthur. 1893. *J. J. Rousseau.* Paris: Hachette.

Clagett, Marshall., ed. 1959. *The Science of Mechanics in the Middle Ages.* Madison: University of Wisconsin Press.

Clark, Ronald W. 1971. *Einstein; the Life and Times.* New York: World Publishing Co.

Corell, Alice F. 1927. *The Contribution of Verhaeren to Modern French Lyric Poetry.* The University of Buffalo Studies, vol. 5, no. 2. Monographs in Romance Languages, no. 1. Buffalo.

Dante. 1970–75. *The Divine Comedy.* Translated by Charles S. Singleton. 3 vols. in 6. Princeton: Princeton University Press.

de Santillana, Giorgio. 1953. 'Léonard et ceux qu'il n'a pas lus'. In *Léonard de Vinci et l'expérience scientifique au XVI. siècle*, pp. 43–59. Presses universitaires de France. See also de Santillana 1968.

de Santillana, Giorgio. 1968. 'Leonardo, "Man without Letters".' In *Reflections on Men and Ideas*, pp. 1–19. Cambridge, Mass.: M.I.T. Press. Translation of de Santillana 1953.

de Ste. Croix, Geoffrey E. M. 1981. *The Class Struggle in the Ancient Greek World from the Archaic Age to the Arab Conquests.* Ithaca: N.Y.: Cornell University Press.

Descartes, René. 1954. *Discourse on Method.* In *Philosophical Writings*, translated and edited by G. E. M. Anscombe and P. T. Geach, pp. 5–57. Edinburgh: Nelson.

Descartes, René. 1961. *Les pages immortelles de Descartes.* Selected and annotated by Paul Valéry. Paris: Buchet/Chastel. (Originally published 1941.)

Descartes, René. 1969. *Correspondance, avril 1622–février 1638.* Vol. 1 of *Oeuvres*, published by Charles Adam and Paul Tannery. Paris: Vrin, 1964 –. (Originally published, Paris: Cerf, 1897–1910.)

Descartes, René. 1972. *Correspondance, juillet 1643–avril 1647.* Vol. 4 of *Oeuvres*, published by Charles Adam and Paul Tannery. Paris: Vrin, 1964 –. (Originally published, Paris: Cerf, 1897–1910.)

Dostoevsky, Fedor M. 1930. 'Tri rasskaza Edgara Poe.' In *Polnoe sobranie khudozhestvennykh proizvedenii*, edited by B. V. Tomashevskii and K. I. Khalabayev (Moscow—Leningrad: Gosizdat, 1926—30), vol. 13, pp. 523—524.

Dostoevsky, Fedor M. 1966. *The Brothers Karamazov*. Translated by David Magarshack. 2 vols. Harmondsworth, Middlesex: Penguin Books.

Dostoevsky, Fedor M. 1968. *Crime and Punishment*. Translated by David Magarshack. Harmondsworth, Middlesex: Penguin Books.

Ehrenberg, Il'ia. 1966. *Post-War Years, 1945—1954*. Vol. 6 of *Men, Years-Life*. London: MacGibbon and Kee.

Einstein, Albert. 1905. "Ist die Trägheit eines Körpers von seinem Energieinhalt abhängig?" *Annalen der Physik* **17**, 639—641.

Einstein, Albert. 1932. 'Bemerkungen ueber den Wandel der Problemstellungen in der Theoretischen Physik'. In *Contributions to the Medical Sciences in Honor of Dr. Emanuel Libman* . . . , pp. 363—364. New York: International Press.

Einstein, Albert. 1934. *The World as I See It*. New York: Covici Friede.

Einstein, Albert. 1936a. 'Physics and Reality'. *Journal of the Franklin Institute* **221**, 349—382. Translation of Einstein 1936b.

Einstein, Albert. 1936b. 'Physik und Realität'. *Journal of the Franklin Institute* **221**, 313—348. See also Einstein 1936a.

Einstein, Albert. 1956. *Lettres à Maurice Solovine*. Paris: Gauthier-Villars.

Einstein, Albert. 1969. *Albert Einstein und Hedwig und Max Born: Briefwechsel, 1916—1955*. Munich: Nymphenburger Verlagshandlung. See also Einstein 1971.

Einstein, Albert. 1971. *The Born-Einstein Letters; Correspondence between Albert Einstein and Max and Hedwig Born from 1916 to 1955*. Translated by Irene Born. New York: Walker. Translation of Einstein 1969.

Engels, Friedrich. 1954. *Dialectics of Nature*. Moscow: Foreign Languages Publishing House. (Also published in London, by Lawrence and Wishart, 1955.)

Epicurus. [1926] 1979. *Epicurus, the Extant Remains*. Translated by Cyril Bailey. Westport, Conn.: Hyperion Press.

Fermi, Enrico. 1932. 'L'etat actuel de la physique du noyau atomique'. In *Comptes rendus du Congrès international d'électricité*, Paris, 1932. Edited under the direction of Robert de Valbreuze. 13 vols. Vol. 2: Première section: *Science de l'électricité* . . . , pp. 789—807.

Feuerbach, Ludwig. 1966. *Principles of the Philosophy of the Future*. Translated by Manfred H. Vogel. Indianapolis: Bobbs-Merrill.

Feuerbach, Ludwig. 1969. *Geschichte der neuern Philosophie; Darstellung, Entwicklung und Kritik der Leibnizschen Philosophie*. Berlin: Akademie Verlag. (*Gesammelte Werke*, vol. 3).

Fischer, Kuno. 1887. *History of Modern Philosophy. Descartes and His School*. Translated by J. P. Gordy. New York: Scribner's.

Frege, Gottlob. 1962. *Grundgesetze der Arithmetik*. 2nd ed. 2 vols in 1. Hildesheim: Olm.

Fridman, Aleksandr A. 1965. *Mir kak prostranstvo i vremia* [The world as space and time]. 2nd ed. Moscow: Nauka.

Furley, David J. 1967. *Two Studies in the Greek Atomists: Study I, Indivisible Magnitudes; Study II, Aristotle and Epicurus on Voluntary Action*. Princeton, N.J.: Princeton University Press.

Gaidenko, Piama P. 1970. *Tragediia estetizma*. Moscow: Iskusstvo.

Galileo. 1890–1909. *Le opere ...* Edizione nazionale. 20 vols. in 21. Florence: Barbèra.

Galileo. 1892. *Il sidereus nuncius e le scritture ad esso attinenti*. Florence: Barbèra. (*Le opere*, vol. 3, pt. 1).

Galileo. 1894. 'Risposta alle opposizioni di L. delle Colombe e di V. di Grazia contro al trattato Delle cose che stanno su l'acqua o che in quella si muovono'. In *Delle cose che stanno in su l'acqua o che in quella si muovono*, pp. 449–671. Florence: Barbèra. (*Le opere*, vol. 4).

Galileo. 1897. *I due massimi sistemi del mondo: Dialogo sopra i due massimi sistemi del mondo*. Florence: Barbèra. (*Le opere*, vol. 7). See also Galileo 1967.

Galileo. 1898. *Le nuove scienze: Discorsi e dimostrazioni matematiche intorno a due nuove scienze*. Florence: Barbèra. (*Le opere*, vol. 8). See also Galileo 1974.

Galileo. 1967. *Dialogue Concerning the Two Chief World Systems — Ptolemaic and Copernican*. Translated by Stillman Drake. 2nd ed. Berkeley: University of California Press. Translation of Galileo 1897.

Galileo. 1974. *Two New Sciences ...* Translated by Stillman Drake. Madison: University of Wisconsin Press. Translation of Galileo 1898.

Gillispie, Charles C., ed. 1970–1980. *Dictionary of Scientific Biography*. 16 vols. New York: Scribner.

Gilson, Étienne. 1948. *L'être et l'essence*. Paris: Vrin.

Gilson, Étienne. 1951. *Études sur le rôle de la pensée médiévale dans la formation du système cartésien*. Paris: Vrin.

Gombrich, Ernst. 1972. 'Raphael's *Stanza della Segnatura* and the Nature of Its Symbolism'. In *Symbolic Images*. London: Phaidon.

Hegel, Georg W. F. 1936. *Dokumente zu Hegels Entwicklung*. Edited by Johannes Hoffmeister. Stuttgart: Frommann.

Hegel, Georg W. F. 1939–41. *La phénoménologie de l'esprit*. Translated by Jean Hyppolite. 2 vols. Paris: Aubier, Éditions Montaigne.

Hegel, Georg W. F. 1967. *Philosophy of Right*. Translated by T. M. Knox. Oxford: Clarendon Press. (Reprint of 1952 ed.)

Hegel, Georg W. F. 1969. *Science of Logic*. Translated by A. V. Miller. London: Allen and Unwin; New York: Humanities Press.

Hegel, Georg W. F. 1969–81. *Briefe von und an Hegel*. Edited by Johannes Hoffmeister. 4 vols. in 5. 3rd ed. Hamburg: Meiner.

Hegel, Georg W. F. 1974. *Lectures on the History of Philosophy*. Translated by E. S. Haldane and Frances H. Simson. 3 vols. London: Routledge and Kegan Paul; New York: Humanities Press. (Reprint of London, 1892–1896 ed.)

Hegel, Georg W. F. 1975. *Logic*. Translated by William Wallace. Oxford: Clarendon Press.

Hegel, Georg W. F. 1981. *Phenomenology of Spirit*. Translated by A. V. Miller. Oxford: Oxford University Press. Reprint of 1977 edition.

Heine, Heinrich. 1854. *Lutezia; Berichte über Politik, Kunst und Volksleben*. 2 vols. Hamburg: Hoffmann und Campe. (*Vermischte Schriften*, vols. 2–3.) See below for later German edition (Heine 1974); French editions (Heine 1855, 1977); English translation (Heine 1893).

Heine, Heinrich, 1855. *Lutèce; lettres sur la vie politique, artistique et sociale de la France*. Paris: Michel Lévy.

Heine, Heinrich. 1892. *Florentine Nights* . . . Translated by Charles G. Leland. (*Works*, vol. 1.) Translation of Heine 1979.

Heine, Heinrich. 1893. *Lutetia*. Translated by Charles G. Leland. Vol. 2 of *French Affairs: Letters from Paris*. London: Heinemann. (*Works*, vol. 8.)

Heine, Heinrich. 1970—. *Werke, Briefwechsel, Lebenszeugnisse*. Säkularausgabe. 31 vols to date. Berlin: Akademie-Verlag.

Heine, Heinrich. 1974. *Lutezia; Berichte über Politik, Kunst und Volksleben*. Edited by Lucienne Netter. Berlin: Akademie-Verlag. (*Werke* . . . , vol. 11.)

Heine, Heinrich. 1977. *Lutèce*. Edited by Jacques Voisine. Berlin: Akademie-Verlag. (*Werke* . . . , vol. 19.)

Heine, Heinrich. 1979. *Florentinische Nächte*. In *Prosa, 1836–1840*, edited by Fritz Mende, pp. 7—53. Berlin: Akademie-Verlag. (*Werke* . . . , vol. 9). First published as volume 3 of *Der Salon* (Hamburg: Hoffmann u. Campe, 1837).

Helmholtz, Hermann. 1847. *Über die Erhaltung der Kraft: eine physikalische Abhandlung*. Berlin: Reimer. See also Helmholtz 1971.

Helmholtz, Hermann. 1971. *The Conservation of Force: A Physical Memoir*. In *Selected Writings of Hermann von Helmholtz*, edited by Russell Kahl, pp. 3—55. Middletown, Conn.: Wesleyan University Press. Translation of Helmholtz 1847.

Hilbert, David. 1899. *Grundlagen der Geometrie*. Leipzig: Teubner. See also Hilbert 1971.

Hilbert, David. 1915—1916. 'Die Grundlagen der Physik'. In *Nachrichten von der königlichen Gesellschaft der Wissenschaft zu Göttingen*, 20 November 1915 (p. 395 of 1915 vol.) and 23 December 1916 (pp. 53—76 of 1917 vol.). Reprinted in *Mathematischen Annalen* **92** (1924) 1-32; and in Hilbert's *Gesammelte Abhandlungen*. 3 vols. Vol. 3: pp. 258–289. Berlin: Springer.

Hilbert, David. 1971. *Foundations of Geometry*. Translated by Leo Unger. 2nd ed. La Salle, Ill.: Open Court. See also Hilbert 1899.

Hingley, Ronald. 1978. *Dostoevsky, His Life and Work*. New York: Scribner's.

Hyppolite, Jean. 1953. *Logique et existence; essai sur la logique de Hegel*. Paris: Presses universitaires de France.

Hyppolite, Jean. 1955. *Études sur Marx et Hegel*. Paris: Rivière.

Ihering, Rudolf von. 1909. *Scherz und Ernst in der Jurisprudenz. Eine Weihnachtsgabe für das juristische Publikum*. 10th ed. Leipzig: Breitkopf und Härtel.

Infeld, Leopold. 1941. *Quest, the Evolution of a Scientist*. New York: Doubleday, Doran and Co.

Jackson, Robert L. 1958. *Dostoevsky's Underground Man in Russian Literature*. The Hague: Mouton.

Kant, Immanuel. 1970. *Metaphysical Foundations of Natural Science*. Translated by James Ellington. Indianapolis: Bobbs-Merrill.

Kepler, Johann. 1941. 'De fundamentis astrologiae certioribus' (1602). In *Gesammelte Werke* (Munich: Beck, 1937—), vol. 4, pp. 9—35.

Kepler, Johann. 1965. *Conversation with Galileo's Sidereal Messenger*. Translated by Edward Rosen. New York: Johnson Reprint.

Kierkegaard, Søren A. 1941. *Concluding Unscientific Postscript.* Translated by David F. Swenson and Walter Lowrie. Princeton: Princeton University Press, for the American Scandinavian Foundation.

Klein, Felix. 1893a. 'A Comparative Review of Recent Researches in Geometry'. *N.Y. Math. Soc. Bull.* **2**, 215—249. See also Klein 1893b.

Klein, Felix. 1893b. 'Vergleichende Betrachtungen über neuere geometrische Forschungen' (1872). *Math. Ann.* **43**, 63—100; reprinted in Klein's *Gesammelte mathematische Abhandlungen.* 2 vols. Vol. 1, pp. 460—497. Berlin: Springer, 1921—1923.

Kline, Morris. 1972. *Mathematical Thought from Ancient to Modern Times.* New York: Oxford University Press.

Koyré, Alexandre. 1957. *From the Closed World to the Infinite Universe* Baltimore: Johns Hopkins Press.

Kroner, Richard. 1921—24. *Von Kant bis Hegel.* 2 vols. Tübingen: Mohr.

Kuznetsov, Boris G. 1962. *Einshtein.* Moscow: Akad. nauk. See also Kuznetsov 1965.

Kuznetsov, Boris G. 1965. *Einstein.* Translated by Vladimir Talmy. Progress Publishers. Translation of Kuznetsov 1962.

Kuznetsov, Boris G. 1966a. 'Complementarity and Relativity'. *Philosophy of Science* **33**, 199—209.

Kuznetsov, Boris G. 1966b. 'Maks Plank i perekhod ot klassicheskoi fiziki k kvantovoi'. Appendix to Planck 1966, pp. 233—245.

Kuznetsov, Boris G. 1967. 'Spinoza et Einstein'. *Revue de Synthèse* **88,** 3rd series, nos. 45—46, pp. 31—52.

Kuznetsov, Boris G. 1970a. *Dzhordano Bruno i genezis klassicheskoi nauki.* Moscow: Nauka.

Kuznetsov, Boris G. 1970b. *Etiudy ob Einshteine.* Moscow: Nauka.

Kuznetsov, Boris G. 1972a. *Einshtein. Zhizn', smert', bessmertie.* Moscow: Nauka.

Kuznetsov, Boris G. 1972b. *Einstein and Dostoevsky.* Translated by Vladimir Talmy. London: Hutchinson.

Leibniz, Gottfried W. 1840. *Opera philosophica.* Edited by Johannes Erdmann. 2 vols. in 1. Berlin: Eichler.

Leibniz, Gottfried W. 1877. Letter from Leibniz to Queen Sophie-Charlotte, 22 April 1702. In *Die Werke von Leibniz gemäss seinem handschriftlichen Nachlass in der Königlichen Bibliothek zu Hannover,* edited by Onno Klopp (Hanover: Klindworth's Verlag, 1864—84), vol. 10, pp. 143—145.

Leibniz, Gottfried W. 1898. *The Monadology and Other Philosophical Writings.* Translated by Robert Latta. Oxford: Clarendon Press.

Leibniz, Gottfried W. 1951. 'Whether the Essence of a Body Consists in Extension'. In *Selections,* edited by Philip P. Wiener, pp. 100—102. New York: Scribner.

Leibniz, Gottfried W. 1969. *The Controversy between Leibniz and Clarke, 1715—1716.* In *Philosophical Papers and Letters,* translated and edited by Leroy E. Loemker, pp. 675—721. 2nd ed. Dordrecht: D. Reidel.

Lenin, Vladimir I. 1961. 'Aristotle. *The Metaphysics*'. In *Philosophical Notebooks,* pp. 367—374. London: Lawrence and Wishart. (*Collected Works,* vol. 38). Translation of Lenin 1963.

Lenin, Vladimir I. 1963. 'Aristotel'. *Metafizika . . .*' In *Filosofiskie tetradi,* pp. 325—332. Moscow: Gospolitizdat. (*Polnoe sobranie sochinenii* (5th ed.), vol. 29). See also Lenin 1961.

Leonardo da Vinci. 1956. *Treatise on Painting — 'Codex urbinas latinus 1270'*. Translated by A. Philip McMahon. 2 vols. Princeton: Princeton University Press.
Lessing, Gotthold. 1893. *Laokoon oder über die Grenzen der Malerei und Poesie*. In *Sämtliche Schriften*, edited by Karl Lachmann, vol. 9, pp. 3—177. Stuttgart: Göschen.
Lucretius. 1923—24. *De rerum natura*. 2 vols. Berlin: Weidmann. Preface by Albert Einstein, vol. 2, pp. vi[a]—vi[b].
Lucretius. 1945—47. *De rerum natura*. Translated into Russian by Theodore Petrovski. 2 vols. Leningrad: Academy of Sciences of the U.S.S.R.
Lucretius. 1975. *De rerum natura*. Cambridge, Mass.: Harvard University Press.
Luporini, Cesare. 1953. *La mente di Leonardo*. Florence: Sansoni.
Marx, Karl. 1913. *A Contribution to the Critique of Political Economy*. Translated by N. I. Stone. Chicago: Kerr.
Marx, Karl. 1927—35. 'Differenz der demokritischen und epikureischen Naturphilosophie . . .' In Karl Marx, Friedrich Engels: *Historisch-kritische Gesamtausgabe, Werke, Schriften, Briefe* (Frankfurt a. Main: Marx-Engels-Archiv), section 1, vol. 1, part 1, pp. 1—144. See also Marx 1975—.
Marx, Karl. 1964. *Economic and Philosophic Manuscripts of 1844*. Edited by Dirk J. Struik. Translated by Martin Milligan. New York: International Publishers.
Marx, Karl. 1975—. 'On the Difference between the Democritean and Epicurean Philosophy of Nature'. In Karl Marx, Friedrich Engels: *Collected Works*, vol. 1, pp. 25—107. New York: International Publishers. Translation of Marx 1927—35.
Maxwell, James C. 1890. *The Scientific Papers* . . . Edited by W. D. Niven. 2 vols. Cambridge: University Press.
Maxwell, James C. 1954. *A Treatise on Electricity and Magnetism* (1891). Edited by J. J. Thomson. 3rd ed. 2 vols. New York: Dover.
Meyerson, Émile. 1921. *De l'explication dans les sciences*. 2 vols. Paris: Payot.
Meyerson, Émile. 1930. *Identity and Reality*. Translated by Kate Loewenberg. London: Allen and Unwin; New York: Macmillan.
Meyerson, Émile. 1931. *Du cheminement de la pensée* . . . 3 vols. Paris: Alcan.
Meyerson, Émile. 1936. *Essais*. With a Preface by Louis de Broglie. Paris: Vrin.
Michel, Henri. 1962. *La cosmologie de Giordano Bruno*. Paris: Hermann. See also Michel 1973.
Michel, Henri. 1973. *The Cosmology of Giordano Bruno*. Translated by R. E. W. Maddison. Paris: Hermann; Ithaca, N.Y.: Cornell University Press. Translation of Michel 1962.
Mikhailovsky, Nikolai K. 1882. 'Zhestokii talant'. A review of volumes 2 and 3 of Dostoevsky's *Polnoe sobranie sochinenii* (St. Petersburg, 1882). In *Otechestvennia zapiski* (St. Petersburg) **264**, no. 2, pp. 81—123; 242—274. Reprinted in Mikhailovsky's *Sochineniia* (St. Petersburg: Russkoe bogatstvo, 1896—97), vol. 5 (1897), pp. 1—78. Also reprinted in Belkin 1956, pp. 306—385. See also Mikhailovsky 1978.
Mikhailovsky, Nikolai K. 1978. *Dostoevsky — A Cruel Talent*. Translated by Spencer Cadmus. Ann Arbor, Michigan: Ardis. Translation of Mikhailovsky 1882.
Moscovici, Serge. 1956. 'A propos de quelques travaux d'Adam Smith sur l'histoire et la philosophie des sciences'. *Revue d'histoire des sciences* **9**, 1–20.
Muratov, Pavel P. 1924. *Obrazy Italii*. 3 vols. Berlin: Grzhebina.
Murdoch, John E. 1960. 'Thomas Bradwardine: Geometry and the Continuum in the Fourteenth Century'. In *Actes du IXe Congrès International d'Histoire des Sciences*,

Barcelona-Madrid, 1—7 September 1959. Vol. I, pp. 538—543. Barcelona: Asociación para la Historia de la Ciencia Española; Paris: Hermann.

Newton, Sir Isaac. 1934. *Mathematical Principles of Natural Philosophy* . . . Translated by Andrew Motte and Florian Cajori. Berkeley: University of California Press. Translation of Newton 1972.

Newton, Sir Isaac. 1972. *Philosophiae naturalis principia mathematica*. 3rd ed. Edited by Alexandre Koyré and I. Bernard Cohen with the assistance of Anne Whitman. 2 vols. Cambridge, Mass.: Harvard University Press. See also Newton 1934.

Noether, Emmy. 1918a. 'Invarianten beliebiger Differentialausdrücke'. *Nachrichten von der Königlichen Gesellschaft der Wissenschaften zu Göttingen* (Math.-phys. Kl.), Heft 1, pp. 37—44.

Noether, Emmy, 1918b. 'Invariante Variationsprobleme'. *Nachrichten von der königlichen Gesellschaft der Wissenschaften zu Göttingen* (Math.-phys. Kl.), Heft 2, pp. 235—257.

Pais, Abraham. 1982. *Subtle is the Lord —: the Science and the Life of Albert Einstein*. Oxford; New York: Oxford University Press.

Pallucchini, Rodolfo. 1962. *Giovanni Bellini*. Translated by R. H. Boothroyd. Milan: Martello.

Pascal, Blaise. 1904. *Pensées*. Edited by Léon Brunschvicg. 3 vols. Paris: Hachette.

Pascal, Blaise. 1958. *Pensées*. Edited by L. Lafuma. 2 vols. Paris: Le Club du Meilleur Livre.

Pascal, Blaise. 1966. *Pensées*. Translated by A. J. Krailsheimer. Harmondsworth: Penguin Books.

Patrizi, Francesco. 1587. *Philosophiae, de rerum natura* . . . Ferrara.

Petrovskii, F. A. 'Kommentarii'. In Lucretius 1945—47, vol. II, pp. 297—490.

Planck, Max. 1966. *Edinstvo fizicheskoi kartiny mira*. Moscow: Nauka.

Plekhanov, Georgii V. 1976. 'Bernstein and Materialism'. In *Selected Philosophical Works*, vol. 2, pp. 326—339. Moscow: Progress Publishers.

Pope-Hennessy, Sir John. 1970. *Raphael*. New York: New York University Press.

Raphael Sanzio. 1972. *Il cartone per la Scuola di Atene*. Milan: Silvana.

Redig de Campos, Deoclecio. 1952. *The Stanze of Raphael*. Rome: Del Turco.

Redig de Campos, Deoclecio. 1965. *Raffaello nelle stanze*. Milan: Martello.

Renouvier, Charles B. 1896—97. *Philosophie analytique de l'histoire* . . . 4 vols. Paris: Leroux.

Riemann, Bernhard. 1854. 'Ueber die Hypothesen, welche der Geometrie zu Grunde liegen'. Delivered in 1854 and published in 1867 in *Abhandlungen der Königlichen Gesellschaft der Wissenschaften zu Göttingen* **13**, 133—152. Reprinted in *Gesammelte mathematische Werke und wissenschaftlicher Nachlass*. (2nd ed., 1892). Pp. 272—287. New York: Dover, 1953.

Rosenkranz, Karl. 1844. *Georg Wilhelm Friedrich Hegels Leben*. Berlin: Duncker und Humblot.

Rossi, Paolo. 1981. 'Francesco Patrizi: Heavenly Spheres and Flocks of Cranes'. In *Italian Studies in the Philosophy of Science*, edited by Maria Luisa Dalla Chiara, pp. 363—388. Boston Studies in the Philosophy of Science, vol. 47. Dordrecht; Boston: D. Reidel.

Rousseau, Jean Jacques. n.d. 'A Discourse on the Arts and Sciences'. In *The Social Contract and Discourses*, pp. 125—154. London: Dent; New York: Dutton. Translation of Rousseau 1964.

Rousseau, Jean Jacques. 1964. 'Discours sur les sciences et les arts'. In *Oeuvres complètes*, vol. 3, pp. 3—33. Paris: Gallimard.

Russell, Bertrand. 1901. 'Recent Work on the Principles of Mathematics'. *International Monthly* **4**, 81—101.

Schiller, Johann. 1902. *The Poems of Schiller*. Translated by E. P. Arnold-Forster. New York: Henry Holt.

Schiller, Johann. 1943—. *Werke*. Nationalausgabe. Vol. I, *Gedichte*. Weimar: Böhlau.

Schilpp, Paul A., ed. 1949. *Albert Einstein, Philosopher-Scientist*. Evanston, Ill.: Library of Living Philosophers.

Schmidt, Johann K. 1845. *Der Einzige und sein Eigentum*. Leipzig: Wigand. See also Schmidt 1971.

Schmidt, Johann K. 1971. *Max Stirner: The Ego and His Own*. New York: Harper and Row.

Seelig, Carl, ed. 1956. *Helle Zeit — dunkle Zeit. In memoriam Albert Einstein*. Zürich, Stuttgart, Vienna: Europa-Verlag.

Simplicius, of Cilicia. 196—. *In Aristotelis Physicorum libros octo commentaria*. Edited by Hermann Diels. 2 vols. Berlin: Reimer. (Reprint of 1882—95 ed.)

Slive, Seymour. 1970—74. *Frans Hals*. 3 vols. London: Phaidon.

Smoluchowski, Marian. 1907. 'Kinetyczna teorya opalescencyi gazów w stanie krytycznym oraz innych zjawisk pokrewnych [Théorie cinétique de l'opalescence des gaz à l'état critique et de certains phénomènes corrélatifs]'. *Bulletin International de l'Académie des Sciences de Cracovie* (Classe des Sciences Mathématiques et Naturelles), pp. 1057—1075.

Spinoza, Benedictus de. 1802—03. *Opera quae supersunt omnia*. 2 vols. Jena.

Spinoza, Benedictus de. 1925. *Opera*. Edited by Carl Gebhardt. 4 vols. Heidelberg: Winter.

Spinoza, Benedictus de. 1966. *The Correspondence of Spinoza*. Translated and edited by A. Wolf. London: Cass.

Spinoza, Benedictus de. 1982. *The Ethics and Selected Letters*. Edited by Seymour Feldman. Translated by Samuel Shirley. Indianapolis: Hackett.

Starkie, Enid. 1927. *Les sources du lyrisme dans la poésie d'Emile Verhaeren*. Paris: Boccard.

Stirner, Max, see Schmidt, Johann K.

Tamm, I. E. 1956. 'A. Einstein i sovremennaia fizika'. *Uspeki fizicheskikh nauk* **59**, 1.

Taylor, Charles. 1975. *Hegel*. Cambridge, Eng.; New York: Cambridge University Press.

Teilhard de Chardin, Pierre. 1961. *The Phenomenon of Man*. Translated by Bernard Wall. New York: Harper.

Tolstoy, Aleksei N. 1953. *Bleak Morning*. Translated by Ivy and Tatiana Litvinova. Volume 3 of *Ordeal; a Trilogy*. Moscow: Foreign Languages Publishing House.

Turgenev, Ivan. 1970. 'A Conversation'. In *Dream Tales and Prose Poems*, pp. 242—244. Translated by Constance Garnett. New York: AMS Press. (*Novels*, vol. 10. Reprint of London, 1897 ed.)

Usener, H. 1963. *Epicurea*. Rome: 'L'Erma' di Bretschneider. Reprint of 1887 ed.

Valéry, Paul. 1938. *Les divers essais sur Léonard de Vinci*. Paris: Ed. de la N.R.F.

Valéry, Paul. 1956—. *Collected Works*. Edited by Jackson Mathews. Bollingen Series, XLV. 15 vols. New York: Pantheon Books; Princeton: Princeton University Press.

Valéry, Paul. 1958. 'Poetry and Abstract Thought'. In *The Art of Poetry*, translated by Denise Folliot, pp. 52—81. New York: Pantheon. (*Collected Works*, vol. 7.)

Valéry, Paul. 1968. *Masters and Friends*. Translated by Martin Turnell. Princeton: Princeton University Press. (*Collected Works*, vol. 9.)

Valéry, Paul. 1968–70. *Oeuvres*. Edited by J. Hytier. 2 vols. Paris: Gallimard.

Valéry, Paul. 1970. *Analects*. Translated by Stuart Gilbert. Princeton: Princeton University Press. (*Collected Works*, vol. 14.)

Valéry, Paul. 1972. *Leonardo, Poe, Mallarmé*. Translated by Malcolm Cowley and James R. Lawler. Princeton: Princeton University Press. (*Collected Works*, vol. 8.)

Verhaeren, Émile. 1912–33. *Oeuvres*. 9 vols. Paris: Mercure de France.

Wahl, Jean. 1929. *Le malheur de la conscience dans la philosophie de Hegel*. Paris: Rieder.

Wahl, Jean. 1931. 'Les revues philosophiques allemandes et le centenaire de Hegel'. *Revue philosophique* **56**, 448–451.

Zubov, Vasilii P. 1960. 'Traktat Bradvardina "O kontinuume".' *Istoriko-matematicheskie issledovaniia* **13**, 385–440.

Zubov, Vasilii P. 1965. *Razvitie atomisticheskikh predstavlenii do nachala XIX veka*. Moscow: Nauka.

Zubov, Vasilii P. 1968. *Leonardo da Vinci*. Translated by David H. Kraus. Cambridge, Mass.: Harvard University Press.

Zweig, Stefan. 1914. *Emile Verhaeren*. Translated by Jethro Bithell. Boston, New York: Houghton Mifflin.

INDEX OF NAMES

BOSTON STUDIES IN THE PHILOSOPHY OF SCIENCE

1. Marx W. Wartofsky (ed.), *Proceedings of the Boston Colloquium for the Philosophy of Science 1961-1962.* 1963.
2. Robert S. Cohen and Marx W. Wartofsky (eds.), *In Honor of Philipp Frank.* 1965.
3. Robert S. Cohen and Marx W. Wartofsky (eds.), *Proceedings of the Boston Colloquium for the Philosophy of Science 1964-1966. In Memory of Norwood Russell Hanson.* 1967.
4. Robert S. Cohen and Marx W. Wartofsky (eds.), *Proceedings of the Boston Colloquium for the Philosophy of Science 1966-1968.* 1969.
5. Robert S. Cohen and Marx W. Wartofsky (eds.), *Proceedings of the Boston Colloquium for the Philosophy of Science 1966-1968.* 1969.
6. Robert S. Cohen and Raymond J. Seeger (eds.), *Ernst Mach: Physicist and Philosopher.* 1970.
7. Milic Čapek, *Bergson and Modern Physics.* 1971.
8. Roger C. Buck and Robert S. Cohen (eds.), *PSA 1970. In Memory of Rudolf Carnap.* 1971.
9. A. A. Zinov'ev, *Foundations of the Logical Theory of Scientific Knowledge (Complex Logic).* (Revised and enlarged English edition with an appendix by G. A. Smirnov, E. A. Sidorenka, A. M. Fedina, and L. A. Bobrova.) 1973.
10. Ladislav Tondl, *Scientific Procedures.* 1973.
11. R. J. Seeger and Robert S. Cohen (eds.), *Philosophical Foundations of Science.* 1974.
12. Adolf Grünbaum, *Philosophical Problems of Space and Time.* (Second, enlarged edition.) 1973.
13. Robert S. Cohen and Marx W. Wartofsky (eds.), *Logical and Epistemological Studies in Contemporary Physics.* 1973.
14. Robert S. Cohen and Marx W. Wartofsky (eds.), *Methodological and Historical Essays in the Natural and Social Sciences. Proceedings of the Boston Colloquium for the Philosophy of Science 1969-1972.* 1974.
15. Robert S. Cohen, J. J. Stachel and Marx W. Wartofsky (eds.), *For Dirk Struik. Scientific, Historical and Political Essays in Honor of Dirk Struik.* 1974.
16. Norman Geschwind, *Selected Papers on Language and the Brain.* 1974.
17. B. G. Kuznetsov, *Reason and Being.* 1987.
18. Peter Mittelstaedt, *Philosophical Problems of Modern Physics.* 1976.
19. Henry Mehlberg, *Time, Causality, and the Quantum Theory* (2 vols.). 1980.
20. Kenneth F. Schaffner and Robert S. Cohen (eds.), *Proceedings of the 1972 Biennial Meeting, Philosophy of Science Association.* 1974.
21. R. S. Cohen and J. J. Stachel (eds.), *Selected Papers of Léon Rosenfeld.* 1978.

22. Milic Čapek (ed.), *The Concepts of Space and Time. Their Structure and Their Development.* 1976.
23. Marjorie Grene, *The Understanding of Nature. Essays in the Philosophy of Biology.* 1974.
24. Don Ihde, *Technics and Praxis. A Philosophy of Technology.* 1978.
25. Jaakko Hintikka and Unto Remes. *The Method of Analysis. Its Geometrical Origin and Its General Significance.* 1974.
26. John Emery Murdoch and Edith Dudley Sylla, *The Cultural Context of Medieval Learning.* 1975.
27. Marjorie Grene and Everett Mendelsohn (eds.), *Topics in the Philosophy of Biology.* 1976.
28. Joseph Agassi, *Science in Flux.* 1975.
29. Jerzy J. Wiatr (ed.), *Polish Essays in the Methodology of the Social Sciences.* 1979.
30. Peter Janich, *Protophysics of Time.* 1985.
31. Robert S. Cohen and Marx W. Wartofsky (eds.), *Language, Logic, and Method.* 1983.
32. R. S. Cohen, C. A. Hooker, A. C. Michalos, and J. W. van Evra (eds.), *PSA 1974: Proceedings of the 1974 Biennial Meeting of the Philosophy of Science Association.* 1976.
33. Gerald Holton and William Blanpied (eds.), *Science and Its Public: The Changing Relationship.* 1976.
34. Mirko D. Grmek (ed.), *On Scientific Discovery.* 1980.
35. Stefan Amsterdamski, *Between Experience and Metaphysics. Philosophical Problems of the Evolution of Science.* 1975.
36. Mihailo Marković and Gajo Petrović, *Praxis. Yugoslav Essays in the Philosophy and Methodology of the Social Sciences.* 1979.
37. Hermann von Helmholtz, *Epistemological Writings. The Paul Hertz/Moritz Schlick Centenary Edition of 1921 with Notes and Commentary by the Editors.* (Newly translated by Malcolm F. Lowe. Edited, with an Introduction and Bibliography, by Robert S. Cohen and Yehuda Elkana.) 1977.
38. R. M. Martin, *Pragmatics, Truth, and Language.* 1979.
39. R. S. Cohen, P. K. Feyerabend, and M. W. Wartofsky (eds.), *Essays in Memory of Imre Lakatos.* 1976.
42. Humberto R. Maturana and Francisco J. Varela, *Autopoiesis and Cognition. The Realization of the Living.* 1980.
43. A. Kasher (ed.), *Language in Focus: Foundations, Methods and Systems. Essays Dedicated to Yehoshua Bar-Hillel.* 1976.
44. Trân Duc Thao, *Investigations into the Origin of Language and Consciousness.* (Translated by Daniel J. Herman and Robert L. Armstrong; edited by Carolyn R. Fawcett and Robert S. Cohen.) 1984.
46. Peter L. Kapitza, *Experiment, Theory, Practice.* 1980.
47. Maria L. Dalla Chiara (ed.), *Italian Studies in the Philosophy of Science.* 1980.
48. Marx W. Wartofsky, *Models: Representation and the Scientific Understanding.* 1979.
49. Trân Duc Thao, *Phenomenology and Dialectical Materialism.* 1985.
50. Yehuda Fried and Joseph Agassi, *Paranoia: A Study in Diagnosis.* 1976.
51. Kurt H. Wolff, *Surrender and Catch: Experience and Inquiry Today.* 1976.
52. Karel Kosík, *Dialectics of the Concrete.* 1976.
53. Nelson Goodman, *The Structure of Appearance.* (Third edition.) 1977.

54. Herbert A. Simon, *Models of Discovery and Other Topics in the Methods of Science.* 1977.
55. Morris Lazerowitz, *The Language of Philosophy. Freud and Wittgenstein.* 1977.
56. Thomas Nickles (ed.), *Scientific Discovery, Logic, and Rationality.* 1980.
57. Joseph Margolis, *Persons and Minds. The Prospects of Nonreductive Materialism.* 1977.
59. Gerard Radnitzky and Gunnar Andersson (eds.), *The Structure and Development of Science.* 1979.
60. Thomas Nickles (ed.), *Scientific Discovery: Case Studies.* 1980.
61. Maurice A. Finocchiaro, *Galileo and the Art of Reasoning.* 1980.
62. William A. Wallace, *Prelude to Galileo.* 1981.
63. Friedrich Rapp, *Analytical Philosophy of Technology.* 1981.
64. Robert S. Cohen and Marx W. Wartofsky (eds.), *Hegel and the Sciences.* 1984.
65. Joseph Agassi, *Science and Society.* 1981.
66. Ladislav Tondl, *Problems of Semantics.* 1981.
67. Joseph Agassi and Robert S. Cohen (eds.), *Scientific Philosophy Today.* 1982.
68. Władysław Krajewski (ed.), *Polish Essays in the Philosophy of the Natural Sciences.* 1982.
69. James H. Fetzer, *Scientific Knowledge.* 1981.
70. Stephen Grossberg, *Studies of Mind and Brain.* 1982.
71. Robert S. Cohen and Marx W. Wartofsky (eds.), *Epistemology, Methodology, and the Social Sciences.* 1983.
72. Karel Berka, *Measurement.* 1983.
73. G. L. Pandit, *The Structure and Growth of Scientific Knowledge.* 1983.
74. A. A. Zinov'ev, *Logical Physics.* 1983.
75. Gilles-Gaston Granger, *Formal Thought and the Sciences of Man.* 1983.
76. R. S. Cohen and L. Laudan (eds.), *Physics, Philosophy and Psychoanalysis.* 1983.
77. G. Böhme et al., *Finalization in Science,* ed. by W. Schäfer. 1983.
78. D. Shapere, *Reason and the Search for Knowledge.* 1983.
79. G. Andersson, *Rationality in Science and Politics.* 1984.
80. P. T. Durbin and F. Rapp, *Philosophy and Technology.* 1984.
81. M. Marković, *Dialectical Theory of Meaning.* 1984.
82. R. S. Cohen and M. W. Wartofsky, *Physical Sciences and History of Physics.* 1984.
83. E. Meyerson, *The Relativistic Deduction.* 1985.
84. R. S. Cohen and M. W. Wartofsky, *Methodology, Metaphysics and the History of Sciences.* 1984.
85. György Tamás, *The Logic of Categories.* 1985.
86. Sergio L. de C. Fernandes, *Foundations of Objective Knowledge.* 1985.
87. Robert S. Cohen and Thomas Schnelle (eds.), *Cognition and Fact.* 1985.
88. Gideon Freudenthal, *Atom and Individual in the Age of Newton.* 1985.
89. A. Donagan, A. N. Perovich, Jr., and M. V. Wedin (eds.), *Human Nature and Natural Knowledge.* 1985.
90. C. Mitcham and A. Huning (eds.), *Philosophy and Technology II.* 1986.
91. M. Grene and D. Nails (eds.), *Spinoza and the Sciences.* 1986.
92. S. P. Turner, *The Search for a Methodology of Social Science.* 1986.
93. I. C. Jarvie, *Thinking About Society: Theory and Practice.* 1986.
94. Edna Ullmann-Margalit (ed.), *The Kaleidoscope of Science.* 1986.
95. Edna Ullmann-Margalit (ed.), *The Prism of Science.* 1986.
96. G. Markus, *Language and Production.* 1986.
97. F. Amrine, F. J. Zucker, and H. Wheeler (eds.), *Goethe and the Sciences: A Reappraisal.* 1987.

GPSR Compliance

The European Union's (EU) General Product Safety Regulation (GPSR) is a set of rules that requires consumer products to be safe and our obligations to ensure this.

If you have any concerns about our products, you can contact us on ProductSafety@springernature.com

In case Publisher is established outside the EU, the EU authorized representative is:

Springer Nature Customer Service Center GmbH
Europaplatz 3
69115 Heidelberg, Germany

Batch number: 10761588

Printed by Printforce, the Netherlands